ADVANCED AMATEUR ASTRONOMY

About the author

Gerald North is an experienced amateur astronomer with a special interest in transient phenomena on the Moon. He has contributed observations to the Lunar Section of the British Astronomical Association since 1979, and has been a guest observer at the Royal Greenwich Observatory. He likes the hands-on approach to astronomy, and has built much of his own equipment. This practical experience is shared in his book, *Mastering Astronomy* (1988) and is further developed in this second edition of his acclaimed *Advanced Amateur Astronomy* (first edition 1991).

Cover illustration

Imaging Comet Hyakutake at Chiricahua National Monument, Arizona, USA, from an altitude of 2100 m (7000 ft). © Glenn LeDrew 1996, used with permission.

ADVANCED AMATEUR ASTRONOMY

GERALD NORTH BSc.

PUBLISHED BY THE PRESS SYNDICATE OF THE UNIVERSITY OF CAMBRIDGE
The Pitt Building, Trumpington Street, Cambridge CB2 1RP, United Kingdom

CAMBRIDGE UNIVERSITY PRESS
The Edinburgh Building, Cambridge, CB2 2RU, United Kingdom
40 West 20th Street, New York, NY 10011–4211, USA
10 Stamford Road, Oakleigh, Melbourne 3166, Australia

First published by Edinburgh University Press 1991
Second edition published by Cambridge University Press 1997

Printed in the United Kingdom at the University Press, Cambridge

Typeset in 10/13 Monotype Times

A catalogue record for this book is available from the British Library

Library of Congress Cataloguing in Publication data

North, Gerald.
Advanced Amateur Astronomy/Gerald North.
p. cm.
Includes index.
ISBN 0-521-57407-2. – ISBN 0-521-57430-7 (pbk.)
1. Astronomy. 2. Astronomy–Amateur's manuals. 3. Telescopes.
I. Title.
QB45.N67 1997
520–dc21 96-51100 CIP

ISBN 0 521 57407 2 hardback
ISBN 0 521 57430 7 paperback

To the memory of Mr F. W. (Bill) Peters
astronomer, gentleman, and friend

Contents

Preface to the second edition

The few years since the first edition of this book was published have seen some remarkable changes. Chief amongst these must be the blossoming of the field of electronic imaging in the arena of amateur astronomy. Consequently, I have included a special new chapter on this subject, as well as expanding on some of the more traditional subjects. I have brought the whole of the text up to date, including a greatly expanded section on the sources of further information.

I hope that you will enjoy reading this book and find it useful. Above all I hope that you, like me, continue to pursue astronomy *for the love of it!*

Bexhill-on-sea
30 April, 1996.

Gerald North

Preface to the first edition

There are currently many books on the market which cater for the novice amateur astronomer. However astronomy, like almost everything else, has become much more sophisticated in recent years. Many observers wish to go on beyond the stage of just looking, or making simple observations of the celestial bodies. There are only something like ten thousand professional astronomers, worldwide. Consequently the opportunities are there for amateurs to make a particularly valuable contribution in astronomy, as in no other science. Even if one has no real wish to advance knowledge, astronomy is a hobby which becomes so much more enjoyable with just a little increase in effort and in sophistication of approach.

At present there is a real paucity of books which cater for the more advanced amateur astronomer. This book was written with the aim of filling that gap. What about that term - *amateur*? It strikes up connotations of slap-dash and trifling work carried out by someone very inexpert. I use that term many times in this book but never, **never, NEVER** in that context. Instead, I use it in the sense of its more exact meaning - *someone who cultivates an art or study for the love of it.*

I hope that you enjoy this book and find it useful. In order to make it of manageable (and affordable) size I have had to deal with some subjects extremely tersely and one or two areas I have had to leave out altogether. However, I have been able to include all the major areas of work likely to be of interest to most readers. In addition I recommend reading further in your chosen specialist fields and to this end I have included an extensive list of references in Chapter 17.

I have enjoyed pursuing my interest in astronomy for over two decades. I hope to continue doing so for a lot longer. Above all I hope that you, like me, will pursue this fascinating subject - *for the love of it.*

Bexhill-on-sea
23 February 1990.

Gerald North

Acknowledgements

The following people have contributed examples of their work for inclusion in this book: Mr Ron Arbour, Dr Tony Cook, the late Mr Jack Ells, Mr Peter Ells, Commander Henry Hatfield, Mr Nick James, Mr Dave Jackson, Mr Andrew Johnson, Mr Martin Mobberley, Mr Terry Platt, the late Mr F. W. (Bill) Peters, Mr Rob Moseley, Mr Peter Strugnell, Mr Eric Strach, and Mr John Watson. I have also been allowed to reproduce information and data from the following companies: Kodak Ltd., Ilford Ltd, EEV Ltd., Hamamatsu Ltd. I offer my grateful thanks to them all.

The first edition of this book was published by Edinburgh University Press and I would like to thank Mrs Vivian C. Bone and her staff for their excellent work which made it the success it was. Of course, my grateful thanks also go to Dr Simon Mitton and his staff at Cambridge University Press for taking this book on and for *their* excellent work on it.

I would also like to use this opportunity to offer my special thanks to the Director and staff of the former Royal Greenwhich Observatory, at Herstmonceux, for allowing me, as a guest observer, the privilege of using their resources and telescopes between January 1985 and March 1990.

1
Telescope optics

This is a book on advanced techniques in amateur astronomy, not a textbook on optics. However, the amateur astronomer should understand the basics of optical theory as applied to telescopes. How else might he/she make sensible choices as to the equipment to be purchased or constructed? The characteristics and performance of the telescope and its auxiliary equipment depend heavily on design. Aperture, focal length, focal ratio, image scale, resolving power, image brightness, image contrast, magnification and diffraction pattern structure are just some of the interrelated factors of crucial importance. The purpose of this chapter is to provide a summary of the optical matters relevant to the needs of the telescope user.

Focal length and image scale

Before considering specific types of telescopes let us take the imaginary case of a single, perfect, converging lens forming an image of a distant object. Further, imagine that the object is a point source, such as a star. With reference to Figure 1.1(a), the rays from the star will arrive at the lens virtually parallel. The line perpendicular to the plane of the lens and passing through its centre is known as the *optical axis*. If, as is shown in the diagram, the optical axis of the lens is aligned in the direction of the star, then all the arriving rays will be parallel to the optical axis.

If the lens is truly perfect then all the rays from the star will be brought together at a common point after passing through the lens. A screen placed at this point would show a focused image of the star. This position is known as the *principal focus* of the lens. The distance from the centre of the lens to the principal focus is known as the *focal length*. Since the optical axis of the lens is aligned to the direction of the star, the focused image is, itself, formed on the optical axis.

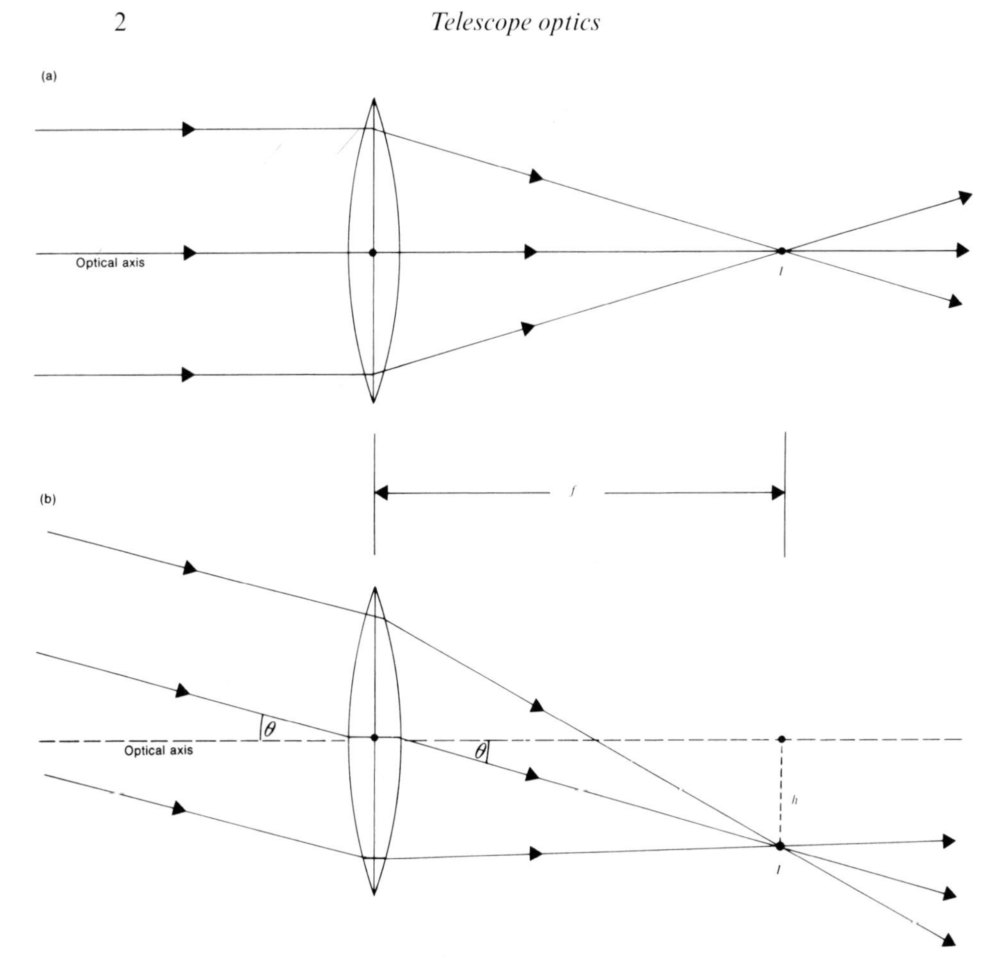

Figure 1.1. Light from a star being focused by an optically perfect lens, of focal length f. In (a) the optical axis of the lens is aligned to the direction of the star. After refraction through the lens the light achieves a focus at a point I. In (a) this point is on the optical axis. In (b) the arriving starlight makes an angle θ to the optical axis. In this case the point of focus is on the focal plane of the lens but is a distance h from the optical axis.

Figure 1.1(b) illustrates the case of the starlight entering the lens at an angle to the optical axis. Notice that the ray passing through the centre of the lens makes the same angle, θ, with the optical axis both before and after refraction by it. In this case the image is formed a distance h from the optical axis. In both (a) and (b) the focal length of the lens is f. The plane defined by the positions of sharp focus for distant point objects at varying angles to the optical axis is known as the *focal plane* of the lens.

For small values of θ we can make the approximation that:

$$\theta_{\text{radians}} = h/f,$$

where θ is measured in radians (see Figure 1.2 for an explanation of radian measure).

Expressing θ in arcseconds:

$$\theta_{\text{arcsec}} = 206\,265 \times h/f$$

From which:

$$\theta_{\text{arcsec}}/h = 206\,265/f$$

Now, θ_{arcsec}/h is the *image scale* – the linear distance in the focal plane that corresponds to a given angular distance in the sky/distant object.

$$\therefore \textbf{Image scale} = \mathbf{206\,265}/\boldsymbol{f}$$

If the focal length of the lens is measured in millimetres, then the image scale is in arcseconds per millimetre. For example, suppose that the diameter of the Moon subtends an angle of 2000 arcseconds. If the lens has a focal length of 1000 mm then the image scale in its focal plane is, to 3 significant figures, 206 arcseconds per millimetre. The focused image of the Moon will, in this case, be just under 10 mm across.

The image scale in the focal plane actually varies with the distance from the optical axis, though only by a negligible amount for small values of θ.

Aperture, focal ratio and light grasp

The *aperture* (diameter), focal length and *focal ratio* of our hypothetical lens are related by:

$$\text{Focal ratio} = \text{focal length/aperture},$$

the aperture and the focal length being expressed in the same units.

The light collected by the lens is proportional to its area, and is thus proportional to the square of its aperture:

$$\text{Light grasp} \propto (\text{aperture})^2$$

Diffraction and resolution

It is an oversimplification to say that a lens, even a perfect one, will form a truly point image of a point object. In practice the image will consist of a

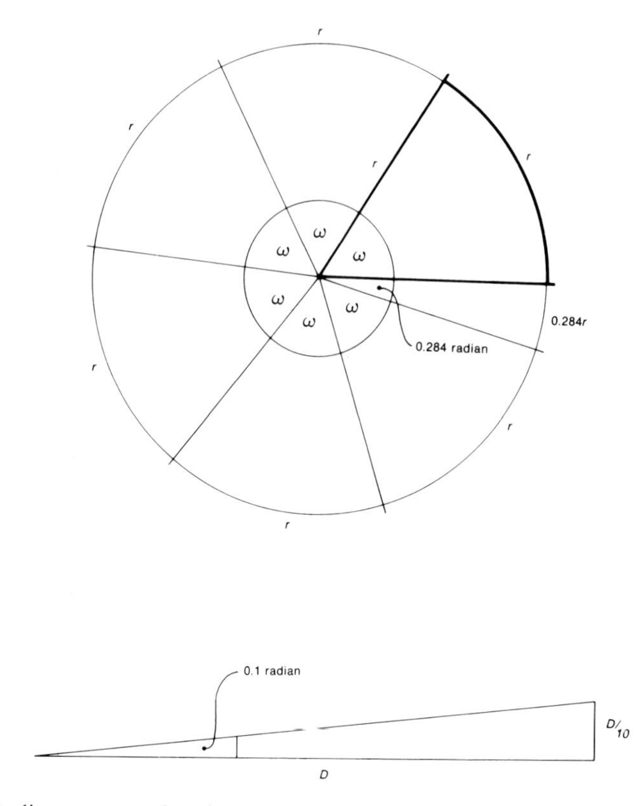

Figure 1.2. Radian measure. Consider a circle of radius r. If an arc of length equal to r is drawn along the circumference, the angle that the arc subtends from the centre of the circle is defined to be 1 radian. This is illustrated in (a) by the boldly outlined sector. In this diagram the angle ω is equal to 1 radian. Since the circumference of a circle is equal to 2π (6.284) times its radius, it follows that there are 2π radians in 360°. Hence:

$$1 \textit{ radian} = 360°/2\pi = 57°.29578$$
$$= 206\,265 \textit{ arcseconds}.$$

In general, if the arc length is l and the radius of the arc is R, then the angle, θ, is given by:

$$\theta \text{ (in radians)} = l/R,$$
$$\text{or } \theta \text{ (in arcseconds)} = 206\,265\ l/R$$

Diagram (b) shows the concept of the radian applied to smaller angles. In this case the angle subtended by a length *D/10* (notice that the distiction between a straight line and an arc can be ignored for small angles) from a distance *D* is equal to 0.1 radian.

diffraction pattern, an effect caused by the wave nature of the light passing through the restricted aperture of the lens.

The diffraction pattern that our hypothetical lens might form of a star is represented, at a very large scale, in Figure 1.3. It consists of a central bright disc, the *Airy disc*, surrounded by concentric rings of decreasing brightness. The diffraction pattern of a star produced by a perfect lens has 84 per cent of the light energy contained in the Airy disc. The remainder of the light energy is shared amongst the more diffuse rings.

The diameter of the diffraction pattern is inversely proportional to the diameter of the lens producing it. Hence **larger** apertures produce **smaller** diffraction patterns. This effect is of crucial importance when the lens is used to image fine details. For instance, consider the case of two close stars being imaged by a lens of given aperture. The size of the aperture sets the sizes of the diffraction pattern images of each star. If the stars are too close together their diffraction patterns will merge and the stars will appear as one. In order to resolve the two stars as separate a lens of larger aperture must be used. Figure 1.4 (a), (b) and (c) illustrates the cases of two stars which are not resolved as separate, just resolved, and easily resolved, respectively.

The relationship between the limiting resolution of a lens and its aperture is expressed mathematically by the equation:

$$R = 206\,265 \times \frac{1.22\lambda}{d}$$

where R is the minimum angular separation of a pair of stars in order to be just resolved, measured in arcseconds; λ is the wavelength of the light being imaged and d is the diameter of the lens. Both λ and d are measured in the same units, for instance both in metres, or both in centimetres, etc.

You should notice that the resolution of the lens is wavelength dependent. The wavelength of light to which the eye is most sensitive is taken as 5.5×10^{-7}m (This corresponds to the yellow-green part of the spectrum). A perfect lens of 0.1m aperture should then be able to resolve two stars separated by 1.4 arcseconds when used visually.

Another formula for resolving power at visual wavelengths is that derived empirically by Dawes:

$$R = 4.56/D,$$

where R is the minimum resolvable separation of two stars in arcseconds, as before, and D is the aperture expressed in inches (1 inch = 25.4 mm). Dawes' formula predicts a minimum separation for resolution about 20 per

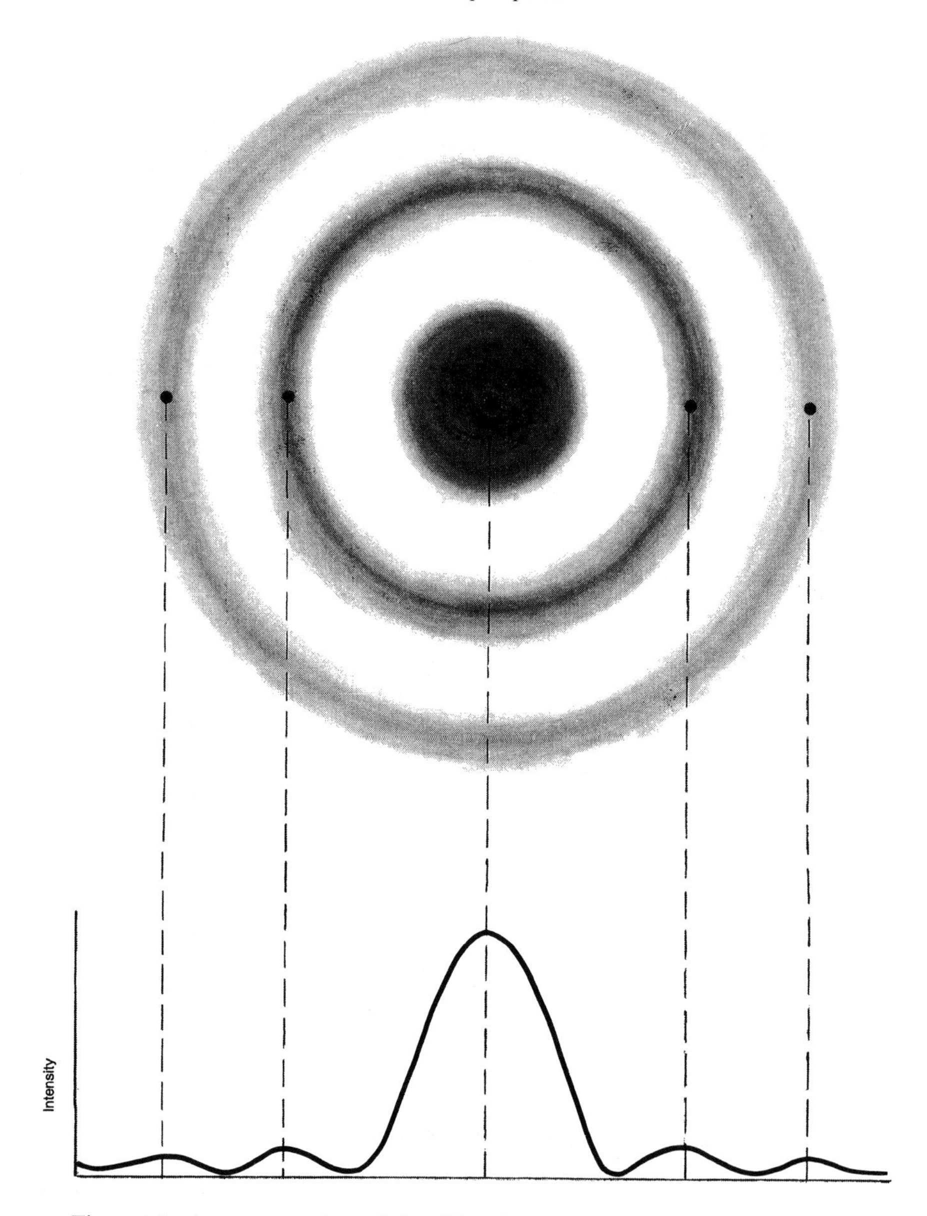

Figure 1.3. A representation of the diffraction pattern of a point source as seen through a limited aperture. The graph below illustrates the variation of the intensity of the light across the diameter of the diffraction pattern.

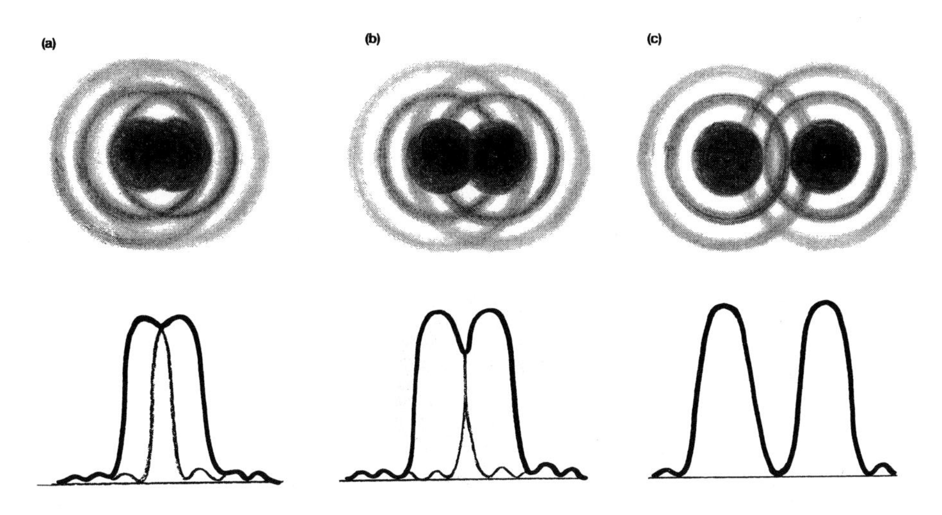

Figure 1.4. The overlapping diffraction patterns produced by two stars as seen through a given aperture. In (a) the stars are too close together to be resolved as separate. In (b) the stars are just resolvable and in (c) the stars are easily resolved.

cent closer than that of the mathematically derived formula but double star observers have found it to be more accurate in practice. Strictly, Dawes' formula applies to a pair of double stars of the sixth magnitude. Brighter stars, and those of unequal brightness, are less easily resolved.

The image of an extended object (anything which does not appear as a point), such as the surface of the Moon or a planet, can be thought of as a series of overlapping diffraction patterns. Here, too, the smaller diffraction patterns produced by larger apertures enable finer detail to be resolved. The situation has been compared to viewing a mosaic made of tiles. The finest details that can be seen in the mosaic are the size of the individual tiles. Making the mosaic from smaller tiles allows finer details to be represented.

The simple refractor

Figure 1.5(a) illustrates the optical principles of the simple refractor. Here the object glass and eyepiece are each represented as single lenses. The single-arrowed light rays originate from the top of some distant object and the double-arrowed light rays originate from the bottom of it. The object glass forms an inverted image of the object at position *I*. This is the object glass' focal plane.

When in normal adjustment, the eyepiece's focal plane is also made to coincide with position *I*. The rays then emerge from the eyepiece in parallel bundles. Notice that all the single-arrowed rays are parallel after passing through the eyepiece. The same is true for the other set of rays shown. An observer with normal vision can most comfortably view the image formed by the telescope when the rays emerge from the eyepiece in parallel bundles.

All the rays that pass through the object glass also pass through the plane at a position I'. This is variously known as the *Ramsden disc*, the *eye ring*, or, most popularly, the *exit pupil*. The exit pupil is an image of the object glass formed by the eyepiece (see Figure 1.6). It's significance is that all the rays collected by the object glass pass through this position and so it is the best location for the pupil of the observer's eye. If the observer were to place his/her eye far from the exit pupil then much of the light collected by the object glass would not pass through his/her eye pupil, and so would be wasted.

The magnification produced by the telescope can best be understood by referring to Figure 1.5(b), which shows only the relevant light rays. Notice that the object subtends an angle θ_1 from the telescope objective. It produces an image of height h at position *I*. h, θ_1 and the focal length of the object glass (f_o) are related by the equation:

$$\theta_1 = h/f_o.$$

(Strictly, this equation is only accurate if θ_1 is a small angle). The observer views this image through the eyepiece. Since the focal length of the eyepiece (f_e) is less than that of the object glass, the image subtends a larger angle, θ_2, from the eyepiece:

$$\theta_2 = h/f_e.$$

Dividing the two equations gives:

$$\theta_2/\theta_1 = f_o/f_e.$$

The telescope has increased the apparent angular size of the object from θ_1 to θ_2. θ_2/θ_1 is the *angular magnification, m*, produced by the telescope. Hence the well known equation for the magnification of the telescope in terms of the focal lengths of its object glass and eyepiece:

$$m = f_o/f_e.$$

Similar reasoning would show that the diameter of the exit pupil produced by the telescope is also linked to the magnification, as well as to the aperture of the telescope:

$$m = \text{aperture/diameter of exit pupil},$$

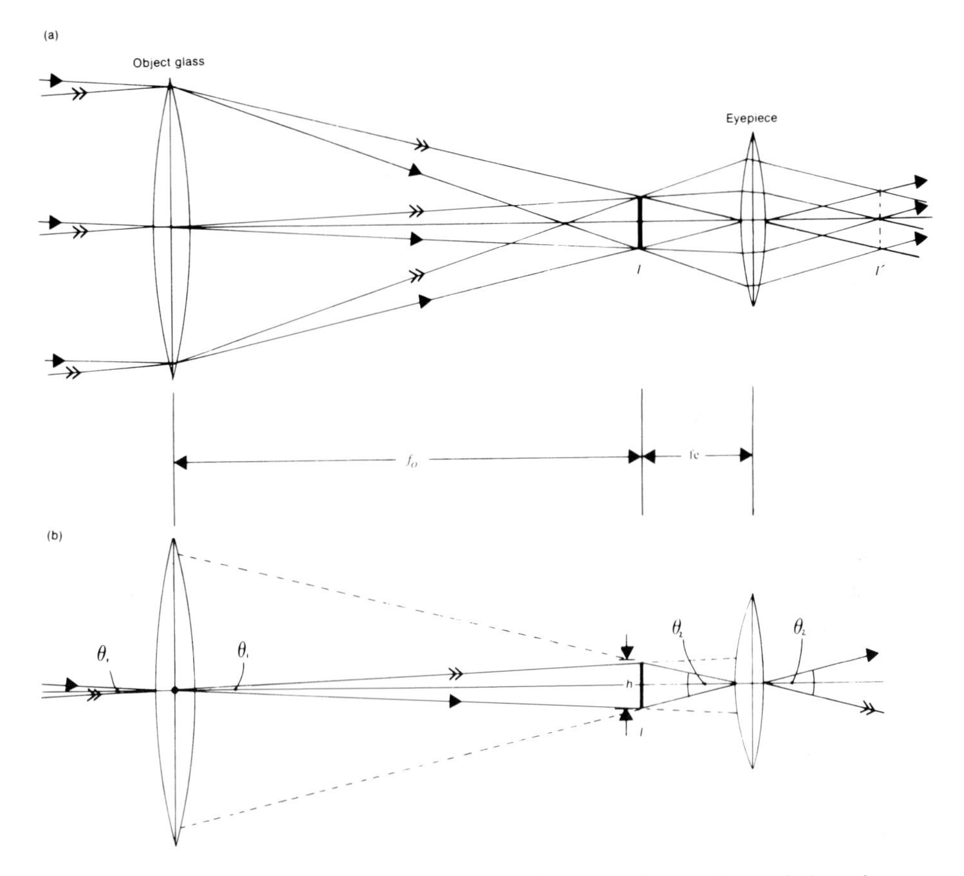

Figure 1.5. The simple refracting telescope. *I* denotes the positions of the coincident focal planes of the object glass and the eyepiece. In (a) *I′* marks the position of the exit pupil. (b) shows only the rays necessary for the explanation of the magnification the telescope produces (see text for details).

where the aperture and exit pupil diameter are measured in the same units. The exit pupil diameter is important – if it is larger than the observer's eye pupil then not all of the light collected by the object glass will enter his/her eye.

The iris of a youth's eye will open to about 8 millimetres diameter, when fully dark-adapted. In older people the maximum size of the opening decreases. Studies have shown that it averages 6.5 mm diameter for a thirty year old and further decreases with advancing years at the rate of about 0.5 mm per decade. The size of the eye pupil puts a lower limit on the magnification that can be used on a telescope without wasting any of the light gathered by its objective. If a particular

Figure 1.6. The bright image of the primary mirror of the author's telescope can be clearly seen in the eyelens of the eyepiece. This image is the exit pupil, the best position for the observer's eye in order to receive all the light collected by the mirror. The image is actually centred on the optical axis but is formed a short distance in front of the eyelens, which is why it appears displaced to the upper left in this oblique view.

observer's dark-adapted eye has a pupil diameter of 6 mm then any magnification less than the aperture in millimetres divided by 6 (this is roughly equal to ×4 per inch of aperture) will produce an exit pupil larger than 6 mm across. Some of the precious light gathered by the telescope's objective will then be wasted.

Optical aberrations

Any practical optical system will have its imperfections. The errors differ both in magnitude and type for different systems, depending upon their design and accuracy of manufacture.

The possible optical errors can be broadly grouped into *chromatic aberration* and five *seidal aberrations*. The following notes briefly detail these aberrations, using a single lens as an example.

Chromatic aberration

This occurs where the path of light rays through an optical system is wavelength dependent. For instance integrated light (light composed of a mixture of wavelengths) passing through a triangular glass prism is *dispersed* such that the shorter wavelengths are refracted through the largest angles (Figure 1.7(a)).

The shape of a lens in cross-section can be thought of as a series of triangular prisms, each of differing apex angle. For a bi-convex (converging) lens all the bases of the imaginary prisms face the optical axis. In the case of a bi-concave (diverging) lens all the bases face away from the optical axis. Figures 1.7(b) and (c) show how this leads to the red and violet rays (which are the extremes of the visible spectrum) reaching differing foci. All of the other colours (not shown) reach foci in intermediate positions.

Chromatic aberration is manifest in two distinct forms, though both have the same root cause. The spread of focal positions along the optical axis is known as *longitudinal chromatic aberration*. Since the light of only one colour can be brought to sharp focus at once, the out-of-focus images formed in the other colours are superimposed on the one in-focus image. The result is that the best image that can be formed is somewhat softened and lacking in resolution.

The situation for rays passing through the lens at an angle to the optical axis is shown in Figure 1.8. In this case the best focus images for each colour, being formed at different distances from the lens, are of differing sizes. The result is that the light–dark boundaries in the best-focused image are fringed with colour. This is *lateral chromatic aberration*.

If the lens/lens system is used to image monochromatic light (light composed of only one wavelength) then the image cannot suffer from the effects of chromatic aberration. However it can suffer from one, or a combination, of the five seidal aberrations. These are *spherical aberration, astigmatism, coma, distortion* and *curvature of field*.

Spherical aberration

Illustrated for a lens in Figure 1.9(a), this aberration causes a general softening of the focus of the image. If the rays incident on the lens were parallel then they should all intersect at a given point on the focal plane. Since the surfaces of the lens are spherical in form, the rays from its outer zones are brought to a focus closer to it than those from the inner zones. The best image that can be formed of a point object is a blurred disc, the so-called 'circle of least confusion'.

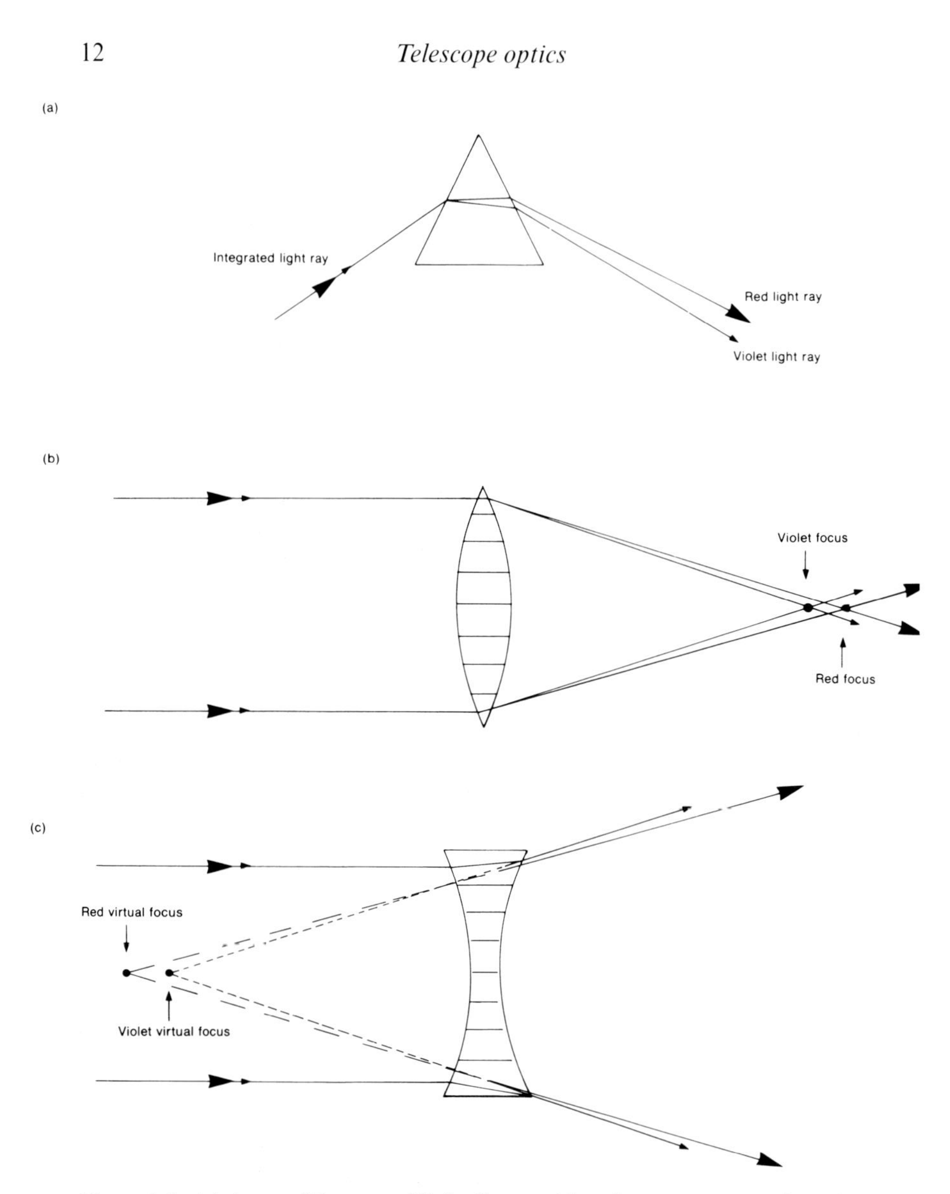

Figure 1.7. (a) A ray of integrated light dispersed into its component colours by a triangular glass prism. (b) How this leads to the generation of chromatic aberration by a single converging lens. (c) The generation of chromatic aberration by a single diverging lens.

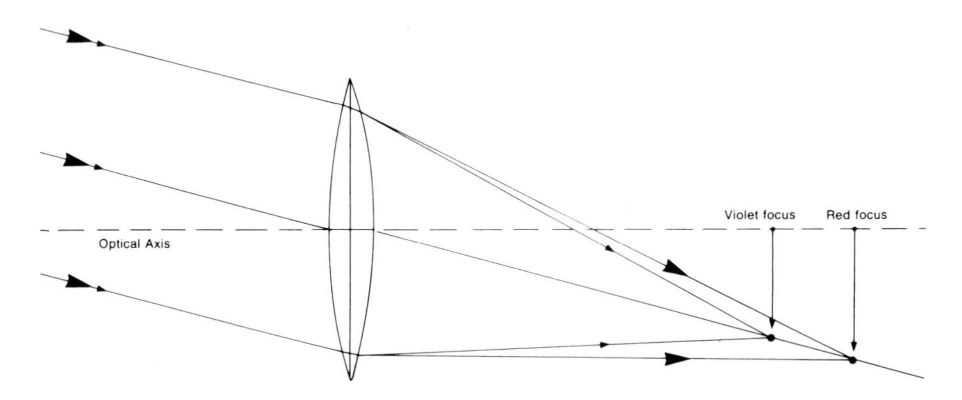

Figure 1.8. Lateral chromatic aberration. Different colours are brought to focus at differing distances from the lens. Hence the different coloured images of an object are of differing sizes. The result is that the best-focused image (usually the yellow–green, as this is the colour to which the eye is most sensitive) is fringed with false colours.

Astigmatism

This occurs when an imaging system brings sets of rays that are in planes to one another (eg. the tangential and sagittal planes) to differing focal positions along the optical axis (see Figure 1.9(b)). A star forms a good test of the presence of this aberration in a telescope. At the best focus a 'circle of least confusion' is formed but the intrafocal and extrafocal images are distorted as shown in Figure 1.9(c).

Coma

This is a complex aberration, akin to astigmatism. Rays that enter the optical system at an angle to the optical axis produce a series of radially displaced and overlapping disc images. Each disc image increases in size (and decreases in brightness) with distance from the optical axis. The result is that a star image that is not in the centre of the field of view develops a fan-shaped tail, pointing away from the centre of the field of view. Figure 1.10(a) illustrates the effect of severe coma on a hypothetical field of stars.

Distortion

This arises when the magnification delivered by an optical system varies with radial distance from the centre of the field of view. If the

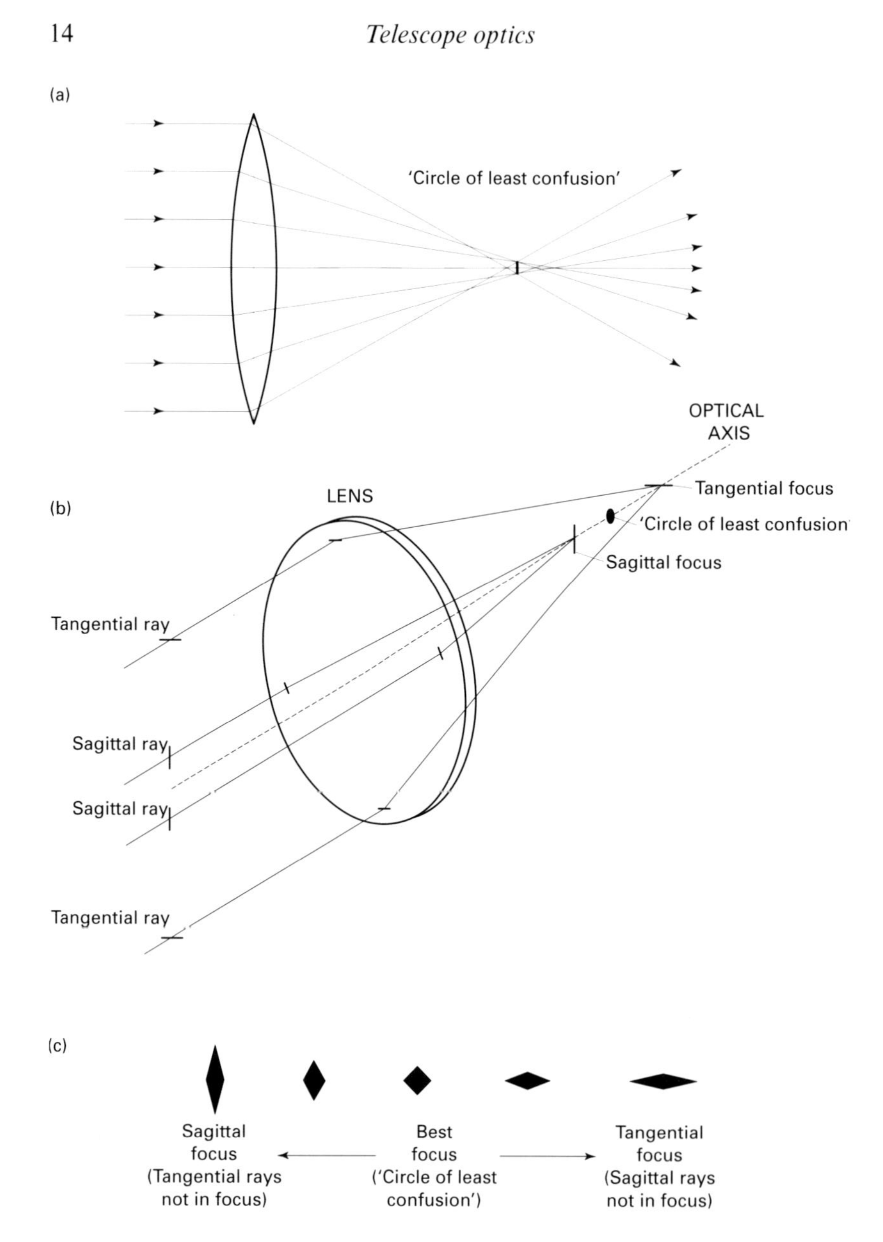

Figure 1.9. (a) Spherical aberration. The smallest image that can be formed of a point object is a blurred disc, known as a 'circle of least confusion'. (b) Astigmatism. In the case illustrated the tangential rays (those in the vertical plane) passing through the lens are brought to a focus further from it than the sagittal rays (those in the horizontal plane). (c) The effect of astigmatism on the focused image of a star. At the sagittal focus the tangential rays are out of focus, whilst the opposite is true at the tangential focus. The best-focused image that can be achieved is a 'circle of least confusion'.

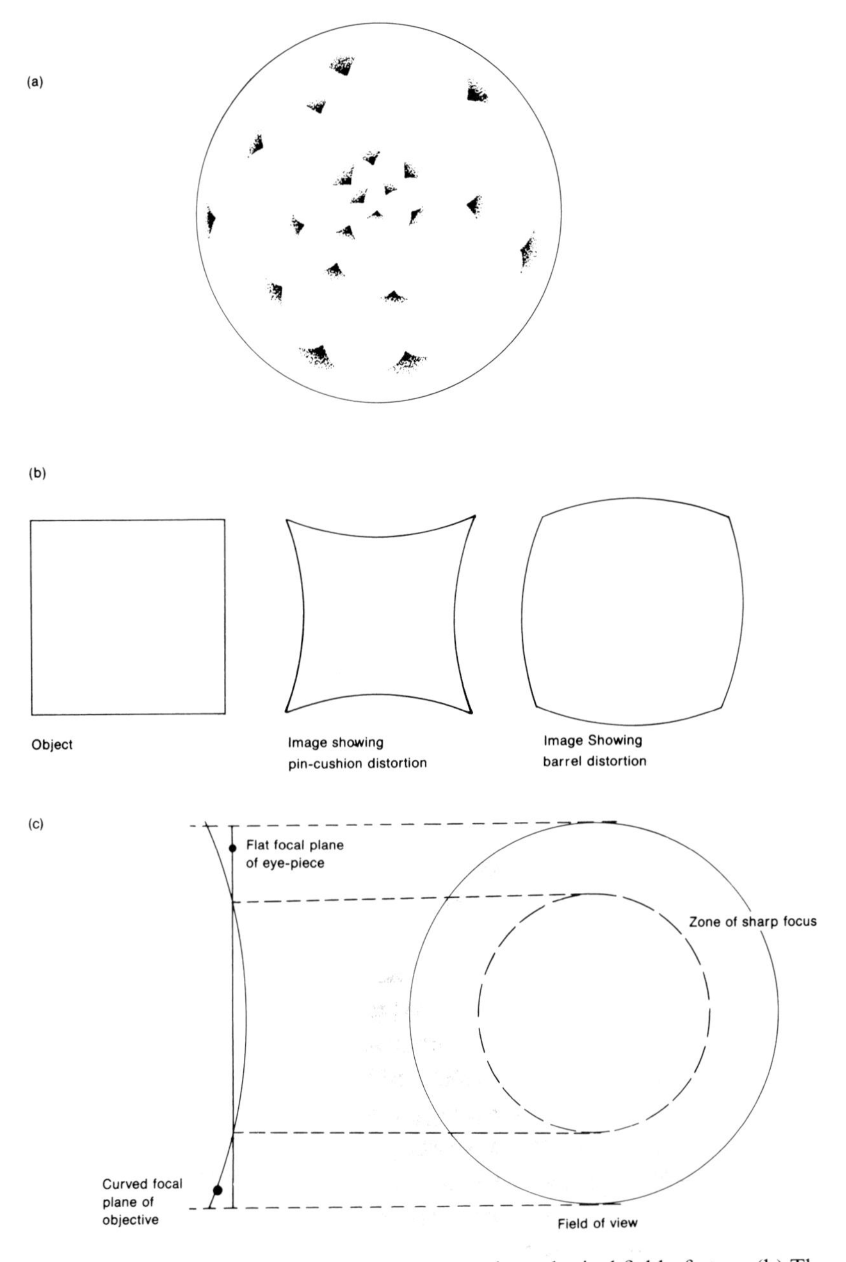

Figure 1.10. (a) The effect of severe coma on a hypothetical field of stars. (b) The square at left represents an object to be imaged. The representations to the right and the middle show the images as they would be affected by barrel distortion and pin-cushion distortion, respectively. (c) Curvature of field. In the example shown the focal plane of the telescope's objective is curved but that of its eyepiece is flat. The result is that only one zone of the field of view can be brought to a sharp focus at once.

magnification increases with distance then pin-cushion distortion results. The opposite effect produces barrel distortion (see Figure 1.10(b)).

Curvature of field

Here the focal plane of an optical system is curved instead of being flat. As Figure 1.10(c) illustrates, images in only one zone of the field of view can be in sharp focus at once.

Practical telescopes: the refractor

The chromatic aberration inherent in the simple refractor can only be rendered tolerable by making its object glass of large focal ratio (the rays arriving at the focal plane then all make small angles to the optical axis and the amount of dispersion is small). As a rough guide the focal ratio has to be at least 50 times the aperture in inches (125 times the aperture in centimetres). A 6-inch (152 mm) lens would have to have a focal ratio of at least $f/300$ and, so, a focal length of over 45 metres! Fortunately the chromatic aberration can be partially cured by making the lens of two, or sometimes three, components each of a different type of glass.

Different types of glass differ in their *refractive index* and their *dispersive power*. In very crude terms these quantities represent the amount of bending of light rays produced by the glass and the difference in the amount of bending of light of different wavelengths, respectively. A lens made of a glass of high refractive index will have a shorter focal length than a lens of identical shape but of lower refractive index. Similarly, the spread in focal lengths for light of differing colours will be greater for the lens of higher dispersive power, even if the average refractive indices and the shapes of the lenses are the same.

A typical two-element refractor objective consists of a biconvex lens (usually crown glass) with a plano-concave lens of a material of higher dispersive power (usually flint glass) situated just behind (see Figure 1.11(a)). The crown glass component is strongly converging and the flint glass component is weakly diverging. However the flint glass lens produces the same amount of dispersion as the crown lens but the spread of colours is opposite (refer to Figure 1.7 (b) and (c)). The net result is an object glass that is less strongly converging (and so longer focal length) than the crown glass component alone but has most of its chromatic aberration cancelled out.

In practice the *achromatic* object glass so formed has a minimum focus for one particular colour with focal positions for light of the other colours

(a)

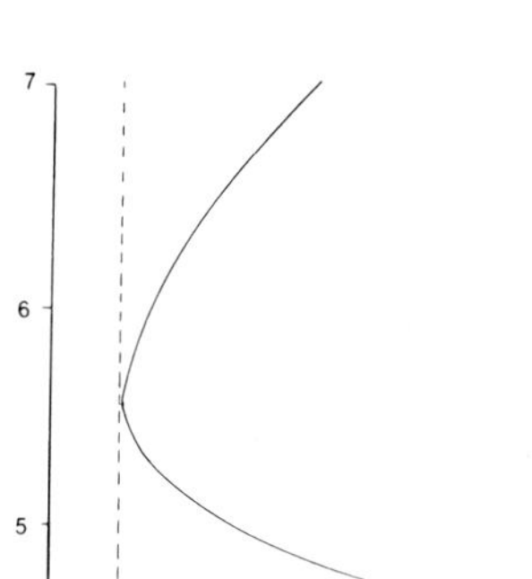

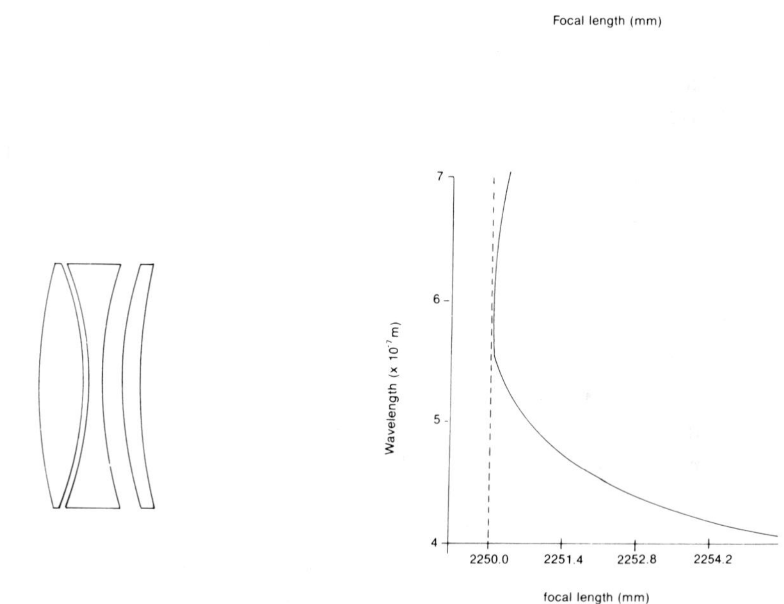

Figure 1.11. (a) The achromatic object glass. A typical lens arrangement is shown on the left. The left-hand component is usually made of a crown glass and the right-hand component is usually made of a flint. The accompanying graph shows the spread of focal positions for light ranging from deep violet (at 4×10^{-7} m) to deep red (at 7×10^{-7} m), for a 150 mm aperture *f*/15 lens. (b) The apochromatic object glass. Variants exist but in this common type the left-hand and middle components are usually made of different types of flint glass, with the other made of a crown glass. The graph shows the spread of focal positions for a typical apochromatic object glass of the same aperture and focal ratio as the lens in (a). Notice how the graph has a much larger region close to the minimum focus, though the correction for violet light is still rather poor.

extending further from the lens (shown in Figure 1.11(a)). Where the object glass is for visual use the lens designer chooses the types of glass and the curves on the lens components such that the wavelength at the minimum focus closely matches that of maximum sensitivity of the observer's eye (around 5×10^{-7}m). The spread of colours (known as the *secondary spectrum*), though much reduced from that produced by a single lens of the same aperture, imposes a lower limit on the focal ratio of an achromatic doublet. In addition, the lens designer has to ensure that the seidal aberrations are minimised when computing the correct curves for the lens components.

Over the years I have used a number of refractors both large and small. Based upon my experience I would say that, for a telescope which is to be used for planetary observation, the focal ratio of a two-element achromat ought to be at the very least 1.3 times the aperture in inches (3.3 times the aperture in centimetres) in order that the secondary spectrum should not be too obtrusive. The secondary spectrum normally reveals itself as a haze of out-of-focus colour (usually blue or purple) surrounding the focused image. This causes an overall reduction in the visibility of faint markings seen against a bright background (such as the markings on the surface of a planet).

However, what is tolerable depends very much on the use to which the telescope is put. I have observed with the 12¾-inch (324 mm), *f*/16.4, 'Mertz' refractor (the guide telescope for the Thompson 26-inch astrographic refractor at Herstmonceux) and seen lunar craters fringed with yellow, with the black interior shadows filled in with a delicate blue haze. Yet I have looked at double stars through the Greenwhich 28-inch (710 mm) refractor and seen little trace of any halo and that object glass is an *f*/11.9!

Three-element *apochromatic* objectives generally produce a secondary spectrum around 11 per cent of the extent of their two-element counterparts (Figure 1.11(b)) and they are becoming quite popular with amateurs, despite their high cost. Achromatic objectives larger than 3-inch aperture are rarely manufactured with focal ratio's of less than *f*/12. Currently there are a number of commercial apochromatic refractors of 4 to 7 inches aperture with focal ratios of *f*/5 to *f*/9. Using modern types of glass, opticians have been able to manufacture objectives (the so-called *semi-apochromats*) which, though two-element, are a considerable improvement on the early designs.

Roland Christen in the U.S.A. has pioneered the use of a triplet corrector lens, placed about a third of the way inside the focus of a two-element objective, which produces a residual secondary spectrum at the final focus even smaller than that produced by an apochromatic triplet. He calls the resulting telescope a 'tri-space refractor'. This novel design is described in the October 1985 *Sky & Telescope* magazine, page 375.

Practical telescopes: the reflector

As far as optics go, those which produce images of the highest possible contrast and the finest resolution are the ideal for lunar and planetary observation. If the seidal wavefront error at the focal plane of a telescope is as great as 1/4 of the wavelength of visible light then the potential resolution of the telescope is still that given by theory. However, the same cannot be said of the apparent contrast of faint details seen against a bright background. All other things being equal, the telescope considered would have to be very roughly half again the aperture of an optically perfect telescope in order to show low contrast planetary markings as well.

The manufacturing tolerances in the surfaces of the components of an object glass to produce this wavefront error are rather less exacting than is the case for the mirrors of a reflecting telescope. For a two-mirror reflecting telescope a 1/4 wavelength error at the focal plane means that the error at each reflection is limited to no more than 1/8 wave (we are here assuming that the errors add, it is possible that they could cancel out). Worse, the wavefront error produced at each reflection is **twice** the innacuracy in its surface. In other words, for a 1/4 wave error at the focal plane each reflecting surface should be accurate to better than 1/16 wavelength (about 30 nm). Even then, the performance of the telescope is still much less than perfect.

Telescopes are commonly advertised with '1/8 wave' and '1/16 wave' optics. It is rarely stated whether this figure refers to the wavefront error at the focal plane or the accuracy of the surfaces of the optics (it is usually the latter, though the focal plane error is the one that really counts). Even then, there are a numbers of ways this figure can be calculated (maximum deviation from perfection, average deviation from perfection, root mean square (r.m.s.) error, etc). The r.m.s value ought to be no more than 0.08 wavelength as measured at the focal plane to ensure reasonable optical quality.

Also, manufacturers are sometimes coy about revealing the wavelength of light against which these accuracies are measured. After all, the wavelength of red light is more than 50 per cent greater than that of violet light! The 1/4 wave peak–peak or the 1/12 wave r.m.s. values that are desired are for a wavelength of 5×10^{-7} m (500nm) which is in the yellow-green part of the spectrum.

Obviously the manufacturers claims should be very carefully weighed up before settling on the purchase of a particular telescope. Personally, I would always put optical quality ahead of sheer size when choosing a telescope on a limited budget.

One of the chief advantages of the reflecting telescope over the refractor is the absence of any residual chromatic aberration (except that in the eyepiece, which should be minor). Another advantage is the cost. A given outlay will secure a good quality reflecting telescope of at least twice the aperture of the refractor. There are a number of designs of reflecting telescope. The type most commonly found in amateur hands is the Newtonian reflector. The vast majority of amateur telescopes are of this form, with the Cassegrain reflector as the next most popular. We will consider each of these types in turn.

The Newtonian reflector

The basic layout is shown in Figure 1.12(a). For a 6-inch (152 mm) $f/10$ mirror the difference between a spheriod and a paraboloid (the desired shape) is sufficiently small that the former (which is much easier to make) will give an image of adequate quality. Somewhat larger focal ratios are need for larger mirrors and so a 10-inch (254 mm) spheroidal mirror would need to be at least $f/12$. Most amateur-sized reflecting telescopes have primary mirrors in the range $f/4$ to $f/8$ and so their surfaces are made into paraboloids of revolution.

Referring to Figure 1.12(a), if it were not for the presence of the secondary mirror, which has an optically flat surface inclined at 45° to the optical axis of the primary mirror, the image would be formed at position I'. The intersection point of the primary's optical axis and the surface of the secondary mirror is placed a distance a inside the primary focus. The final focus, at I, is this distance from the intersection point but is now in an accessible position outside the telescope tube.

In order to deliver all the rays from the primary mirror to the centre of the focal plane, the secondary mirror has to have a width (the diameter of the minor axis for an elliptical flat) of at least aD/f, where D and f are the diameter and focal length of the primary mirror, respectively (each of these quantities measured in common units). If the focal plane image was viewed with an eyepiece the centre of the field of view would be fully illuminated but the image would be a little less bright away from the centre. In order to obviate this *vignetting* of the rays from the primary mirror the secondary mirror has to be made larger. To fully illuminate the focal

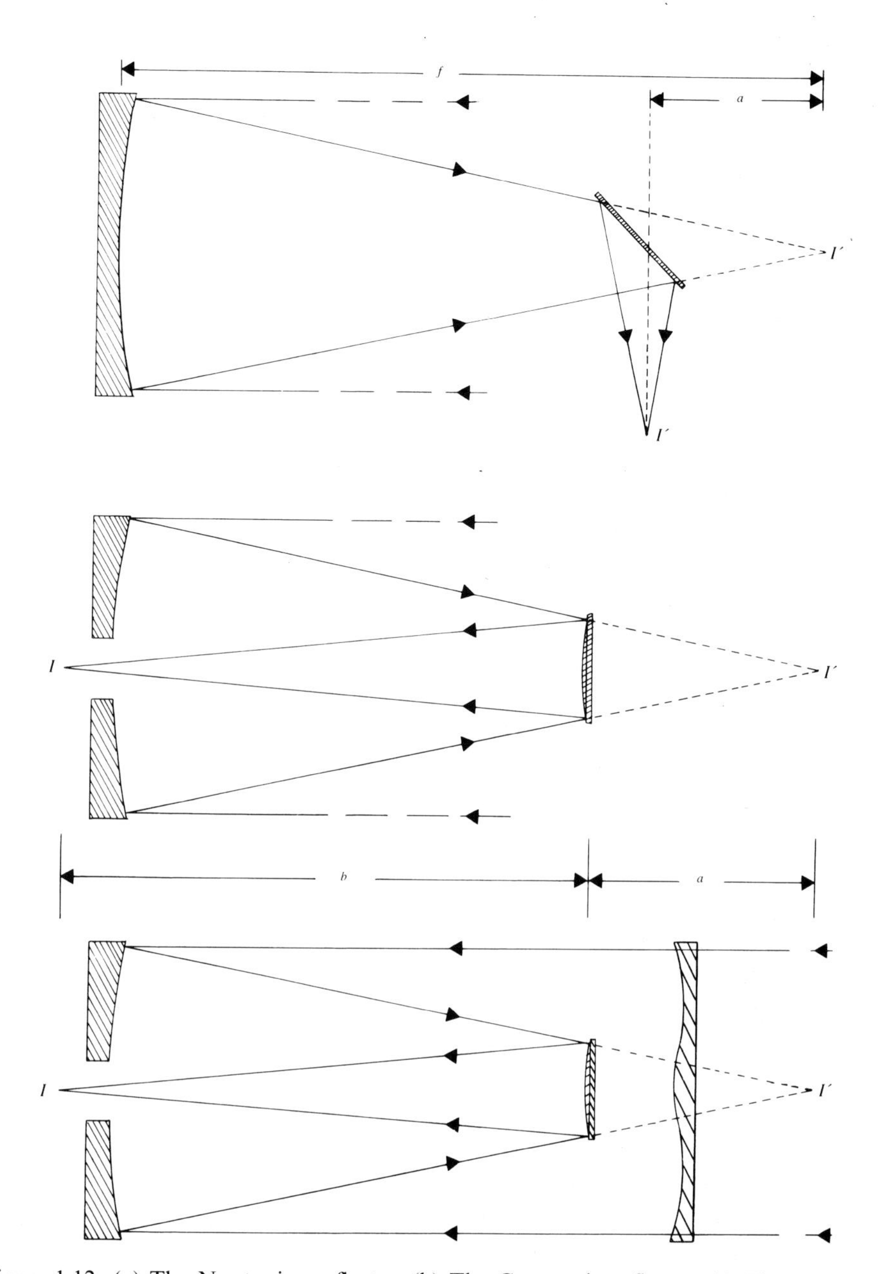

Figure 1.12. (a) The Newtonian reflector. (b) The Cassegrain reflector. (c) The Catadioptric telescope. The curve on the corrector plate has been much exaggerated in this diagram for clarity. Since the corrector plate has negligible optical power the formulae for calculating the effective focal ratio and the size and focal length of the secondary mirror are identical to that for the Cassegrain reflector. Notice that the corrector plate, though of 'Schmidt type', is placed close to the focus of the primary mirror and not twice the focal length from it, as in the case of the true Schmidt camera.

plane across a diameter d, the secondary mirror should have a width w given by:

$$w = d + \frac{a(D-d)}{f}$$

Commercial Newtonian reflectors are made with an elliptical secondary mirror which, typically, has a minor-axis diameter 25 per cent of the diameter of the primary. The diameter across which the field is fully illuminated is usually only a couple of centimetres (and could even be less than a centimetre for the smallest reflectors!). This vignetting is not usually troublesome when using the telescope for Moon or planet observation but could be important when using a wide-field eyepiece for variable star work. Comparison stars near the edges of the field would then be dimmed. Errors could similarly arise from brightness measurements made from a film exposed at the focal plane, not all of the film frame being fully illuminated.

Unfortunately increasing the size of the secondary mirror also has its drawbacks. One obvious disadvantage is that it reduces the light grasp of the primary mirror, though the obstruction is only 1/16 for a secondary which is a quarter of the width of the primary mirror. More important to the planetary observer is that the presence of the secondary (and its spider-support system) modifies the diffraction pattern structure. A perfect, *unsilhouetted* (unobstructed), optical aperture would produce a diffraction pattern with 84 per cent of the light energy concentrated in the Airy disc and 7 per cent of the energy in the first diffraction ring (with smaller amounts in each subsequent ring). The situation for different sized central obstructions is given in Table 1.1.

The effect of the central obstruction on actual resolution is minimal. However, larger secondaries do have an effect on the visibility of low contrast markings seen against a bright background. As a very rough guide, a central obstruction of 30 per cent of the diameter of the primary mirror causes a drop in contrast equivalent to an unobstructed aperture with 1/4 wavefront errors.

Owing to this effect, in the size range of amateur telescopes, a Newtonian reflector's aperture has to be about half as large again as that of a refractor in order to be as good at showing faint planetary markings. The qualifications here are that the refractor's secondary spectrum is minimal, the reflector's secondary mirror is the usual 25 per cent of the primary mirror diameter and both telescopes have high quality optics. The effect of the secondary mirror support vanes is a little different. These produce narrow spikes extending from the Airy disc.

Table 1.1. *The effect of a central obstruction on the diffraction pattern produced by a telescope (see text for details)*

Percentage central obstruction	Percentage of energy in Airy disc	Percentage of energy in 1st ring
0	84	7
5	83	8
10	82	9
15	80	11
20	76	14
25	73	17
30	68	22
35	64	26
40	58	30
45	53	33
50	48	35

A Newtonian reflector with an accurately parabolised mirror can produce nearly perfect images on axis (at the centre of the field of view) but the off-axis images suffer from coma. This limits the useful field of view, even if the secondary mirror is large enough to provide full illumination of the field, unless a correcting lens is installed. What is tolerable depends on the type of observation being conducted. For focal plane astrophotography (see Chapters 4 and 5) and low power visual observation I have found that the maximum useful field diameter, in degrees, is roughly given by:

$$\text{Maximum field diameter} = 450/D,$$

where D is the aperture of the telescope in millimetres. The focal length of my own 18¼-inch (464 mm) Newtonian reflector is 2.59 m. At the focal plane the image scale is 80 arcseconds mm^{-1} and the maximum useable field (about 1°) covers 45 mm. This is exactly the extent of the field fully illuminated by the secondary mirror ($a = 432$ mm, $w = 114$ mm, using the previous equation).

The Cassegrain reflector

In the classical form of this instrument the primary mirror is paraboloidal in figure and the secondary mirror has a hyperboloidal convex surface. Variants exist, notably the Dall–Kirkham and Ritchey–Crétien types which have differently shaped reflecting surfaces (for instance the Dall–Kirkham telescope has a prolate ellipsoidal primary mirror and a spherical secondary).

Figure 1.12(b) shows how the secondary mirror intercepts the rays collected by the primary mirror a distance a before they would reach a focus at I'. The secondary mirror then reduces the convergence of the rays and sends then axially back down the tube, a distance b, where they pass through a hole in the primary mirror to an accessible focus beyond the bottom end of the telescope tube (position I).

In reducing the convergence of the rays, the secondary mirror has the property of multiplying the effective focal length (and hence the effective focal ratio) of the primary mirror by a factor b/a. The effective focal length, E.F.L., and effective focal ratio, E.F.R., of the telescope are then given by:

$$\text{E.F.L.} = fb/a,$$

and

$$\text{E.F.R.} = Fb/a,$$

where f and F are the focal length and focal ratio of the primary mirror, respectively. Just as for the Newtonian reflector and the refractor, the final image of the Cassegrain reflector is 'astronomically normal', that is inverted.

The minimum diameter, w, of the secondary mirror needed to fully illuminate the centre of the field of view is Da/f, where D is the diameter of the primary mirror and f is its focal length. In order to fully illuminate the field at the focal plane out to a diameter d the diameter of the secondary mirror is given by:

$$w = a\left(\frac{D}{f} + \frac{d(f-a)}{bf}\right)$$

As an example let us consider a 400 mm aperture Cassegrain telescope with an $f/4$ primary mirror (1600 mm focal length) and an E.F.R. of $f/20$. Let us suppose that the focus is formed at a distance of 200 mm behind the surface of the primary mirror (Therefore $a+b=1800$ mm). The ratio $a{:}b$ must be 1:5 and so $a=300$ mm and $b=1500$ mm. If we desire that a 40 mm diameter patch of the focal plane be fully illuminated then the above equation predicts that the diameter of the secondary mirror should be 81.5 mm.

The focal length of the Cassegrain telescope's secondary mirror, f_s, can be calculated from:

$$f_s = \frac{ab}{(a-b)}$$

The secondary's focal length will come out as a negative number in the above equation, indicating that the mirror is convex. Using the foregoing data, our hypothetical 400 mm $f/20$ Cassegrain reflector would have to have a secondary mirror of focal length −375 mm.

The longer effective focal length inherent in the Cassegrain design is an advantage to the lunar and planetary observer since high magnifications can be achieved with long focal length eyepieces (which are much easier to use than their shorter focal length counterparts) and the image scale being much larger (fewer arcseconds per millimetre) is an advantage when attempting high resolution photography.

However, the large image scale is a disadvantage for the observer wanting a larger field of view. The barrel size of the eyepiece effectively limits the size of the field stop (eyepieces are dealt with in the next chapter) and this puts a limit to the maximum diameter of field that can be observed. If the field stop has a diameter of 40 mm then a field of diameter 0°.9 could be observed if this eyepiece were plugged into my 18¼-inch, $f/5.6$, Newtonian reflector. The same eyepiece would deliver a field of only 0°.3 diameter if plugged into our hypothetical 400 mm aperture $f/20$ Cassegrain telescope (though the magnification delivered would be higher). By virtue of its large effective focal ratio, the off-axis coma is negligible across the illuminated field of the Cassegrain telescope.

Although the Newtonian and Cassegrain telescopes are the most commonly owned types amongst amateur astronomers, a few practitioners have other types of reflecting telescope. However, these can only be custom-made by a telescope manufacturer (which would be very expensive), or constructed by the home doer (with rather less ease than the building of a Newtonian reflector). Of particular interest are the off-axis designs which obviate the need for silhouetted optics, such as the tri-schiefspiegler. The references given in Chapter 17 will allow interested readers to follow up these other designs.

Practical telescopes: the catadioptric

Telescopes which use both refracting and reflecting optics for their major parts are known as *catadioptric*. Originally the development of this type of telescope was pioneered by Bernard Schmidt. He set out to produce a wide-field camera of low focal ratio.

He succeeded by replacing the parabolic mirror of the reflecting telescope with a spherical mirror. This cured the trouble with outfield coma but, of course, introduced spherical aberration. In order to remedy this, he introduced a specially shaped corrector plate placed at twice the focal length of the mirror. The focal plane, half way between the mirror and the corrector plate, was curved but was almost totally free of both spherical aberration and coma. The fact that the corrector plate had negligible optical power meant that the amount of chromatic aberration it introduced was itself negligible (at least in the smaller sizes of instrument). Photographic plates, specially deformed to fit the focal surface, could then be placed at the focal plane to image very large portions of the sky. One of the largest examples of the *Schmidt camera*, the 1.2 m aperture on Mount Palomar, can photograph 6°×6° on 14-inch×14-inch photographic plates.

A number of telescope manufacturers have produced catadioptric telescopes for the amateur market. The most common type is the Schmidt–Cassegrain, illustrated in Figure 1.12(c). These telescopes, with an effective focal ratio of *f*/10, are very nicely made and are very compact and so easily portable. Despite being rather expensive, they are very popular.

However, I must say that the use of the spherical mirror and corrector plate is only truly necessary where a large field of view is being imaged. The *f*/10 effective focal ratio obviates this and so a conventional Cassegrain configuration would be cheaper. In addition, the corrector plate is rather difficult to manufacture with any great accuracy (one of its surfaces is flat, the other is a special shape called a quaternary) and so the final assembled telescope would probably not perform quite as well as a good quality Cassegrain of the same size and focal ratio. I have read test reports which seem to confirm this conclusion.

On the other hand, there is something to be said for an optical system which is totally enclosed. Certainly, the mirrors are protected from tarnishing and the telescope is much less likely to suffer from tube currents (convecting air which disturbs the light rays passing along the tube). However, a conventional Cassegrain or Newtonian reflector could have the open end of its tube closed with a disc of glass (each side of which is optically flat) for rather less total cost than a Schmidt–Cassegrain telescope. Another slight variant is the Maksutov telescope, though in the size range usually encountered there is little to choose between this type and the Schmidt–Cassegrain.

2

Atmosphere, seeing, magnification and eyepieces

The Earth's atmosphere poses a constant problem to the work of the astronomer. Not only does it absorb much of the radiation arriving from space; it also limits the fineness of detail that can be resolved by a telescope.

Atmospheric turbulence

Temperature gradients cause convective cells to develop in the atmosphere from ground level up to a height of about 7 km. These swirling pockets of air of differing density (and so differing refractive index) disturb the light arriving from space. The arriving rays suffer small changes in direction, as a result. The convective cells typically have diameters ranging from 10 to 20 cm. It is the motion of these in the line of sight of an observer that causes the stars to appear to flash and twinkle (or *scintillate*).

The effect that scintillation has on the image formed by a telescope partly depends on the size of that telescope. If it has an aperture not much bigger than the prevalent size of the convective air cells, then the image of a star will remain fairly sharply focused but will perambulate around its mean position in the focal plane. If the aperture of the telescope is much larger, then a number of different convective cells contribute their random effects to the light arriving at the telescope at any given time. The star image will then be less mobile but will be blurred out into a *seeing disc* roughly equal in radius to the amplitude of the motions as seen in the smaller telescope.

Several mountain-site observatories experience sub-arcsecond seeing on the majority of clear nights. The rest of us have to put up with rather poorer seeing conditions. At Herstmonceux, the old site of the Royal Greenwich Observatory, near my Sussex home the seeing has been measured as averaging around two to three arcseconds. Observing from the garden of my home I find that the image perambulations are of this order of magnitude (if anything, larger), whilst I can usually glimpse detail on the 1 arcsecond scale when using my 18¼-inch reflector. On some nights the seeing is considerably worse and the image can perambulate by 20, or even 30, arcseconds. On rare nights of excellent seeing I have glimpsed details at the resolving limit of my telescope.

The quality of the image can change very rapidly during an observing session. This can be due to the telescope and its immediate surroundings changing temperature, or to actual variations in the atmosphere. Even the type of seeing can be variable. Sometimes the image is disturbed by slow undulations, while remaining fairly sharp. At other times the image is steady but defies a sharp focus. More often the image is affected by a combination of high-frequency and low-frequency undulations and general softening.

Observers should always record the seeing conditions and the atmospheric transparency at the time of the observation (and make a note of any changes during the observation period). Over the years various scales have been adopted for registering the seeing conditions. Perhaps the best (certainly the most widely adopted) for general use is the Antoniadi scale. Officially, it runs as follows:

ANT. I Perfect seeing, without a quiver.
ANT. II Slight undulations, with moments of calm lasting several seconds.
ANT. III Moderate seeing, with larger air tremors.
ANT. IV Poor seeing, with constant troublesome undulations.
ANT. V Very bad seeing, scarcely allowing the making of a rough sketch.

The difficulty here is that near perfect seeing for a small telescope may well be very poor for a large one. My own criteria for telescopes larger than 10-inches aperture are:

ANT. I Can consistently see detail finer than 0.5 arcsecond (or a steady and nearly perfect image in a telescope of 10-inches aperture)
ANT. II Can, for most of the time, see detail finer than 0.7 arcsecond.
ANT. III Can, for most of the time, see detail finer than 1 arcsecond.
ANT. IV Can, for most of the time, see detail finer than 1.5 arcsecond.
ANT. V Cannot consistently see any detail finer than 1.5 arcsecond.

Of course, the estimate of the seeing made at the eyepiece can only be rough. In addition, any qualifying statements that pertain to the accuracy of the observation should be noted at the time.

To stop or not to stop?

The fact that atmospheric turbulence may render details smaller than 1 arcsecond invisible may strike some as a very good reason for not investing in anything larger than, say, a 6-inch telescope. However, the larger telescope will always score when it comes to light grasp and the apparent contrast of the viewed image at any given magnification.

A controversial subject is the use of stops on large telescopes, to reduce their effective aperture. Many observers claim that stopping the telescope down to a few inches clear aperture produces a marked improvement in image quality. Others state that using stops can only reduce the quality of the image and certainly never improve it (except where the telescope is poor quality, when stopping down reduces the visibility of the imperfections present in the optical system).

My own experience is that stopping down a telescope can **sometimes** 'clean up' a flaring and confused image. For its use on a reflector, positioning the hole in the diaphragm such that it avoids the central obstruction and spider support vanes then maximises the quality of the image. I have found that stopping my largest telescope down to 6-inches off-axis certainly does seem to sharpen the image and allow faint planetary markings to be seen more easily on the worse nights. A suitable cardboard mask is easy to make and could be kept in readiness for trial on turbulent nights. I would caution that the stop only be used if there is a definite improvement in the image quality. Otherwise the finer details that might be glimpsed in better moments of seeing would remain unobserved, due to the diffraction limit imposed by the stop.

The minimum magnification to see the finest detail

What magnification is needed to show the finest detail a telescope can show? Let us take, as an example to work on, an image that contains detail as fine as 1 arcsecond in angular extent. This image must be magnified so that the finest details in it can be appreciated by the eye of the observer. From experiments I have carried out on myself and a large number of teenage students, I have found that 100 arcseconds is a fair average for the finest detail that can be resolved by the unaided eye. This figure undoubtably becomes larger with age. This implies that a magnification of $\times 100$ is necessary for the observer in our example to see all the detail that the telescope can show. It also implies that there would be little point in the observer switching to a higher magnification.

If the resolution of the telescope is effectively limited to 1 arcsecond (the diffraction limit for a 5-inch telescope) then magnifications greater than

×100 would have little advantage in seeing finer detail irrespective of how large the telescope is. This figure may strike you as being rather low but I have found it to be a typical value from experiments with telescopes of up to 36-inch (0.9 m) aperture. Of course, this figure is the **minimum** magnification to see detail as fine as 1 arcsecond. It is not the maximum that can be used.

When, as a young boy, I began observing I strove for the highest magnifications possible but as my experience increased I found that I could see much more with a smaller, sharper, brighter and more contrasty image. For 'normal' observing of the Moon and planets I now prefer a power of about twice the minimum necessary to see the finest detail under the given conditions. My most used eyepieces are those that deliver powers of ×144 and ×207 on my 18¼-inch reflector. On better than average nights more power is advantageous and under perfect conditions a power of about ×20 per inch of aperture is necessary to see the finest detail (i.e. ×365 with my 18¼-inch telescope). On odd occasions I have preferred to use powers as high as ×432, or even ×576, but those instances have been very few and far between.

Of course, the magnification you would prefer also depends on your acuity of vision. There is a simple test you can perform to determine the resolving power of your own eye. Fill in two squares on a piece of millimetre-ruled graph paper, leaving one clear square in between. Attach it to a wall at the same height as your eyes above the ground. Back away from the graph paper and measure the distance at which you can only just see the squares as separate. If this distance, measured in millimetres, is D, then the following formula can be used:

$$R = 2 \times 206265/D$$

The 2 is for the 2 millimetre gap between the centres of the filled-in squares on the graph paper. The limiting resolution of your eye, R, is then in arcseconds.

Long- or short-sighted people may keep their glasses on for this test, even if not worn at the telescope, since the correction is effectively made when focusing the telescope eyepiece. Astigmatism is another matter as cylindrical power is necessary to correct for this eye defect – more magnification will be needed at the telescope if correction is not applied by the glasses or a separate cylindrical correction lens.

Eyepiece field of view

In all eyepieces the image quality falls off to some extent with greater distance from the centre of the field of view. Manufacturers mask off the images of unacceptable quality by means of a circular aperture, known as a *field stop*, positioned in the eyepiece's focal plane. As you peer through the eyelens of an eyepiece, the angle through which you would have to swivel your eye to look directly from one edge of the field across the diameter to the other edge is known as the *apparent field* of the eyepiece. Depending upon the design, eyepiece apparent fields lie in the range 25° to 85°.

The other quantity of interest to the observer is the value of the *real field*. This is the actual angular extent of the sky that the observer can see when using a particular eyepiece–telescope combination. A rough figure for the size of the Moon is 32 arcminutes. If the real field of the eyepiece–telescope is bigger than this value then all the Moon can be seen at once. If the real field is smaller then the observer will only be able to see part of the Moon at any one time.

The value of the real field obtained depends upon the values of the apparent field of the eyepiece (fixed by the manufacturer) and the magnification that it produces with the telescope. The relationship is:

$$\text{real field} = \text{apparent field/ magnification.}$$

The values of real and apparent field should be expressed in the same units (for instance both in degrees, or both in arcseconds – etc).

One can find the value of the real field directly by timing the passage of a star across the centre of the field of view from one edge to the other, with the telescope drive switched off. This is repeated several times to get an average. The diameter of the real field, in arcseconds, is then given by:

$$\text{real field} = 15t \cos\delta,$$

where t is the time taken for the star to cross the field of view measured in seconds, and δ is the declination of the star measured in degrees (ignore the negative signs of southerly declinations). What are the values of real and apparent field for each of your eyepieces?

Types of telescope eyepiece

Eyepieces for amateur telescopes tend to come in three standard sizes – those which fit drawtubes of diameters 24.5 mm, 1¼-inch (31.7 mm) and 2-inch (50.8 mm). Eyepieces with values of apparent field in the range 35° to 55° are considered medium-field, those smaller than this range are narrow-field and those larger are wide-field eyepieces. For an eyepiece of apparent field 57° the field stop aperture has the same diameter as the focal length of the eyepiece. This means that wide-field eyepieces of long focal length can only be accommodated in wide-barrel bodies. For instance the field stop aperture for a 36 mm focal length eyepiece of 65° apparent field is 41 mm. In this case the manufacturer would construct the eyepiece to fit into a 2-inch diameter drawtube.

Most modern eyepieces have external markings which indicate their type and focal length (eg. H 20 mm, Or 12.5 mm, etc). The eyepiece types most commonly available, with their designations, are: Huygenian, H; Huygens–Mittenzwey, HM; Ramsden, R; Achromatic (or 'special') Ramsden, SR; Kellner, K; Orthoscopic, Or; and the Plössyl (usually named in full).

A couple of eyepiece types which are not commonly available but which are especially suitable for planetary observation are the Monocentric and the Tolles. There is a large variety of wide-field eyepieces available. The most common of these is probably the Erfle but the best is surely the Nagler. The various eyepiece types are detailed in the following sections.

The Huygenian eyepiece

Refractors and small reflectors are often supplied with this type of eyepiece. In its classical form it consists of two plano-convex lenses, arranged with their flat sides facing the observer's eye (Figure 2.1(a)). Although considerable variations exist, it is normal for the eyelens (the lens closest to the eye) to have a focal length about one third of that of the field lens. The separation of the lenses is then twice the focal length of the eyelens. The field stop is placed at the focus of the eyelens, and thus in between the two lenses. This makes a Huygenian a *negative* eyepiece. Most other eyepieces have external focal planes and are known as *positive* eyepieces. Apart from the Huygens–Mittenzwey and Tolles types, all the other eyepieces considered in this chapter are positive.

A positive eyepiece can be used with a filar micrometer, provided its barrel allows the lenses to come close enough to the micrometer wires (when focused the wires will be in the position normally occupied by the field stop). A negative eyepiece, having no external focal plane, cannot be used with a micrometer without the use of an additional relay lens positioned between the wires and the eyepiece.

It has often been said that Huygenian eyepieces cannot be fitted with crosswires but this is not entirely true. Dismantling the eyepiece will allow access to the field stop and crosswires (the finest copper wire is a suitable material) can be glued in position. However, I must add that eypiece dismantling and cleaning should only be attempted if you can maintain scrupulous cleanliness and you know exactly what you are doing.

A typical Huygenian eyepiece will have a relatively narrow-field of view (apparent field about 30°) and field curvature often results in a very marked deterioration of the image quality near the edges of the field of view. This eyepiece is usually fairly well corrected for chromatic aberration, though the amount of spherical aberration is usually large. It is often said that Huygenian eyepieces will give poor results if the focal ratio of the telescope is less than *f*/10. However, the continental Huygenian eyepieces commonly available usually have 'crossed' eyelenses. In other words, biconvex lenses are used instead of the classical plano-convex form. With this modification the Huygenian does work a little better with shorter focal ratio's. I have had satisfactory results with an *f*/8 telescope.

The *eye-relief* (distance between the exit pupil and the eyelens) is usually about 30 per cent of the focal length for the Huygenian type of eyepiece. If the eye-relief is too small the observer will have difficulty in positioning his eye to see the whole of the field of view. I have found that a 6 mm Huygenian is relatively easy to use but a spectacle-wearer would need an eyepiece providing longer eye-relief.

The Huygens–Mittenzwey eyepiece

This is a variation on the Huygenian type, with a converging meniscus field lens and a slightly closer lens spacing than before (see Figure 2.1(b)). It has an apparent field of 45° to 50°, though the gain in field size is at the expense of the corrections for chromatic and spherical aberration. I once acquired a 9 mm focal length HM eyepiece with a 60 mm *f*/15 refractor. It gave tolerable images with this telescope but proved completely useless with my other instruments.

The Ramsden eyepiece

In this type, two plano-convex lenses of equal focal length are placed with their curved sides facing each other (Figure 2.1(c)). For good chromatic and spherical aberration correction the separation of the lenses should be equal to their focal length. The trouble with this is that the eyepiece would have zero eye-relief and that any dust on the field lens would be in sharp focus. In practice a compromise is reached with the lenses separated by two-thirds of their focal length. The resulting eye-relief is then 30 per cent of the focal length and any dust on the field lens is less easily focused. Chromatic and spherical aberration are markedly increased.

On a practical trial I have found that my 1-inch and ½-inch focal length

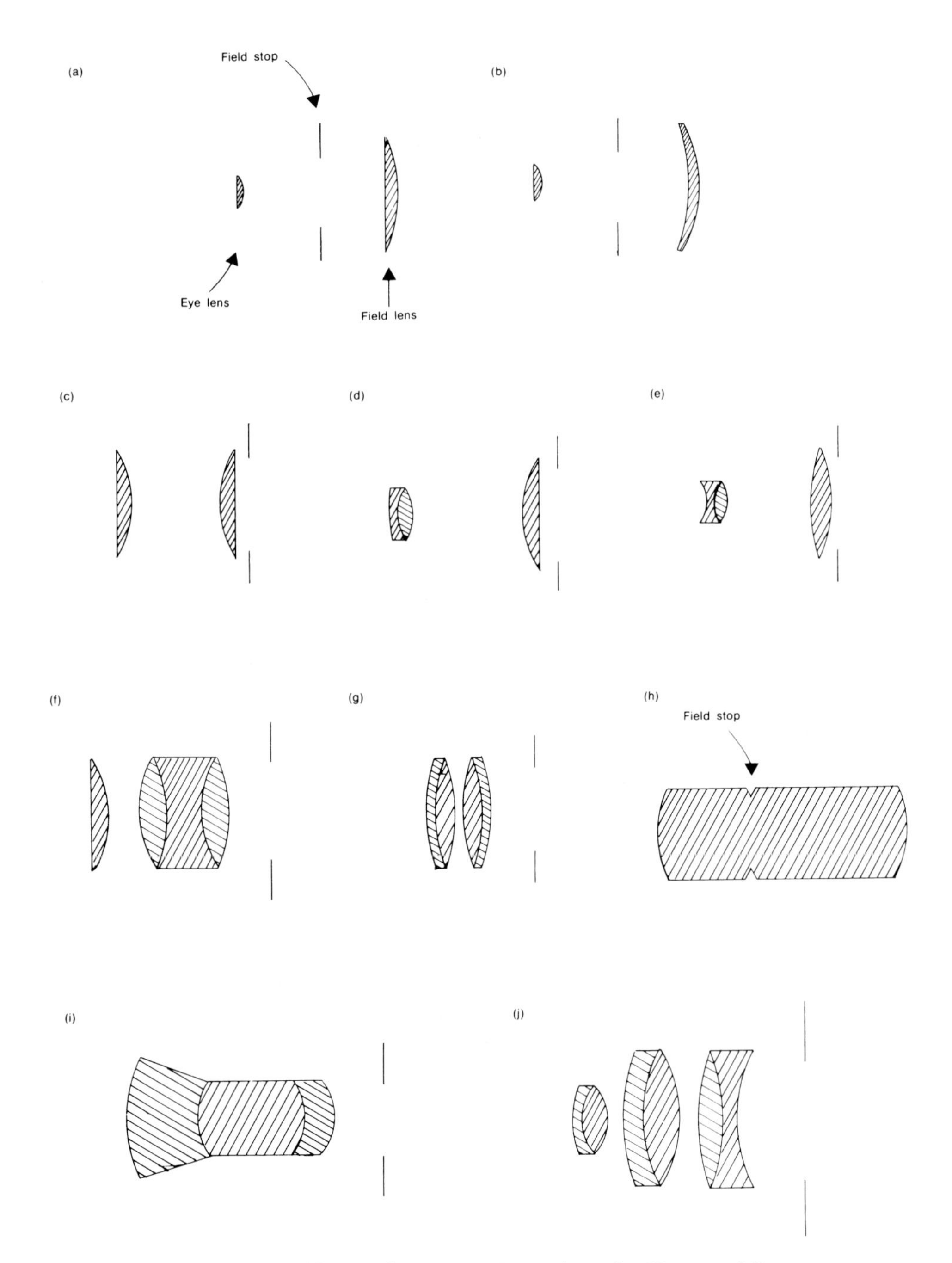

Figure 2.1. Telescope eyepieces: (a) Huygenian; (b) Huygens–Mittenzwey; (c) Ramsden; (d) Achromatic Ramsden; (e) Kellner; (f) Orthoscopic; (g) Plössyl; (h) Tolles; (i) Monocentric; (j) Erfle.

Ramsden eyepieces work extremely poorly with my 18¼-inch, *f*/5.6 reflector. The images they deliver defy sharp focus and are fringed with colour. However, the image is sharpened by the addition of a deep yellow filter. Even with my 6¼-inch reflector (which is an *f*/7.6) the performance leaves much to be desired. It seems that the telescope should have a focal ratio of at least *f*/10 for this type of eyepiece to be used successfully. Ramsden eyepieces usually have apparent fields of about 35°.

The Achromatic Ramsden eyepiece

This is an improvement on the Ramsden type, the eyelens being an achromatic doublet (Figure 2.1(d)). It is thus a three-element eyepiece. Chromatic aberration is much reduced and the apparent field of view is usually larger than the Ramsden type. They can deliver good images with telescopes of focal ratios down to about *f*/7.

The Kellner eyepiece

Shown in Figure 2.1(e), this is an improvement on the Achromatic Ramsden type. The apparent field is similar, at 40° but the corrections for the various aberrations are superior. I find that a 25 mm Kellner eyepiece produces good images with my 18¼-inch telescope but a 12.5 mm delivers a very slightly softer view than does my Orthoscopic of the same focal length. The eye-relief of this eyepiece type is rather small, making it less suitable for high magnifications.

The Orthoscopic eyepiece

This is an excellent general purpose eyepiece. I have a number of this type, ranging from 30 mm to 6 mm focal length and I find that they all deliver very good high contrast views with all of my telescopes. The definition is excellent over the 40° to 45° apparent field of view. The usual form is a single, plano-convex, eyelens and a triplet field lens. Varients exist, some giving wider fields of view. They can be used successfully on telescopes of very low focal ratio, even down to about *f*/4.5. Eye-relief is typically around 75 per cent of the focal length, making them easy to use even in very short focal lengths.

The Plössyl eyepiece

This type of eyepiece is now becoming very popular. The eye-relief is very large (around 80 per cent of the focal length) and the corrections for the various aberrations are virtually as good as for the Orthoscopic. Indeed, the latest generation of Plössyls are, if anything, slightly better than many current commercial Orthoscopics. They deliver good images over apparent fields of 50°–55°. I have a 44 mm focal length eyepiece of this type and can testify to its high quality.

The Tolles eyepiece

This is a single element eyepiece. It is little more than a cylinder of glass with curved ends. In effect, it is a solid form of the Huygenian eyepiece, a groove cut around the glass forming the field stop. Although not easily available nowadays, this type of eyepiece was a firm favourite among planetary specialists. It gives crisp and false colour free images when used on telescopes of focal ratio larger than about *f*/6. The eye-relief of this type of eyepiece is very small, as is the apparent field of view (typically 25°).

The Monocentric eyepiece

Nowadays a rarity, it was another firm favourite with planetary observers. It is a cemented triplet lens system, delivering crisp and high contrast images over an apparent field of similar size to the Tolles but it is usable with telescopes of slightly lower focal ratio. The eye-relief of a Monocentric eyepiece is typically about 80 per cent of its focal length.

Wide-field eyepieces

There are a large number of different designs of wide-field eyepiece. They usually have five or more lens elements and are very much more expensive than the simpler types, though most of them give less critical definition. In effect, quality has been compromised for field size. Another negative factor is that each lens element in an eyepiece adds to the amount of light which is scattered and absorbed, so reducing contrast, or even giving rise to ghost images of bright objects. However, it is true that the anti-reflection coatings (*blooming*) given to lenses tends to negate this effect.

A common wide-field eyepiece design is the *Erfle* (see Figure 2.1(j)). It has six elements and produces an apparent field of about 65° (larger in some variants). The eye-relief is usually around 50 per cent of the focal length. A number of manufacturers market eypieces which are really just 5-element, or 6-element variations of the Erfle. They all have apparent fields of about 65° and give moderate performances with telescopes of focal ratios down to about *f*/5.

A recent development has been the *Nagler* 'Series 1' and 'Series 2' eyepieces, produced by Tele Vue, in America. The Nagler 'Series 1' eyepieces are composed of 7 elements and have an 82° apparent field. They deliver high quality images and are currently manufactured in focal lengths of 4.8 mm, 7 mm, 9 mm and 13 mm. The Nagler 'Series 2' eyepieces are an 8-element design available in focal lengths of 12 mm, 16 mm and 20 mm. The 'Series 2' Naglers have the same awesome apparent fields but improved exit pupil corrections. Both types have larger eye-reliefs than Erfles of the same focal lengths and will cope well with telescopes of focal ratio as low as *f*/4.

Another impressive wide-field design is the 40 mm focal length König 7/70. This is a 7-element eyepiece having an apparent field of 70° and the largest possible real field that can be fitted into a 2-inch barrel. To avoid the wastage of light by an over-large exit pupil, this eyepiece should be used on telescopes of focal ratio no less than *f*/6.

Before closing this section I must mention the Pretoria eyepiece. This is a medium-field (50°) eyepiece of 28 mm focal length which is specifically designed to counteract the outfield coma inherent in low focal ratio Newtonian reflectors. It has a moderate eye-relief and can deliver pin-sharp images over the full field of view. However, the coma correction is optimised for *f*/4 telescopes and so the eyepiece produces images with outfield reverse-coma with telescopes of larger focal ratio.

The Barlow lens

The common Barlow lens is a diverging achromatic doublet which can be used to increase the focal length (and hence the focal ratio) of the telescope. It is usually mounted in its own metal tube (or series of tubes), with the lens at the bottom and an adapter to take eyepieces at the top. In this form the amplification factor is fixed (usually at ×2). Most commercial Barlow lenses are of this type. However, if the separation between the Barlow lens and the eyepiece can be varied, then so can the amplification factor delivered by it (see Figure 2.2). On a note of caution, most Barlow lenses are optimised for just one value of amplification factor and using them otherwise incurs increasing the aberrations introduced by the lens.

A 2× Barlow lens effectively turns an *f*/6 telescope into an *f*/12 one, and so allows the simplest (and cheapest) eyepieces to deliver good images. However, it is also true that the aberrations (particularly chromatic aberration) introduced by the lens itself become increasingly prominent when it is used on telescopes of low focal ratio. I have found that a conventional doublet Barlow lens used on an *f*/6 telescope produces enough lateral chromatic

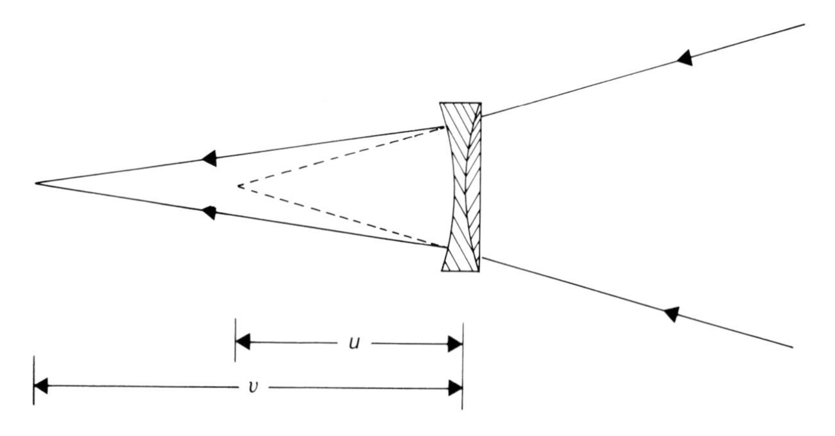

Figure 2.2. The Barlow lens. The lens is positioned at a distance u before the rays from the telescope reach a focus. The Barlow's diverging power causes the rays to be brought to focus at a new position, a distance v from the lens. The amplification factor of the lens is then given by:

$$a = -\frac{v}{u}$$

(The negative sign arises because of an optical convention). An alternative formula is:

$$a = v\left(\frac{1}{u} - \frac{1}{f}\right),$$

where f is the focal length of the Barlow lens (a negative number by convention). The minimum diameter, D, of the lens necessary to avoid vignetting is given by:

$$D = \frac{1}{u}\left(\frac{v}{F} + d\right)$$

where d is the required linear diameter of the fully illuminated field at the focus and F is the focal ratio of the telescope without the Barlow in position.

aberration to be noticeable with eyepieces of focal lengths shorter than about 1 cm. A better quality Barlow lens, composed of three or more lens elements, improves matters and allows good images with short focus eyepieces, even with very low focal ratio telescopes. Be careful that the Barlow lens does not vignette the light reaching the focal position. It should have a minimum diameter determine by the equation given in the caption to Figure 2.2

Transmission of optics and atmosphere

Figure 2.3(a) shows the variation of the transmission of the atmosphere with wavelength. The graph represents the transmission at the zenith, as measured from sea level. Figure 2.3(b) shows the variation of transmission (strictly, just

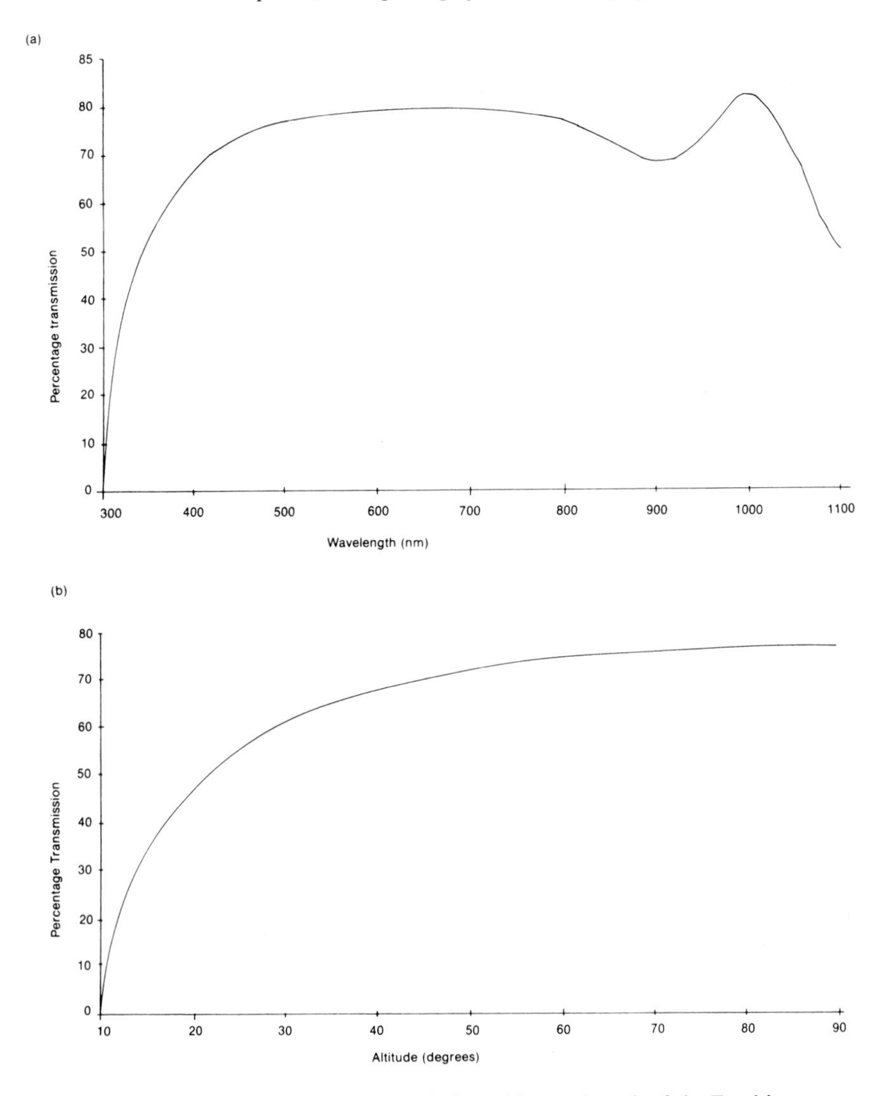

Figure 2.3. (a) The variation of transmission with wavelength of the Earth's atmosphere at the zenith. This is a plot that would be obtained on a very good night from measurements made close to sea level. The transmission at the extremes of wavelength are particularly affected by changes in conditions, hence this plot can only be used as a guide. (b) The variation of transmission of the Earth's atmosphere with altitude (angle above the horizon), as would be measured from a sea level site on a good night. In practice, the prevailing conditions may cause large deviations from this plot, particularly for low values of altitude. Also, this graph only strictly represents the changes in transmission for visual wavelengths (around 5.5×10^{-7}m).

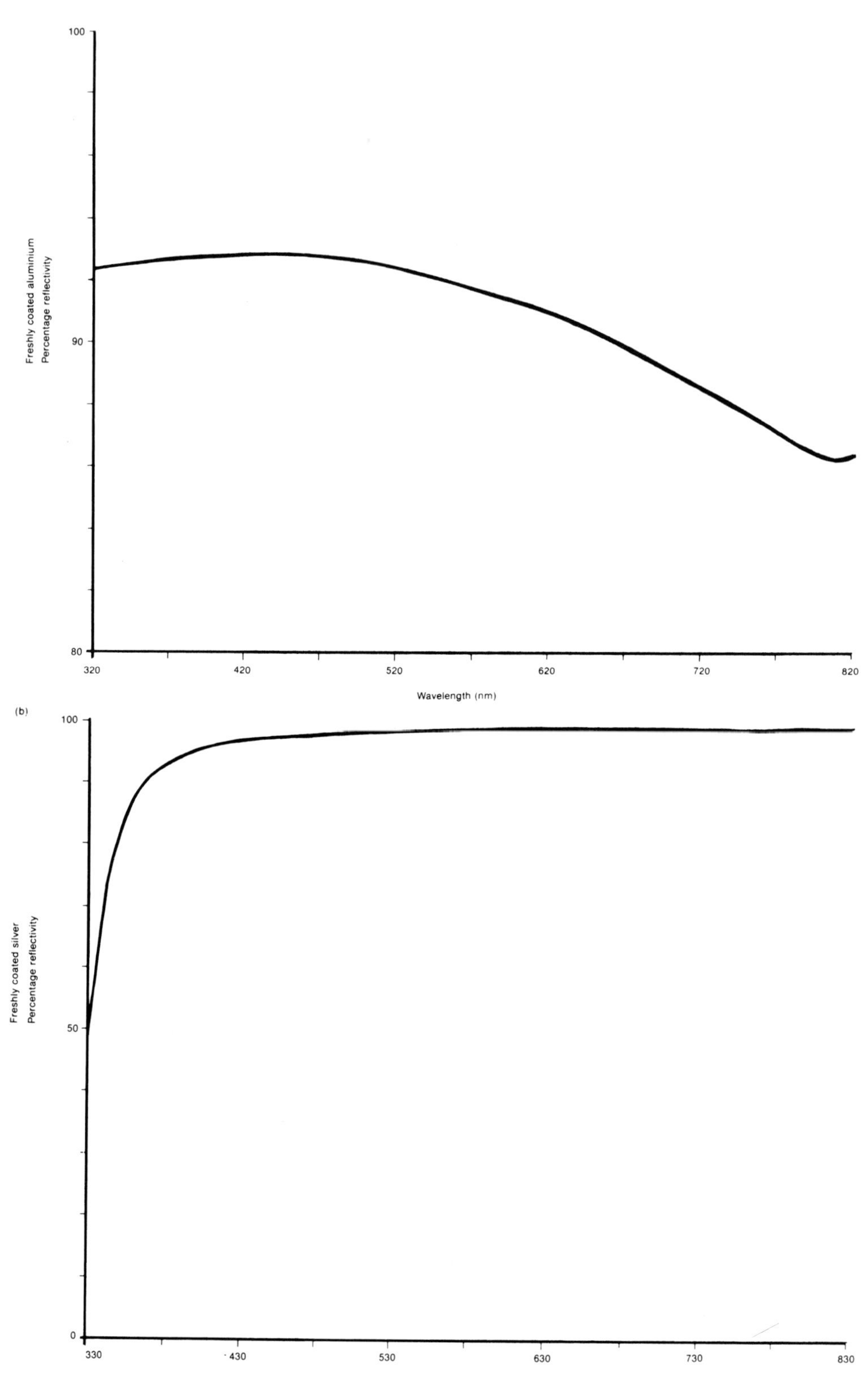

Freshly coated aluminium
Percentage reflectivity
100
90
80
320
420
520
620
720
820
Wavelength (nm)
(b)
Freshly coated silver
Percentage reflectivity
100
50
0
330
430
530
630
730
830
Wavelength (nm)

for a wavelength of 5.5×10^{-7}m) with altitude above the horizon. Both graphs represent measurements taken on an excellent night, with stars of magnitude 6.5 being visible at the zenith to a keen-eyed observer.

Of course, the light energy reaching the focal plane of the telescope also depends upon its own efficiency. Figure 2.4 shows how the percentage of incident light reflected at freshly vacuum-deposited coatings of aluminium and silver varies with wavelength. The process of vacuum-coating the reflective film with a quartz overcoat is highly desirable to give it a long effective life. Aluminium is very reactive to salt and acids, forming a grey tarnish. Silver is much less reactive to these pollutants but it is particularly prone to oxidation when damp, the resulting tarnish being coloured brown.

Some amateurs have their telescope mirrors resilvered by local domestic-mirror making firms for cheapness. These firms almost invariably silver their mirrors by chemical deposition. In this case the graph of reflectivity versus wavelength will resemble that of the vacuum-deposited silver, though the processes differ and so I cannot provide a definitive graph. It should be noted that these firms either leave the silver film uncoated, or else deposit a thickness of varnish over its surface.

The transmission graphs of glasses are rather more problematical, since such a large variety of types are used in telescopes and the auxiliary equipment. However, it is fair to say that most glasses absorb about 1 per cent of the incident light energy per centimetre of thickness for wavelengths between 500 nm (5×10^{-7}m) and 700 nm. This figure only slightly increases for longer wavelengths in the range of interest to the amateur astronomer (up to 1100 nm for those using CCDs). Glasses vary widely in their transmissions at wavelengths shorter than 500 nm, generally being rather more opaque. However, most absorb less than 6 per cent per centimetre, even at 360 nm. At wavelengths shorter than this most glasses absorb heavily.

Also, all uncoated air–glass lens surfaces will reflect away about 4 per cent of the light falling on them. A single-layer blooming reduces the reflection to less than 2 per cent over the visual range (400 nm to 700 nm) and multicoating reduces this even further. When calculating the transmission efficiency of an optical system remember to take into account **all** the optical components. Also allow for any obstructions in the light path, such as the secondary mirror of the reflector.

Figure 2.4. Graphs of the percentage reflectivity of vacuum-deposited aluminium and silver films, versus wavelength. Note the poor reflectivity of the silver film in the near ultraviolet part of the spectrum. Reproduced with kind permission of Dave Jackson of the Royal Greenwich Observatory.

3

Telescope hardware and adjustments

My advice to anybody buying a telescope is to invest in quality rather than sheer size. This applies to the telescope's mounting as well as to its optics. One is able to do so much more with a telescope that is firmly mounted and has an accurate drive.

Nowadays fewer and fewer amateurs make their own telescope optics. However, an ever increasing trend is to install professionally made optics in a home-made mounting. Most telescope manufacturers supply telescope parts, as well as the completed instruments. The home constructor with limited workshop facilities can then make most of the telescope, apart from the optics and one or two of the more 'difficult bits'.

This chapter details some of the major aspects of telescope hardware applicable to the advanced amateur astronomer. Basic knowledge is assumed. Owing to the variety of equipment design and materials in use, no specific contructional details can be given (this subject would require a large book of its own) but the reader is referred to the listing of books and articles given in Chapter 17. In addition, certain aspects of telescope equipment are considered in other relevant chapters of this book. For instance, guide-scopes for astrophotography are dealt with in Chapter 5.

Telescope tubes and baffles

A telescope's tube should maintain the optics in their correct spatial relationship whatever its orientation. It should also be difficult to induce vibrations within it. Further, any induced vibrations should be rapidly damped out. Satisfying these conditions allows for a large number of designs of telescope tube.

Figure 3.1 shows my own 18¼-inch (464 mm) *f*/5.6 Newtonian reflector. The wires that criss-cross its open framework tube are attached to threaded

Figure 3.1. The author's 18¼-inch Newtonian reflector, formerly at Seaford.

rods near the tube's lower end. These keep certain parts of the tube in tension and others in compression. This makes it much more resistant to vibration than would otherwise be the case. By the same token, this also allows for a light-weight design. The tube of my telescope, once stripped of its fittings, has a mass of about 50lb (22kg). In its turn, this saving of weight allows rigidity to be obtained more easily in the mounting.

Figure 3.2 (a) shows Ron Arbour's 16-inch (406 mm) Newtonian reflector with its original tube. He has built a replacement (Figure 3.2(b)) for it, based on the 'Serrurièr truss' design and has found this to be much more rigid. This tube design is the basis of most large professional telescopes and is to be highly recommended.

All astronomers agree on the need for rigidity. However, one point about which many amateurs disagree is whether a reflector's tube should be 'solid' or open frame-work. Unless it is long in comparison to its diameter, rigidity for solid tubes is achieved with much less weight than is the case for open-tubes. Another factor to take into account is baffling against extraneous light. A solid tube is very efficient in this respect if its inside surface is rendered non-reflective. Automobile 'underseal' is particularly good for this, as is 'lampblack' paint. However, should the inside of a solid telescope tube be left at all shiny then considerable trouble will be experienced from light glancing off the tube walls.

(a)

(b)

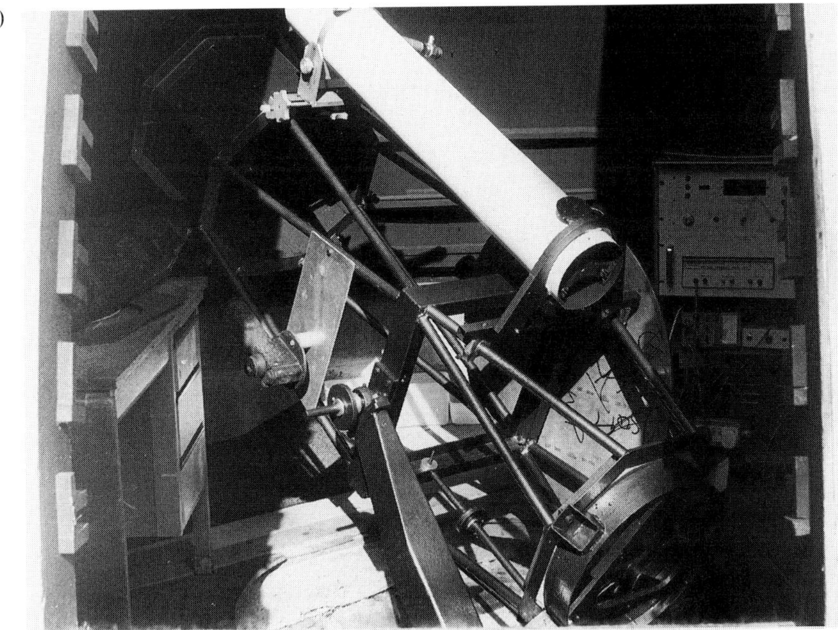

Figure 3.2. (a) [left] Ron Arbour's 16-inch Newtonian reflector with its original tube. (b) [above] The same telescope with its replacement Serrurièr truss tube, which has proved to be much more rigid.

Ideally, the only light that should reach the focal plane of the telescope is that collected by its object glass or primary mirror. Even if there is no artificial light around, it is surprising just how detrimental the glow from the night sky is when it enters the telescope drawtube. With my open-tubed telescope I attach a large black painted card on the side opposite the eyepiece focusing mount when observing faint objects. The secondary mirror of a Newtonian or Cassegrain reflector adds to the shielding, as does the drawtube.

A Cassegrain telescope usually has a 'stove-pipe' baffle extending up through the hole in the primary mirror. However, this additional baffling must not vignette the rays arriving from the outer parts of the mirror. Also, many drawtubes are far too long (especially for low focal ratio telescopes) and drastically reduce the illumination of the outer zone of the field of view. The diameter, *d*, of the fully illuminated field at the focal plane can be calculated using the approximation:

$$d = D - \frac{L}{f}$$

where *D* is the diameter of the drawtube/baffle tube, *L* is the distance of its

(a)

Figure 3.3. (a) [left] The author's $8\frac{1}{2}$-inch Newtonian reflector at Bexhill-on-sea. (b) [above] Notice the access door to the primary mirror cover.

far end from the focal plane and f is the effective focal ratio of the telescope. D, L and d are expressed in the same units, though it does not matter what those units are (f has no units, since it is a ratio).

Many people argue against solid telescope tubes on the grounds of *tube currents*. If the inside surface of the tube is at a different temperature to the air a convection current is set up. If the telescope tube is cooler than the ambient air temperature, the air close to the tube wall is cooled and descends while the displaced air moves up the middle of it. The reverse is the case if the tube is warmer. Similar effects are produced by the optics and other fittings within the tube. Indeed, the temperature lag of a reflector's primary mirror may well be the chief source of trouble. The moving masses of air of differing temperature (and hence differing density and refractive index) disturb the passage of light in the telescope and degrade the image.

Open-tubes do not suffer from convecting air currents flowing along them. Also, the air currents generated by warm optics and fittings pass out directly through the open-tube-work, minimising their deleterous effects. I have an $8\frac{1}{2}$-inch (216 mm) reflector which has a solid tube, but has removeable protective covers over the mirrors. A door is positioned close to the primary mirror and on the upper (north) face of the tube to allow access to its covering 'lid' (see Figure 3.3). I have found that on many nights the

Figure 3.4. Martin Mobberley's 14-inch Cassegrain–Newtonian reflector. Note its partially closed tube.

images are rather poor with the door closed. The effect is similar to astigmatism and the intrafocal and extrafocal images (see later in the chapter) betray the obvious evidence of a tube current. Opening the door always produces a marked improvement within a few seconds because it allows the convecting warm air from the mirror to escape directly, rather than being funnelled up the tube to produce the worst effect.

Another plus-point of open-tubes is that they are less easily buffeted by wind. However, it is also true that the optics are much more prone to dewing, or even frosting over, in an open tube. In addition, the currents of air generated by the observer's body can have a much more serious effect than can tube currents. Using a CCD-video system, Martin Mobberley has found this to be the case with his 14-inch (356 mm) Newtonian–Cassegrain reflector. The image quality as recorded on video took a distinct turn for the worse when he approached the telescope. Consequently he has partially closed in the previously open-framework tube (see Figure 3.4). He subsequently had his largest telescope, a 19-inch (490 mm) Newtonian reflector built with a solid tube, though the driving reason behind this was to have good light baffling in a heavily light-polluted site (at a different location to his 14-inch telescope).

Where does all this get us? Clearly, both sorts of telescope tube have their pros and cons. The best compromise does seem to be to have the telescope tube partly closed and partly open. Closing the tube at the eyepiece end is obviously most important. I am planning to rebuild my largest telescope. One of the improvements I shall make is to construct a partially closed tube.

The mounting of the optical components

A reflecting telescope's mirrors should be mounted in a way which allows them to be firmly held without distorting the glass. The pressure bearing radially on a mirror's rim should be not much more than that due to its own weight when the telescope is trained on a low-altitude and so its mirror is resting on its edge. It is particularly important that any retaining clips must not tightly clamp down on the front surface of the mirror. Always allow for differential thermal expansion/contraction of the cell with respect to the glass. If the glass is at all pinched in its cell the image quality will suffer. In addition, the primary mirror of a reflector should be properly supported on its rear side.

For mirrors with a thickness:diameter ratio of about 1:6 to 1:8 and diameters up to about 20 cm a three-point support system is adequate, the mirror's weight being equally supported by three pads when the telescope is pointed at the zenith. For larger diameter mirrors, especially those which are proportionately thinner compared to their diameters, a more elaborate support system is desirable. Figure 3.5 illustrates a nine-point flotation system, commonly used for medium and large-sized reflecting telescopes.

Once again, each of the pads bears an equal portion of the weight of the mirror when the telescope is pointed at the zenith. The same principle can be extended to provide 18-point or 27-point flotation. Some telescopes have even more elaborate mirror support systems. Some use weighted cantilevers which alter their effect in a way that maintains the correct pressure on the glass whatever the orientation of the telescope. At the other extreme, some home-doers mount their telescope mirrors on nothing more elaborate than a double thickness of stout carpet, or even a layer of 'Bubble-Wrap'! Apparently this can work satisfactorily even for large diameter mirrors, though I cannot speak with any experience on this point. The primary mirror of my own largest telescope is mounted on a nine-point flotation system.

The cell of a refractor's object glass is usually very much more simple, the lenses only being held round the edge. Nonetheless, they must not be held so tightly that they are distorted. Unless of diameter above about

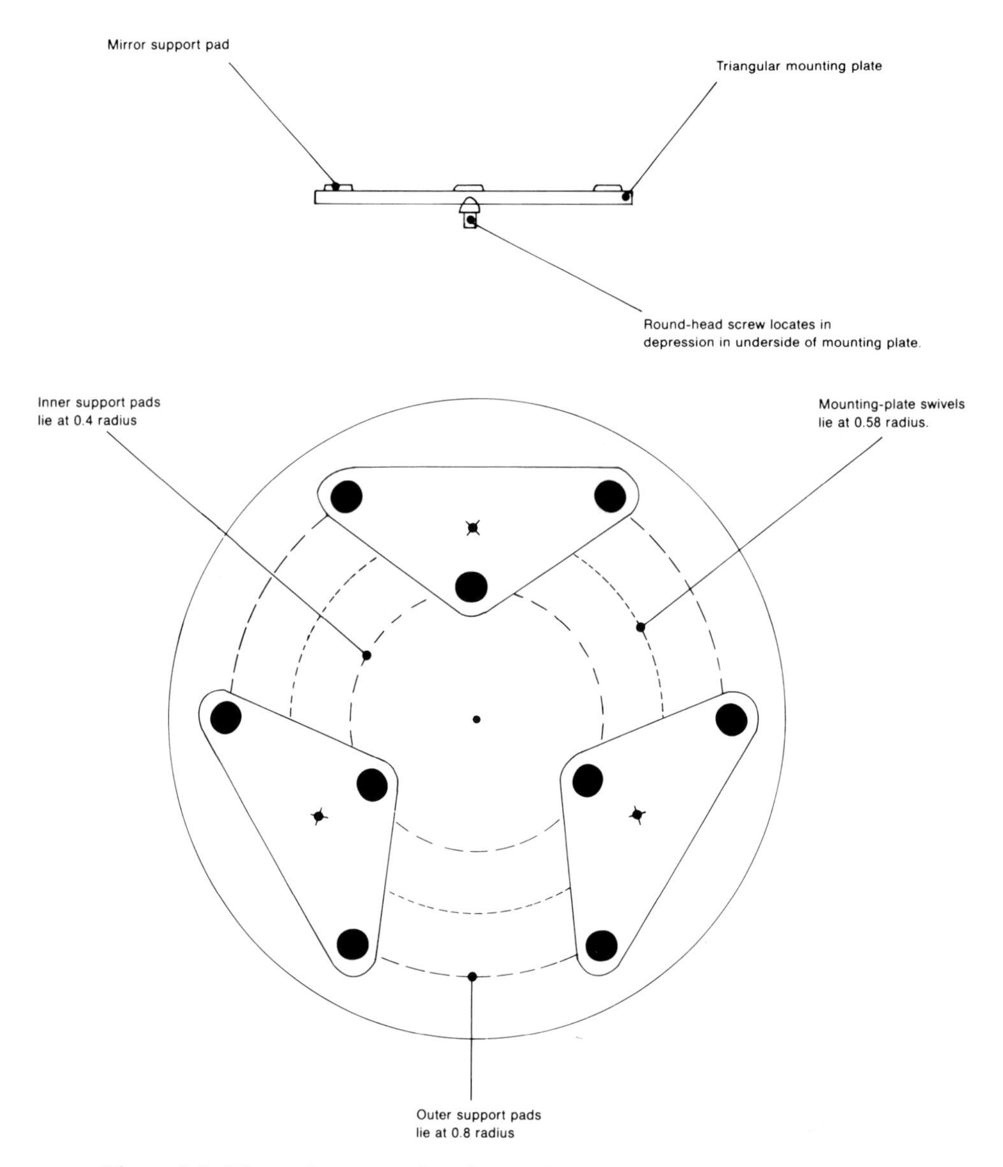

Figure 3.5. Nine-point suspension for a telescope primary mirror. The upper diagram shows an edge-on view of one of the triangular mounting plates. The round head of a large screw locates in a depression drilled to the underside of the plate. It is held in place by its own weight and that of the primary mirror (in addition the mirror is held in its cell by retaining clips. These prevent the mirror from falling out should the telescope tube becomes upended). The screws, one for each plate, are fixed into the base plate of the mirror cell. These form swivels. Each plate can rock slightly on its swivel to ensure even support exerted by the nine pads. With the pads and swivels located at the mirror-radii shown in the lower plan view, each pad bears an equal portion of the weight of the mirror. Slightly pliant materials, such as cork, nylon, or linoleum are suitable for the pads. These should be firmly bonded to the mounting plates.

13 cm, there is usually no provision for adjustment to the squaring on to the telescope tube. Larger refractors usually have three pairs of screws. In each pair one screw tends to pull the cell one way, the other producing the opposite effect. Adjustment is effected by loosening one of the screws and tightening the other. The mirror cells of reflectors are also mounted into the telescope tube in a way which allows the tilt to be adjusted at three points for collimation.

Adjustments to the optics

I have personally encountered a number of cases of telescopes performing poorly, merely due to the misalignment of the optics. Some telescope owners are loathe to touch the adjustments for fear of making the situation worse. In fact, if a proper method is adopted for collimating the optics, the telescope can be made to perform at its best fairly easily.

Collimating a Newtonian reflector of focal ratio larger than f/6

The first step is to make a 'dummy eyepiece'. This is really no more than a plug which fits into the telescope drawtube and which has a small hole (1.5 mm diameter is best) drilled exactly centre. Its function is to steer your eye onto the axis of the drawtube. A 35 mm film canister should fit into a 1-inch drawtube and will serve the purpose admirably, once a small hole is drilled through the bottom of it. For the moment we will assume that the axis of the drawtube is exactly perpendicular to the telescope tube. It certainly should be if the telescope has been commercially manufactured.

The following procedure is followed with the telescope pointed at a light-coloured wall, or the sky. The secondary mirror mounting will probably have adjustments that will allow it to rotate and to move laterally up and down the telescope tube. Using these the secondary mirror is positioned so that its outer edge is concentric with the visible edge of the telescope drawtube (see Figure 3.6(a)). Racking out the drawtube so that the apparent diameter of the secondary mirror is not much smaller than that of the edge of the drawtube will help you to be more precise. At this stage most of the primary mirror should be visible as a reflection in the secondary, though it is unlikely to be properly centred.

Next adjust the **tilt** and rotation of the secondary mirror until the reflection of the primary mirror is concentric within it (Figure 3.6(b)). As before, rack the drawtube in or out as necessary so that the reflection of the primary mirror nearly fills the secondary mirror in order to help you to be

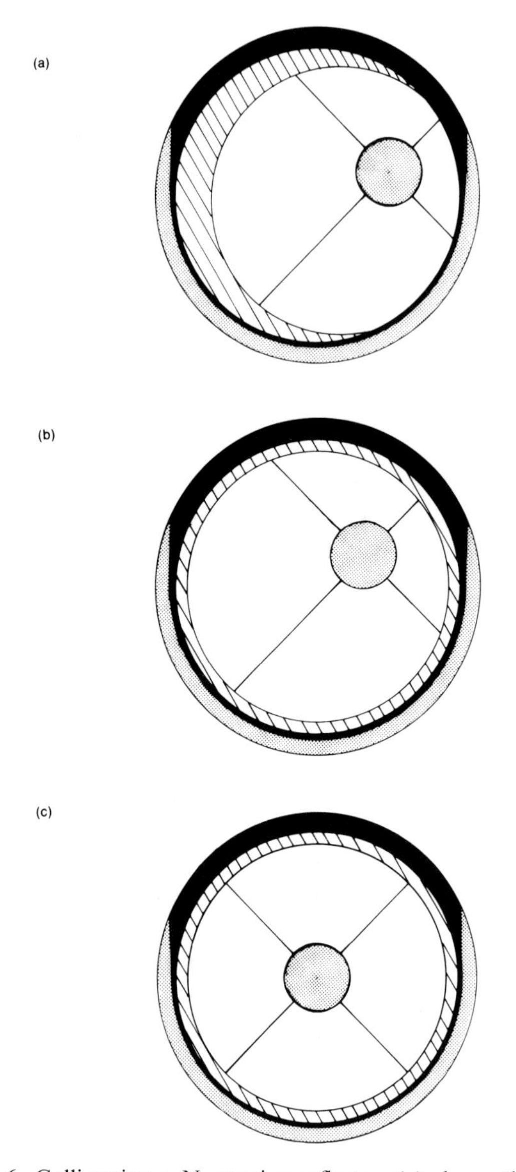

Figure 3.6. Collimating a Newtonian reflector. (a) shows the view through the 'dummy eyepiece' after adjusting the position of the secondary mirror to make its visible edge concentric with the edge of the drawtube. (b) shows the view after fine-adjusting the tilt and rotation of the secondary mirror, to make the reflection of the primary mirror concentric within it. (c) shows the view after adjusting the tilt of the primary mirror, to make the reflection of the secondary mirror-mount concentric.

precise. Finally, and still looking through the dummy eyepiece, adjust the tilt of the primary mirror until the reflection of the secondary mirror in the primary mirror is centred (Figure 3.6(c)). If necessary, further refine the alignment by running through the whole sequence again. Your telescope is now adequately collimated.

Collimating a Newtonian reflector of focal ratio smaller than f/6

Lower focal ratio Newtonian telescopes require a slightly refined approach to collimation if the very best results are to be achieved. The exact centre of the primary mirror is carefully determined by measurement and a small spot is marked on its surface at this position. A spirit marker will do to make the mark. Check that the primary mirror is centred in the telescope tube to better than a couple of millimetres (do this by simply measuring the gap between the edge of the mirror and the tube wall). If no means of adjustment are provided you may have to resort to using cardboard shims around the edge of the mirror (but be careful that these shims do not squeeze the mirror tightly).

With the secondary mirror exactly centred on the axis of the telescope tube the area of full illumination of the focal plane will **not** be exactly centred on the axis of the drawtube. The mirror ought to be slightly off-set away from the telescope drawtube and an equal distance towards the primary mirror. However, with telescopes of focal ratios more than *f*/3.5 for a 300 mm aperture, *f*/4 for a 400 mm, or *f*/5 for a 500 mm, the required adjustment is less than 7 mm in each direction. The fall-off of image brightness is quite small for distances up to a few centimetres away from the fully illuminated area of the focal plane. Vignetting from the drawtube is much more likely to dominate. Personally, I would not bother to off-set the secondary mirror.

After going through the collimation steps previously outlined, it is advisable to replace the 'dummy eyepiece' with a carefully made 'sighting tube'. This is a tube of about 17cm length closed at one end with a 'dummy eyepiece', the other end being open but crossed with wires. The wires should cross at the exact centre of the tube. The names and addresses of suppliers of commercially made 'sighting tubes' are given in Chapter 17. Using the 'sighting tube', make any fine adjustments to the secondary mirror that may be necessary so that the spot you placed at the centre of the primary mirror appears to be under the intersection of the crosswires.

The reflection of the open end of the 'sighting tube' should also be visible to you (still looking through the 'sighting tube') in the primary mirror. If necessary, use a shielded bright light held inside the top end of the telescope tube to show up the end of the 'sighting tube'. Try not to get dazzled!. Fine adjust the tilt of the primary mirror until the reflection of the open end of the 'sighting tube' is exactly centred over the dark spot on the primary mirror (and, hence, the intersection of the crosswires). If necessary go once more through the whole procedure to achieve the best possible alignment.

Refinements over the 'sighting tube' are commercially available. These are the 'Chesire eyepiece' and the 'autocollimating eyepiece'. The suppliers of these are listed in Chapter 17 and they are described in an article in the March 1988 issue of *Sky & Telescope* magazine.

Collimating a Cassegrain reflector

The first step is to use a ruler to carefully check that both the primary and the secondary mirrors are centred within the telescope tube to better than a millimetre. Make any necessary adjustments in the same way as previously described for the Newtonian telescope.

Next, insert a 'sighting tube' (a 'dummy eyepiece' is not really accurate enough for collimating a Cassegrain telescope) into the drawtube. Point the telescope at a light-coloured wall, or perhaps the daytime sky. Looking though the 'sighting tube', adjust the tilt of the secondary mirror until the reflection of the primary mirror is exactly centred within it. Then adjust the tilt of the primary mirror until the reflection of the secondary mirror within the primary mirror is exactly centred. The spider-support vanes will then all appear to be of equal length. At this point a mere rotation of the 'sighting tube' in the eyepiece drawtube should be sufficient to make the crosswires at the end of it line up exactly with the reflection of the spider vanes. Adjust as necessary to make this so.

Now the telescope will be quite close to its optimum collimation. Nonetheless, a Cassegrain telescope is rather more finicky than a Newtonian and so the final adjustment should be made by sighting the telescope on a test-star. Plug in an eyepiece that delivers a high power (several hundred). If your telescope does not have a drive use Polaris, otherwise pick any star that appears moderately bright in the telescope. It should have the highest possible altitude, so that the effects of atmospheric turbulence are minimised. Once the star is centred, slightly defocus it while watching for any asymmetry in the image. If the expanding disc of light becomes oval and less bright in one direction, the telescope is still slightly misaligned (it is here assumed that the optics are of excellent quality – a small amount of astigmatism in the image will produce an effect which is hard to distinguish from misalignment).

With the telescope still pointed at the test-star adjust the tilt of the primary mirror such that the image moves in the direction that the out-of-focus star image is most distended and faintest. Only a very minor adjustment should be given. Next adjust the secondary mirror until the image is once again centred. The star image should now be more circular and more evenly illuminated. Continue this step-by-step adjustment until the out-of-focus star image is as circular and as evenly illuminated as possible. Check that it remains so for all positions of the focuser.

Finally, focus the star and move the telescope in order that the star is brought to different positions in the field of view. Accepting that the eyepiece may cause some defocusing near the edge of the field of view, the image of the star ought to remain sharply focused over most of it. If the image remains sharp along one diameter but not in the perpendicular direction, either the drawtube is misaligned or the optics are astigmatic. See the following notes on the alignment of the eyepiece focusing mount.

Collimating a refractor

The procedure is similar to the fine-tuning of a Cassegrain reflector. Select a good 'test-star'. Centre it in the field of a high power eyepiece. Slightly defocus it and watch for any asymmetry in the expanding disc of light. Adjust the tilt of the objective (this will displace the image) until the re-centred image is cured of any asymmetry for focus positions inside and outside the telescope's focal plane. Finally, check for drawtube misalignment or astigmatism, as detailed for the Cassegrain reflector.

Collimating Maksutov and Schmidt–Cassegrain telescopes

These usually have their primary mirrors fixed in position by the manufacturer and only the secondary mirror (or the secondary mirror together with the corrector plate) is adustable. The procedure to follow is that given for the Cassegrain reflector, fine-tuning on a 'test-star' as before. Using a 'Chesire eyepiece', or an 'autocollimation eyepiece' will make life easier. The manufacturers of these collimation aids supply full instructions for their use with different types of telescope.

'Squaring on' the eyepiece focusing mount

If the telescope drawtube is not aligned to the optical axis of the telescope, the focal planes of the eyepiece and the objective will be slightly tilted with respect to each other. These planes may be brought into coincidence only along a line. The image will be in sharp focus only along this line, becoming more blurred with increasing distance from it.

To check the accuracy of aligment of the focusing mount of the Newtonian reflector first remove the secondary mirror to allow a clear view to the other side of the telescope tube. By careful measurement with a ruler determine the spot exactly opposite the focusing mount and mark this position (a small sticky label will do to make the mark). Put in a 'sighting tube'. Does the intersection of the crosswires appear to lay over the mark you made? If not the eyepiece focusing mount needs adjustment. Removing and then replacing the focusing mount with shims appropriately inserted between it and the telescope tube will do if (as is usual) there is no other provision for adjustment.

With a Cassegrain or a Catadioptric telescope the crosswires of the sighting eyepiece should appear to coincide with the centre of the secondary mirror. If not, then correct as before. Similarly with the refractor – the crosswires should appear to intersect at the centre of the object glass. Covering the refractor's objective with a cap which has a small hole exactly centred in it allows for greater precision. The crosswires should then appear to intersect over the hole.

Collimating a 'star diagonal'

The better examples of this device have some provision for adjusting the tilt of the mirror. The procedure is simple, especially if your telescope has a drive. Once your telescope is adequately collimated (including the eyepiece focusing mount), plug a high powered eyepiece into the telescope and point it at a bright star. Make sure the star is centred. Replace the eyepiece with the 'star diagonal' and plug the same eyepiece into this. Refocus. Is the star still centred? If not, adjust the tilt of the diagonal's mirror until it is. Keep changing over from eyepiece to diagonal plus eyepiece, adjusting as necessary, until the star remains centred in both.

Field testing a telescope's optics

The majority of today's amateur astronomers do not make their own telescope optics. Nonetheless, most of us are aware of the basic procedures involved and are familiar with the principles of the Foucault and Ronchi grating tests. I would, however, wager that very few telescope users realise that simple versions of these tests can be used to evaluate the overall accuracy of both figuring and alignment of the assembled telescope's optics with much greater ease than the toiling mirror maker could assess the individual components in his/her workshop!

The easiest method involves using a Ronchi grating. Many optical firms (especially those specialising in telescope optics and telescope making materials – see Chapter 17) will sell these gratings. They are very cheap to buy. If a choice is given, select a grating of at least four lines per millimetre (100 lines per inch), though more lines per millimetre would provide for a more sensitive test.

Simply mount a small piece of the grating over the central hole in a 'dummy eyepiece' and you have all that you need to perform the evaluation. Set on a fairly bright test-star in the normal way and then unplug the normal eyepiece and replace it with the mounted grating. Peering though it, you will see the primary mirror/objective of the telescope flooded with light from the star but crossed with dark stripes. These stripes are effectively greatly magnified images of the grating.

You will find that rotating the grating produces a consequent rotation of the pattern and adjusting the focuser produces a dramatic change in the apparent magnification of the pattern and so the number of dark stripes that cross the field of view. The closer to the focal plane the grating is, the smaller the number of stripes visible. For this test, I recommend adjusting the focuser until four dark stripes are visible, the grating being just intra-focal (inside the telescope's focal plane). Assuming the grating is not faulty or dirty, the image of the bars you should see should be straight and as shown in Figure 3.7(a). Trapped grains of dust within the grating will show up as a jaggedness of the bands.

Any faults in the figuring of the optics or their alignment will be immediately apparent as shown in the other views in Figure 3.7 and are described in the caption accompanying it. Rotating the grating through 180° in several steps will allow all of the optics' radial zones to be evaluated.

The beauty of this test is that the complete optical system of the telescope is tested in field conditions – and this is surely what really counts. Also, the test is applicable to **all** telescopes. Even better this is a *null test*. In other words the appearance of the bands are straight and regular when all is well because the star is at infinity and the grating is then close to the principle focal plane, unlike the much more complicated situation for the mirror maker who works with the light source and tester each close to the radius of curvature.

Mounting a razor blade to half cover the hole in the 'dummy eyepiece' enables a version of the Foucault knife edge test to be performed. Set up in the same way as for the Ronchi test. With your eye close to the hole and peering past the edge of the razor blade you will see the telescope primary

Figure 3.7. Ronchi patterns for a telescope field tested on a star. (a) shows what would be seen for good optics in accurate collimation. (b) shows what would be obtained for a spherically overcorrected system. For instance, a pattern like this might result from the primary mirror of a reflecting telescope being ground too deep at the centre, or perhaps the change is temporary and caused as the mirror is rapidly cooling to the ambient air temperature, in which case the error will disappear as the mirror approaches thermal equilibrium. (c) is the pattern produced by a spherically undercorrected system (in this case if the mirror is cooler than the ambient air temperature that fact ought to be apparent because it will soon become covered in dew!). (d) is produced by the 'turned-edge', a very common error found in reflecting telescopes' primary mirrors. (e) illustrates the result from a telescope possessing zonal errors. Obviously an almost infinite number of other permutations of this are possible. (f) illustrates astigmatism but a similar result also stems from misalignment of the optics.

mirror/objective flooded with light from the test-star. If you move the telescope **very** slightly so that the blade begins to cut off the light you will see a black shadow sweep across the pool of light.

Adjust the focuser until moving the telescope causes the edge of the shadow to become more blurred. Keep adjusting the rack-mount until moving the telescope causes the whole of the mirror to darken evenly. In other words, at the correct position you will not be able to decide whether the shadow sweeps across from the right or from the left when you move the telescope. The razor blade is now coincident with the focal plane of the telescope. Any figuring errors or misalignment of the optics are hugely magnified and thrown into 'pseudo 3-D relief' and become very obvious at the position where the pool of light is half extinguished. If all is well with the telescope the disc of light will appear perfectly flat and smooth and an even shade of grey (and atmospheric turbulence will appear as a moving set of corrugations and swirling patterns superimposed on this).

The knife edge test has the same advantages as the Ronchi test performed at the telescope's final focus and using a test-star. It is true that it is potentially more sensitive than the Ronchi test but it is also true that it is more tricky to perform. I think that most people would prefer the easier option of the Ronchi test.

Finally, one can inspect the focused, as well as the intrafocal and extrafocal (beyond the focus position), star images produced when a high power eyepiece is plugged into the telescope. Obviously one should choose a star of medium brightness as seen through the instrument and it should have the highest possible altitude so that the effects of atmospheric turbulence are minimised. The precise analysis of the appearance of these images is a complex business and deserves a book all of its own – and, indeed, it has been given one. *Star Testing Astronomical Telescopes* is written by H. R. Suiter and was published by Willmann–Bell in 1995. All permutations of errors and telescope faults are discussed with illustrations depicting the appearances produced. The *Sky & Telescope* magazine of March 1995 carries an article highlighting some of Suiter's main results.

I would caution, though, that these star tests are more suitable for telescopes of below about 200 mm aperture because in larger ones the effects of atmospheric turbulence would overwhelm all but the most serious errors on most nights. The Ronchi test is the one that I would recommend for users of larger telescopes. Having said that, I thought that I would include an illustration showing the star test appearances for astigmatism, optical misalignment and the effects of tube currents as these are the most common troubles and are by far the easiest to recognize (see Figure 3.8).

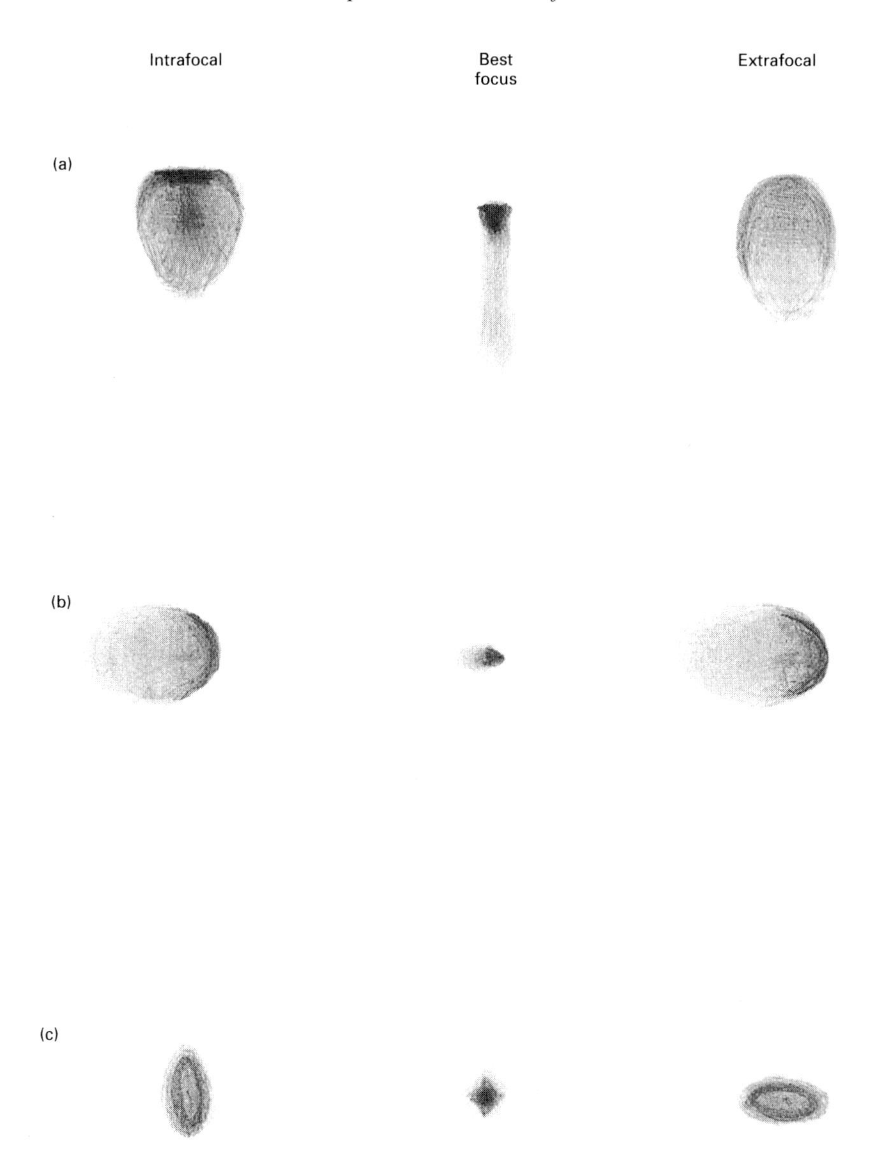

Figure 3.8. Depictions of the telescopic appearances of a star at high magnification, for different focal positions, with each of the following common problems: (a) tube currents. (b) misalignment (note that the orientation of the cometic image does not change either side of the best focus, unlike for astigmatism. (c) astigmatism. These are based on the author's own experiences but agree fairly well with the theoretical results presented in H. R. Suiter's book *Star Testing Astronomical Telescopes*.

Telescope tube accessories

The additional attachments necessary on a telescope tube naturally depend upon the use to which the telescope is put. Finder-telescopes are fairly universal items. The very highest optical quality is not required. Much more important is aperture, which should be as large as possible, especially if you hope to sight the telescope on very faint objects. The eyepiece fitted to a finder ought to deliver a power of about ×4 or ×5 per inch of aperture (×1.6 to ×2 per centimetre) and should have the largest field of view possible.

Crosswires are a distinct help in the finder, though I don't regard them as essential. In fact, they aren't usually of much help unless they are illuminated. The home-doer can fairly easily make serviceable crosswires from very fine copper wire. These should be glued across the field stop of the finder's eyepiece. A small hole can be drilled in the body of the eyepiece, close to the eye-end, and a small red LED fitted through the hole. Make sure that the LED is hidden from view when you look through the eyelens of the eyepiece. You only want to see the dim red glow illuminating the crosswires, not the LED itself. The LED is run from a low-voltage source but don't forget to include a series resistor to limit the current to the maximum of 10mA to 40mA (depending on the LED) specified by the maker. The size of resistor required can be found from:

$$R = \frac{(V-2)}{I_{max}}$$

where V is the supply voltage and I_{max} is the maximum allowable current. Round **up** the answer to the closest standard size of resistor. For instance, if I_{max} for a given LED is 10mA and it is to be run from a 6V battery, select a series resistance of value 470Ω. Adding a 10kΩ linear potentiometer (used as a rheostat) in series provides some degree of variability of the brightness of the illumination. This could be a useful addition – nothing is more frustrating than to have a faint object vanish from view when the crosswire illumination is switched on!

Eyepiece focusing mounts can be manual or electrically driven. If the latter, make sure that two focusing speeds are provided. Sometimes a very large shift of focus is required when changing over eyepieces. Unless you are to wait for an age, the focusing action should then be rapid. However, precise focusing requires that the focusing action is slow – especially when using high powers. If the eyepiece mount is manually adjusted for focus, make sure that its action is smooth and the drawtube does not sloppily shift off-line. See Chapter 5 for comments about attaching cameras to the eyepiece focusing mount.

A particularly useful type of telescope is the Cassegrain–Newtonian, where changing over the secondary mirrors effects the change between the two configurations. Mounting the secondaries on pre-collimated units which are quickly attached to the spider-support avoids collimating the telescope afresh each time the change is made. The secondary mounts each have a pin which locates in a small hole in the spider-support to ensure that the mirror is correctly positioned after each change.

Telescope mountings

Accepting that the telescope mounting must allow the telescope to be pointed almost anywhere over the sky, top of the list of requirements must be rigidity. A telescope that trembles in a breeze and shudders with every touch of the focuser will prove a trial for visual work. It will stop most other forms of observational work altogether. When purchasing beware of elegant, spindly, designs. A telescope mounting should be stocky with thick shafts. If you are building the mounting yourself, build it as massive as you can. Obviously, if your telescope has to be portable then a compromise will have to be reached. In that case a Dobsonian mount may serve you best, even though it is a humble altazimuth. Filling columns, pipes and fork arms with concrete is one way of achieving rigidity, though do make sure that the rest of the mounting is up to carrying the extra weight.

Falling-weight driven mechanisms, d.c. electric motors, a.c. synchronous electric motors (either powered straight from the mains, or via a low output voltage transformer, or a variable frequency oscillator) and pulse-driven stepper motors (driven from a pulse-producing unit, or a computer) have all been successfully used for driving telescopes.

Considerable variety also exists in the way in which the motion is communicated to the telescope. Traditionally, the worm-and-wheel arrangement (see Figure 3.9(a)) has been most popular. Figure 3.9(b) shows a close up view of an interesting alternative: instead of meshing gear teeth, this drive uses the friction between plain rollers. The periodic errors and backlash inherent in gear-driven drives are virtually eliminated. An additional advantage for the DIY telescope builder is that the need for gear cutting (or the expense of purchasing ready-made gears) is avoided. I would refer interested readers to the paper 'An efficient gearless telescope drive' by B. Knight in the December 1989 issue of the *Journal of the British Astronomical Association*. In an article written for the November 1989 *Sky & Telescope* magazine, 'Deep-sky photography without guiding', Ron Arbour describes the gearless drive system used on his own telescope.

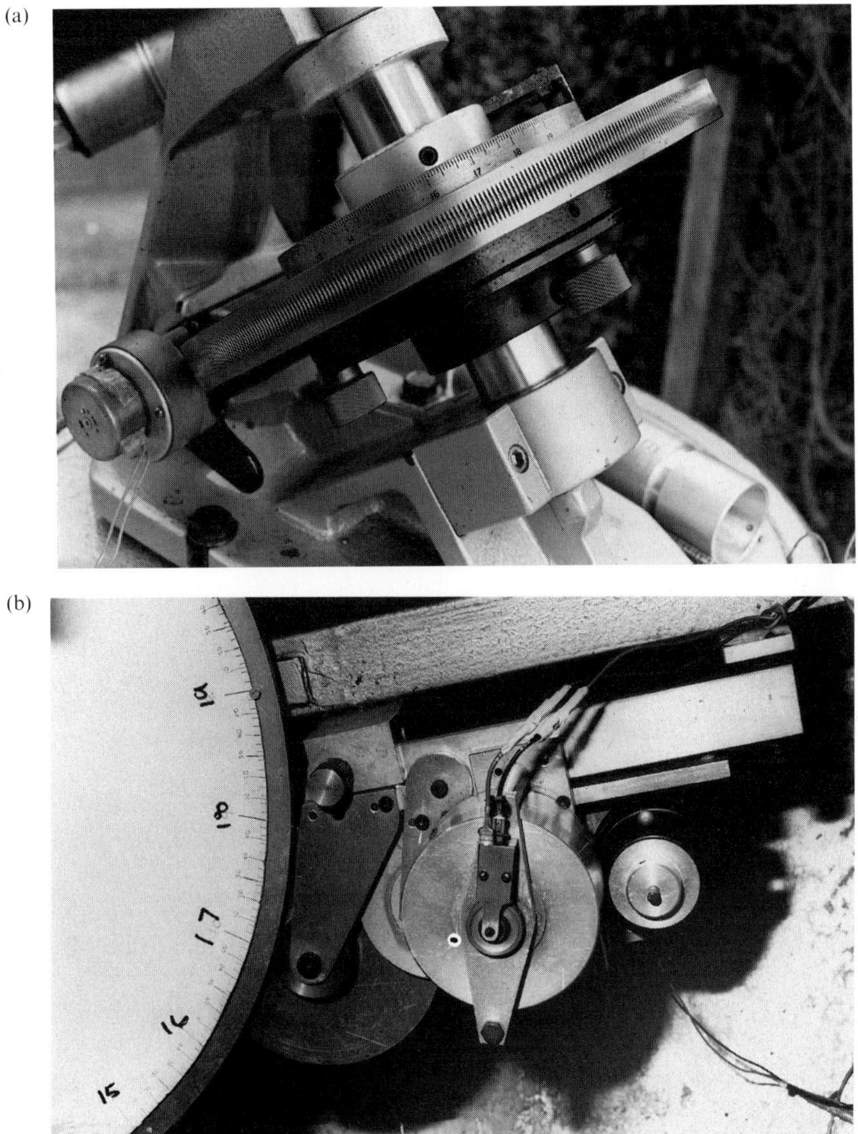

Figure 3.9. (a) The worm-and-wheel drive of Martin Mobberley's 14-inch telescope. (b) The 'friction drive' arrangement of rollers of Ron Arbour's 16-inch telescope.

Figure 3.10. The German equatorial mount of Martin Mobberley's 14-inch telescope. Note the addition of a ruggedly mounted worm-and-wheel drive attached to the declination axis.

For a number of tasks, for instance long exposure astrophotography, provision should be made for making fine adjustments to the right ascension (RA) and declination of the telescope while it tracks. In the case of adjustments in RA the corrections can be applied by slightly altering the speed of the drive motor, or by another motor acting via a differential through the gear train. Another possibility is the use of a threaded rod at the end of a tangent arm. Fine adjustments in declination can also be effected by means of a tangent arm, or a separate worm-and-wheel (see Figure 3.10).

Figures 3.1, 3.2 and 3.3 show fork equatorially mounted telescopes. Figure 3.10 shows a German equatorial mounting. The vast majority of amateur owned telescopes are provided with either of these forms of equatorial mounting. Of the alternative equatorial mountings, those shown in Figure 3.11 are the most common. However, unusual optical configurations may require different mountings to suit.

Telescopes and computers

Several manufacturers (listed in Chapter 17) produce computerised devices that assist in the setting of a telescope on a given object. They

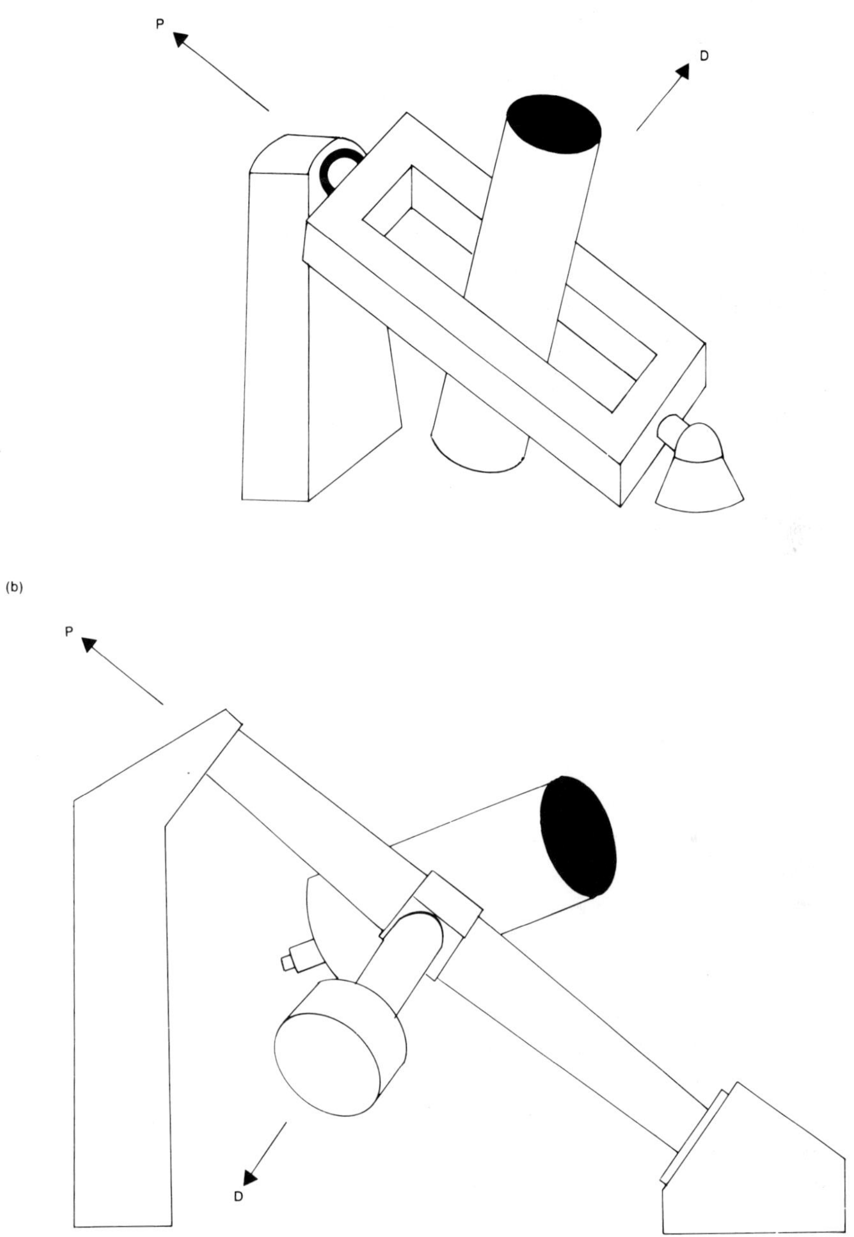

Figure 3.11. The most common alternatives to the fork and German equatorial mountings. In each case P indicates the polar axis and D the declination axis. (a) The English equatorial (sometimes called the 'yoke') mounting. (b) The 'modified English' (or 'cross-axis').

consist of encoders that attach to the telescope's polar and declination axes which are wired to a 'black box' which has its own keyboard and display. The device's memory contains the coordinates (and other details) of thousands of celestial objects. At the beginning of an observing session the user selects two bright stars and points the telescope at each in turn, using the keyboard to 'tell' the black box which star the telescope is pointing at. From then on the observer can key up a specific object and follow the indications given on the display to manually move the telescope to locate the object. Some of these devices have other facilities: the identification of a celestial object that the observer may come across in sweeping the sky, a program to help the observer accurately align the polar axis of the telescope, a readout of the telescope's coordinates, etc.

One step up from computer-aided telescopes are computer-**controlled** telescopes. Some manufacturers produce these (see Chapter 17) but a few amateurs have made their own. The telescope's axes are equipped with stepper motors. These motors have four phase coils. Energising each of these coils in sequence causes the rotor to revolve in a series of small steps. The correct sequence can be generated by a computer and issued as a repeating sequence of binary numbers from its user port. Of course, the computer's user port can only produce currents of up to 20 mA (at 5V), or so. The phase coils of the motor require currents of the order of 1A (and typically at 12V). Hence each output from the user port is used to operate a switching transistor (a Darlington has a sufficiently high gain) which in turn controls the current to the corresponding phase coil.

In essence, the computer produces a rapid stream of pulses which operate the stepper motors and, via gearing, drive the telescope to the correct R.A and declination. The computer can then continue to supply pulses, at a slower rate, to the R.A. motor in order to provide a sidereal drive. The computer can also be programmed to get the telescope to track on any comets or asteroids or any other body which does not move with diurnal motion. As long as the computer can keep track of the pulses it produces (which it can do at a rate of several hundred per second, depending upon the model and the software used) the pointing of the telescope can be as accurate as mechanical considerations allow, certainly to within a small fraction of a degree. Figure 3.12 shows the basic layout of system hardware of a computer-operated telescope.

Of course, it is but a step to dispence with the mechanical complexity and lesser rigidity of an equatorial mounting and follow the lead of the modern professional telescopes and mount the telescope on a driven altazimuth mounting. Since the first edition of this book was published a number of

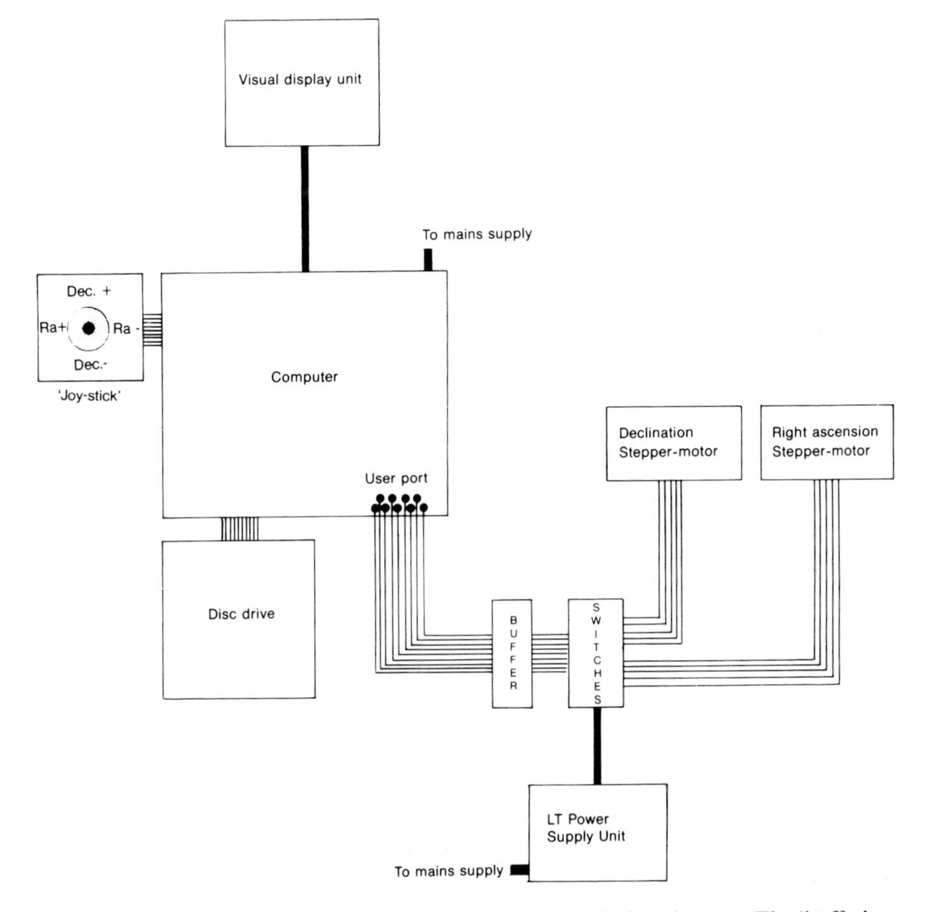

Figure 3.12. Schematic layout for the computer control of a telescope. The 'buffer' is a chip which simply relays the 'off' or 'on' signals from each of the tracks from the user port. Its function is to protect the computer from damage should a fault occur with the motors or the switching. The 'switches' are, in fact, Darlington transistors which connect the appropriate phase coils of the motor to the live output of the L.T. supply. The 'joy-stick' control shown is a possible added feature. It could be used for slewing the telescope (overriding any programmed settings) or for fine adjustment (for instance during photography).

manufacturers have done just that and provided computer-driven altazimuth telescopes for the amateur market. As ever, the listing in Chapter 17 will help. Note that a driven altazimuthly mounted telescope will also have to be provided with a focusing mount which is driven round to compensate for the image rotation inevitably produced if long exposure photography or CCD imaging is desired.

In these brief notes I have had to gloss over many of the details and difficulties that the amateur telescope constructor would face. For instance, it is not easy to provided smooth tracking at the sidereal rate and yet get the motor to operate fast enough to also allow the telescope to slew to its target in a reasonably short time. The motor will also have to 'ramp up' to speed and 'ramp down' to a stop, because of the inertia of the telescope, if pointing accuracy is not to be lost because of 'missed pulses'. Obviously, building a computer-controlled telescope is a job for people who are technically knowledgeable, as is designing the software to operate it. Nonetheless, using a computer to automate the telescope for variable star work (see Chapter 13), or photographing galaxies to patrol for supernovae (Chapter 14), or a host of other observing programs, makes the effort very well worth while. The books and articles listed in Chapter 17 should help.

Aligning an equatorial mount

A good telescope mounting will have provisions for making fine adjustments to the elevation and azimuth of the polar axis. Some have built-in aids to alignment, such as a small telescope within the polar axis. These usually have an eyepiece fitted with a graticule of specially marked circles and divisions. Using these and a chart of the area around the celestial pole, the polar axis can be aligned on the celestial pole to an accuracy of a small fraction of a degree.

Otherwise (and for the really good alignment of telescopes, whether or not they are equipped with this aid), the mounting can be aligned by means of observing the apparent north–south drift in the image of a star with the telescope pointing in specific directions. The following notes detail the procedure.

Set the telescope on a star that is close to the meridian and the Celestial Equator. With the declination axis firmly locked and the telescope drive engaged monitor the apparent north–south drift of the star. Ignore any east–west drift. If you are at all unsure as to the orientation of the image, momentarily move the telescope in declination a small amount in the direction of Polaris. The direction towards which the star appears to move defines **south** in the field of view. Having re-centred the star and locked the declination axis note any north–south drift of the star as the telescope tracks (if it has no drive then simply move the telescope in right ascension after a period of a few minutes in order to re-locate the star, keeping the declination axis firmly locked. Does it come back to the centre of the field of view?). If the star appears to drift to the **south**, the polar axis is pointing a little to the **east** of the true celestial pole. The star will appear to drift to the north if the polar axis is pointing to the west of the celestial pole.

If the star appears to remain centred then either use a crosswire eyepiece or a micrometer in order to make a more precise determination. Failing this, move the telescope so that the star lies on the extreme southern part of the field of view, virtually on the edge of it. If the gap between the star and the southern edge of the field of view increases then the star is clearly moving northwards. However, if the star becomes hidden by the field stop and cannot be recovered by movements in right ascension the star is moving southwards. In this instance, move the telescope so that the star is close to the opposite extreme of the field of view. You will then be able to monitor the rate at which the star moves more precisely. Make any necessary alterations to the altitude adjustment of the polar axis in order to reduce the north–south drift to a minimum. Each time you have made an adjustment, note down how big it was. Then note down its effect (how far north–south the star appears to move in equal intervals of time). In this way you will rapidly be able to bring your telescope's polar axis to the correct azimuth.

Select a star at about 6^h east of the meridian (and preferably within 20° of the Celestial Equator) in order to check the elevation of the polar axis. Use the same procedure as before. If the star drifts to the **north** the elevation is too **high**. A drift to the south indicates that the polar axis elevation is too low. A star which is about 6^h west of the meridian and of similar declination can also be used for assessing the required elevation adjustment. Any errors now produce the opposite effect in the apparent north–south drift of the star. Adjust as necessary to reduce this drifting to a minimum.

Once you are sure that the polar axis alignment of your telescope is correct, you can go ahead and adjust the rate of its drive motor to produce the best east–west tracking possible; but not before – an altitude error in the polar axis produces an east–west drift for objects which is greatest when they are near the meridian. An azimuth error produces an east–west drift that is greatest for object at 6^h east and west of the meridian.

Observatory buildings

A telescope can be designed to be portable. Many have been built onto trailers for hauling over long distances. This may be the only solution for people living under heavily light-polluted skies. Others have been provided with castors so that they can be kept indoors and trundled out into the back garden for an observing session. Of course, for an equatorially mounted telescope this means aligning the polar axis each time (though marking the ground, or fixing clamps to it may help). With telescopes of moderate size it is possible to permanently fix the mounting outside but keep the tube indoors until needed. However, a telescope that is quickly and conveniently set up for observation is one that is most likely to be extensively used. For a large telescope this really means that it should be permanently mounted outdoors.

Home-built telescopes can be contructed to be sufficiently weather-proof not to need any shelter. It is largely a question of design and the materials used. More usually some form of shelter is needed. The simplest form is a large plastic sheet tied into position. It is cheap but far from ideal. Fiddling with wet pieces of string or rope in the dark and hauling a wet plastic sheet about is time consuming and far from being fun. Also, the sheet will undoubtably become brittle and tear, especially in the face of one or two gales, and so will need replacing, maybe every two or three years. I can speak with experience since my present home has a garden which is too small to allow for the building of an observatory. My mother has very kindly cut and sewn together large plastic sheets to make covers which now shield my two main telescopes from the elements.

Telescopes that are not too tall can be housed in a run-off shed. They are very easy for the handyman to construct. They are also relatively cheap. Figure 3.13 shows the run-off shed I contructed for my largest telescope when it was sited at my parents' home in Seaford. For cheapness I used ordinary hardboard sheets (rather than the more expensive weather-board) nailed onto 2″×2″ softwood. The framing was treated with wood preservative and the hardboard was loaded with emulsion paint. The outside surface of the hardboard was given an additional finishing coat of polyurethane paint. The roof was a sheet of varnished and over-painted marine plyboard and all the corners of the framework were braced with offcuts of marine ply. Brass hinges, a garage door-stay to stop the door flapping when open, a hasp-and-staple and padlock, a lamp and table top mounted at the far end of the shed, two handles attached to the rear end of the shed, and large nylon rollers to act as wheels completed the contruction.

For rails I used two eight metre long planks of wood, well soaked in wood preservative. A commonly used alternative is angle-iron rails set in concrete. After opening the door (which locked in the open position until a cord was pulled to release it), I simply went to the back of the shed and pulled it along the rails in order to leave the telescope exposed in the open air. Run-off sheds are a very popular solution to the problem of housing a telescope. Mine served me very well for the years my telescope was sited at

Figure 3.13. A photograph of the run-off shed that housed the author's telescope at Seaford. The photograph was taken soon after the shed was built. Shortly afterwards wooden rails were laid for the shed to run on.

Figure 3.14. The late Bill Peters' observatory, housing his $8\frac{1}{2}$-inch reflector. The slit is opened by removing two panels and the whole dome rotates on wheels attached below the suspended floor. The dome is fairly easy to push round but anything larger would need to have fixed walls and a rotating top.

Seaford. One advantage is that the telescope quickly cools to the ambient temperature. Of course a disadvantage is that one is observing fully exposed to the chill of the night air – which can be very uncomfortable if a breeze is also blowing. Run-off sheds are particularly useful for automated telescopes. A telescope that moves around on its own either has to have a building that provides no obstruction, or one which moves around with it.

If well made, a dome can provide the ultimate in luxury when observing at the telescope. Figures 3.14, 3.15 and 3.16 show examples of amateur built domes. They can also be bought commercially. The commercial ones are usually made of fibreglass but amateurs can press many materials into use when building their own. If contemplating a dome for your telescope, do make sure that it is plenty big enough. Remember that the inside of your dome will become a black void at night. You only want to see real stars, not those generated by banging your head on the roof!

If your observatory has a suspended floor, make sure that the telescope mounting does not come in contact with it. Every movement you make would then pass tremors into the instrument. The mounting should be fixed to a concrete base and should pass through a clearance hole in the floor. If the only floor is the concrete base of the observatory, you should really attach the mounting of the telescope to a separate column of concrete

Figure 3.15. John Watson's 3 metre diameter dome with a slide-back shutter, housing a 10-inch reflector. The upper section rotates on nylon wheels.

Figure 3.16. Ron Arbour's 3.6 metre diameter dome, housing his 16-inch reflector. The rotating section has a slit which is covered by removable panels. Wheels attached to the dome run in a rail set upon a low breeze-block wall.

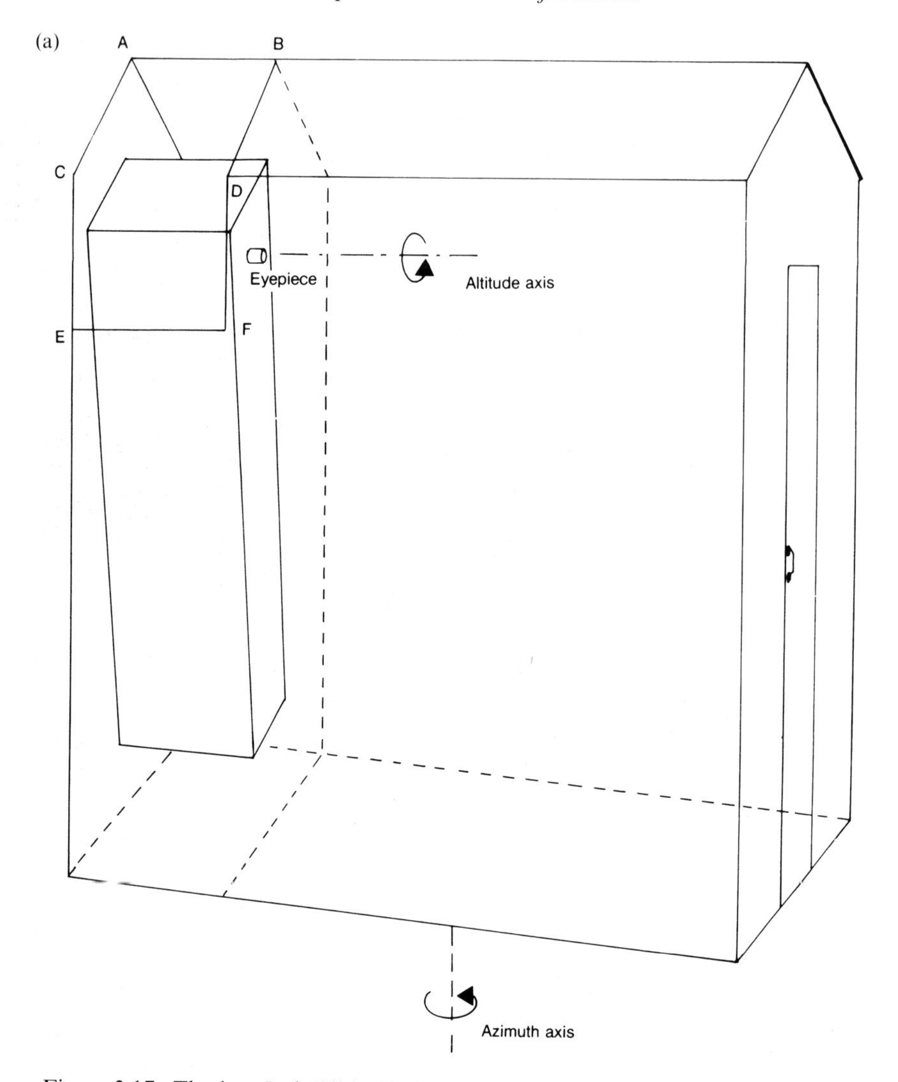

Figure 3.17. The late Jack Ells' 12¾-inch Newtonian reflector was incorporated into his rotating shed. (a) is a schematic showing the manner in which the telescope was confined to its own compartment within the shed. Swing-open shutters covered the sections ABCD and CDEF. The altitude axis of the telescope was set close to the eyepiece, which maintained a virtually fixed position, as a result. The whole shed rotated for motion in azimuth. (b) shows a view of the interior of the observer's cabin. The telescope eyepiece and the eyepiece of a finder-scope can be seen attached to the large, dark coloured, disc incorporating the altitude scale. The cabin could be sealed and heated without disturbing the seeing through the telescope.

(b)

(a)

(b)

Figure 3.18. (a) A model of the base of Jack Ells' rotating observatory, showing the inclined planes that imparted a limited equatorial motion to the shed and telescope. (b) A model of the base-frame of the shed in position. Note the large azimuth scale (illuminated and viewed through a peep-hole in the observer's cabin) and the arrangements for cabling.

(perhaps one which goes down a metre or more) and provide a small clearance between this column and the rest of the floor. The only exception being for a concrete floor of several tons, when its inertia will minimise the effects of vibration. Large wheels or small castors, golf balls or skateboard wheels, all have provided a way of mounting the rotating section of the observatory. Most amateur-sized domes are easily pushed around by hand. However, a surplus washing machine motor can be pressed into service to effect the rotation (but see the following note on safety).

A dome, or any other enclosed observatory, is particularly handy in that lights, furniture, equipment, etc., can be installed and left ready for use. Nonetheless, do beware of the effects of damp and **do make sure that any mains voltage equipment is very thoroughly insulated and earthed**. On that last point, try and use low-voltage equipment wherever possible.

Of course, there are other solutions to the problem of giving a telescope a fixed housing, without necessarily building a rotating dome. A shed where the top section slides back on rails is also quite good, particularly for tall-mounted telescopes, such as refractors, Cassegrains and catadioptrics.

One very clever observatory design, which proved to be highly successful in practice, was that built by the late Jack Ells for his 12¾-inch Newtonian reflector (see Figures 3.17 (a) and (b)). He built his telescope into its own enclosure at one end of a square shed. The telescope pivoted at a point close to the eyepiece and was constrained only to move in altitude. **The whole shed revolved on its base for motion in azimuth**. The only part of the telescope which entered into the main part of the shed was the eyepiece focusing mount (together with the eyepiece of a finder-scope). This section was provided with shelves, lamps, a work surface, a computer – and a heater for the comfort of the observer!

The computer was used to calculate the required azimuth and altitude settings for a given celestial object at a given time (see the Appendix for the appropriate equations) and the observer turned a handle which rotated the shed with respect to a framework built underneath it (a separate control provides fine adjustment). Looking down through a 'peephole' to underneath, the observer sighted on an illuminated azimuth scale. The telescope's altitude was altered by means of another handle (once again, with a separate control for fine adjustment) and the setting was indicated on a circular scale on the wall close to the eyepiece.

With great ingenuity, Mr Ells even provided his telescope/observatory with a limited range of equatorial motion. The lower framework had eight attached wheels. These wheels were set on short planes, each of which was parallel to the equatorial plane (see Figure 3.18 (a) and (b)). A motor driven screw-gear then hauled the shed round at the sidereal rate, forcing the wheels to climb the planes. In fact, Mr Ells provided two motors, one for tracking and the other for rapid slew. The telescope was able to track uninterrupted for 40 minutes before re-winding the drive and re-setting in altitude and azimuth. Of course, to be successful this design has to be extremely well made – and I can vouch for the fact that Mr Ells' observatory was. It was very solidly built and no trouble was ever experienced with observer-generated vibrations transmitting to the telescope. Mr Ells described his observatory in the December 1978 issue of the *Journal of the British Astronomical Association*. His paper is titled: 'The Hill–Poncet heated observatory: A "rocking-type" equatorial'.

4
Astrophotography with the camera

Astrophotography is a natural extension of the visual techniques used by amateur astronomers. However, many observers are put off by the impression that it is a particularly difficult skill to master. The imagined expense and complexity of the required equipment is another deterrent. In fact, with modern films, a home-doer can fairly easily obtain impressive and very useful results using very limited equipment.

Basic equipment

A start can be made with a low cost camera mounted on a tripod and operated by a short cable-release. Cameras come in a variety of types. On some the lenses are permanently fixed in position. Other cameras have lenses which are removable. They either have screw or bayonet fittings to allow the attachment of different lenses in place of the camera's 'standard' lens. Basic astrophotography can be carried out with a camera which is fitted with a fixed lens, though having the facility to remove it is certainly useful for more advanced work. Single-lens reflex (SLR) cameras, having an optical system which allows the operator to see what the camera lens is seeing, are especially convenient and are to be recommended.

The camera will have a range of available shutter speeds. Some have shutter speeds as fast as 1/2000 second but the astrophotographer will find that he/she rarely uses exposures even as short as 1/500 second. Even with a very sensitive film in the camera, the low intensity of light received from most astronomical objects warrents the use of much longer exposures.

It is, for this reason, extremely useful (I would say almost essential) for the camera to have a 'B' setting for exposure. When the exposure selector is set to 'B' the camera shutter remains open all the time the exposure button (or cable release) is pressed. Exposures of seconds, minutes, or even hours can then be made. If you use a cable-release you will not run the risk of shaking the camera during the exposure. Having the camera mounted on a firm tripod is also highly desirable for the same reason. Buy a cable-release of the locking type. They are only slightly more expensive than the non-locking variety and are a great deal more convenient in use. During a long exposure you will then be free to do other things.

Types of film

Of all the film formats currently on the market the most popular world-wide is the 135 (35 mm). 135 films are available in cassettes ready for loading into the camera. Also the home-doer can purchase reels of this film to dispense into his/her own empty cassettes, allowing for greater flexibility and economy of use. In addition, there is the greatest variety of film types available in this format. I would certainly advise any budding astrophotographer to obtain a camera which uses 135 cassettes. Another advantage is that a 35 mm camera (a camera which takes this film format) is much more likely to possess such features as the 'B' exposure setting and the facility to change lenses than the cameras designed for other film sizes.

The types of film available in the 135 format can be broadly grouped into:

(a) Black and white negative film.
(b) Colour negative film.
(c) Colour positive (transparency, or 'colour slide') film.

Black and white positive film used to be easily available but is now little used. All these types are useful to the astrophotographer.

Film speed, granularity and resolution

A film's sensitivity to light is given by its ISO number (ISO stands for International Standards Organisation). The ISO number, properly expressed, consists of two parts – the first is the arithmetic value (equivalent to the old ASA number) and the second is the logarithmic value (equivalent to the old DIN number). As an example, Kodak's Plus-X Pan film is ISO 125/22°. Other scales exist but are now little used. Only the arithmetic values will be quoted in this book. The ISO number is often referred to as the film's 'speed'. Higher speed films are more sensitive to light and have higher ISO numbers. A 125 ISO film is considered to be 'medium speed'. Films of speed rating 50 ISO or less are considered 'slow', whilst those above 300 ISO are considered 'fast'. At the time of writing there are several films rated at 3200 ISO.

Since most of the subjects of the astrophotographer are relatively faint, one might think that low speed films would have little to offer and the highest speed films should always be used. However, there is a penalty in using high speed films. Figure 4.1 is an exposure of the Moon I made, using my 18¼-inch reflector, on 3M Colourslide 1000 film . You will notice that the image is very 'grainy' (try looking at the photograph at a distance, then very closely to see how deleterous the effect of grain is). This is an unfortunate characteristic of high speed films. Another, is that high speed films generally have much lower resolving powers than their less sensitive counterparts.

The resolving power of a photographic emulsion is often expressed in terms of the maximum number of lines per millimetre resolvable. In order to understand this, imagine a grid of alternating black and white lines sharply imaged onto a photographic emulsion. A film of poor resolving power will only clearly show the black and white lines as separate if they are widely spaced. In other words, the grid can only be resolved if it has few black and white lines per millimetre. If a much finer grid (more lines per millimetre) was imaged onto the same film it could not show the black and white lines as separate. A uniform greyness is all that would be seen on the processed film. A film with much superior resolving properties would be needed to show this finer grid.

A high speed film might be expected to have a resolving power of around 40 lines per millimetre. A medium speed film should resolve at least 80 lines per millimetre and a slow film will do better still. However, the way in which the film is processed, as well as the density and contrast levels of the image that is recorded on it, will also have a marked effect. Table 4.1 lists the nominal speed ratings and resolving powers for several commercial films. These figures apply only when the manufacturers instructions for processing are followed. In general, any attempt to increase the effective speed rating of a film during processing will increase its granularity and reduce its effective resolution.

Film exposure characteristics

If we consider a black and white negative film, the blackness of a given image on it after processing depends on three factors:

(a) Its effective speed rating (taking into account any treatment designed to alter its speed rating from its manufactured value).

Figure 4.1. Notice the coarse grain structure in this photograph of the Moon.

Table 4.1. *Nominal speed ratings and resolving powers for a selection of commercially available films*

Film	Manufacturer	ISO	Resolving Power (lines/mm)
Black and White negative film:			
PanF	Ilford	50	60–160
FP4	Ilford	125	50–145
HP5 Plus	Ilford	400	50–125
XP1 400	Ilford	400	50–125
Plus-X Pan	Kodak	125	50–125
Tri-X Pan	Kodak	400	50–125
TP 2415	Kodak	25–200	125–400
2475 Recording Film	Kodak	1000	25–63
T Max 100	Kodak	100	63–200
T Max 400	Kodak	400	50–125
T Max 3200	Kodak	3200	40–125
Colour Negative Film:			
Kodak GOLD 100	Kodak	100	50–100
Kodak GOLD 200	Kodak	200	50–100
Kodak GOLD ULTRA 400	Kodak	400	50–125
Kodak VR 1000	Kodak	1000	40–80
Kodak Ektar 1000	Kodak	1000	40–80
Fuji HR 1600	Fuji	1600	40
Konica SR-V3200	Konica	3200	40
Colour Positive Film:			
Kodachrome 25	Kodak	25	50–125
Kodachrome 64	Kodak	64	50–100
Kodachrome 200	Kodak	200	63–100
Ektachrome 200	Kodak	200	50–125
Ektachrome 400	Kodak	400	40–80
3M Colour Slide 1000	3M	1000	40

Note
These are all manufacturers quoted values. In most cases the resolving power is given as a range of values. The lower figure corresponds to the resolution of the emulsion when imaging a low contrast grid (brightness ratio of 1:1:6). The second figure is the resolution one can expect when imaging a high contrast grid (brightness ratio of 1:1000). Where only one figure is quoted, this corresponds to the resolution of a high contrast grid. Data based on relevant technical publications issued by the manufacturers and, in the case of Ilford, additional information supplied by a company official.

(b) The brightness of the object to be imaged.
(c) The length time that the film was exposed to the light from the object (in other words, the length of time that the camera shutter was opened).

Factors (b) and (c) can be combined to give an 'exposure factor', thus:

$$\text{exposure factor} = \text{brightness} \times \text{time}.$$

If the brightness of the object was halved, then the same 'exposure factor' could be obtained by doubling the length of time that the camera shutter was opened.

The blacker the image on the processed black and white negative film, the less light that will pass through it. If this image were only to pass, for example, one tenth of the light incident on it then we would say that the piece of film containing the image has an *opacity* of 10. If the image were darker, so that it only transmitted one hundredth of the light incident on it, then the opacity would be 100, and so on. The logarithm to base 10 of the opacity of the image on the film we term the *density*. Thus an opacity of 100 is equivalent to a density of 2.

Figure 4.2 shows an idealised plot of density versus Log_{10} (exposure factor) for a typical photographic emulsion of slow to medium speed. Such a plot is known as a *characteristic curve* for the emulsion. It shows how the image density builds with 'exposure factor'. Notice that the plot becomes horizontal at the upper right of the graph. This shows that beyond a certain limit of density continued exposure cannot increase it any further. The emulsion has then reached saturation. Notice also that a given minimum 'exposure factor' is needed before the emulsion responds to any significant degree (lower left of the plot).

Between the plotted points A and B the density is proportional to the logarithm of the 'exposure factor'. In normal use the film is exposed so that the resulting image density values lie in the range corresponding to this portion of the graph. The consequent processed image then shows a given collection of photographed objects in their correct brightness relationships.

The gradient of the straight line portion of the characteristic curve determines the contrast of the emulsion. The steeper it is the greater the change in density of the recorded image for a given change in 'exposure factor', and so the greater is the contrast. In Figure 4.2 θ is the angle the linear portion of the plot makes with a line parallel to the Log_{10} (exposure factor) axis. The tangent of this angle is a number which is used to express the contrast of a given emulsion and is often denoted by γ:

$$\gamma = \tan\theta$$

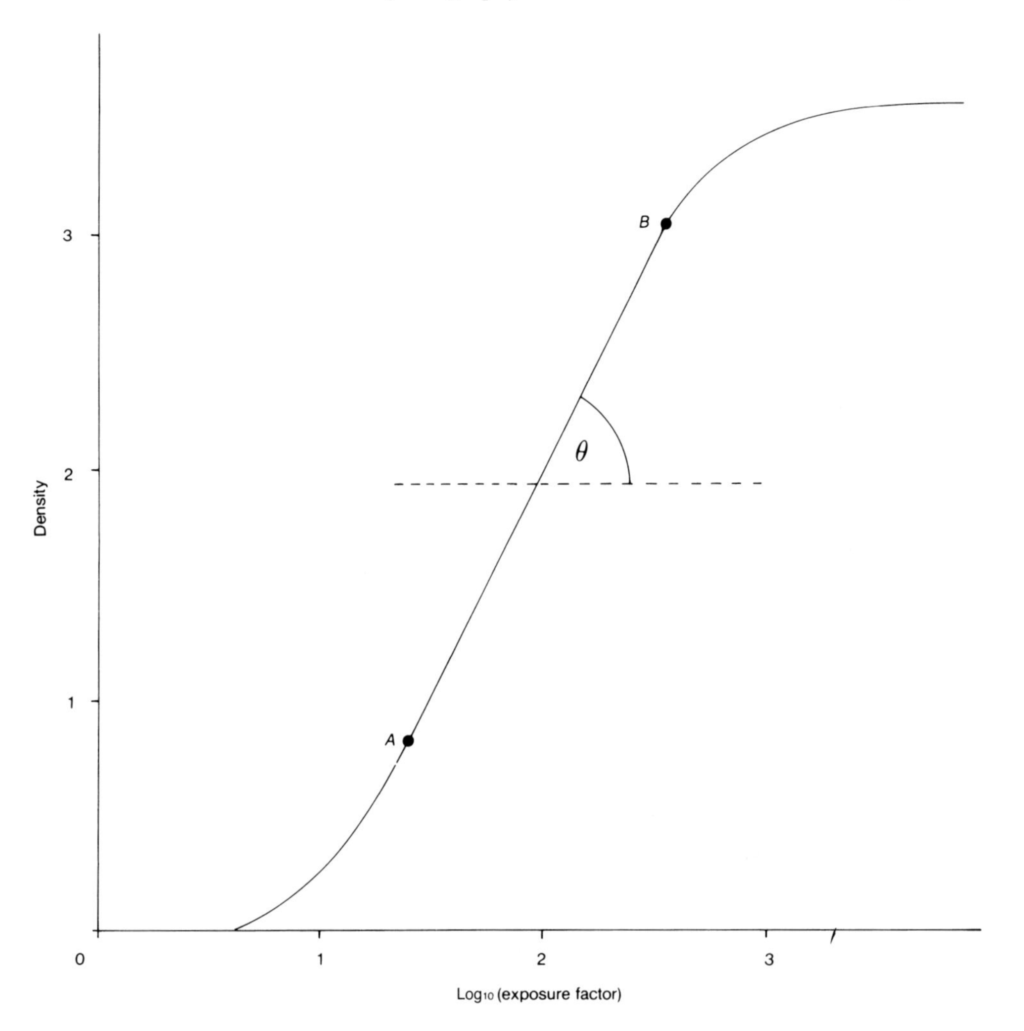

Figure 4.2. The characteristic curve of a typical slow to medium speed photographic emulsion.

In the case shown the angle θ is close to 45° and so γ is close to 1.0. An emulsion of this value of γ will faithfully record a range of brightness levels without unnaturally high or low contrast. Lower values of γ correspond to less contrast and higher values correspond to greater contrast. High speed films tend to have lower values of γ than slower films. Also, the value of γ can be affected by the processing. More about this in the next chapter. Figure 4.3 shows characteristic curves for some commercially available films.

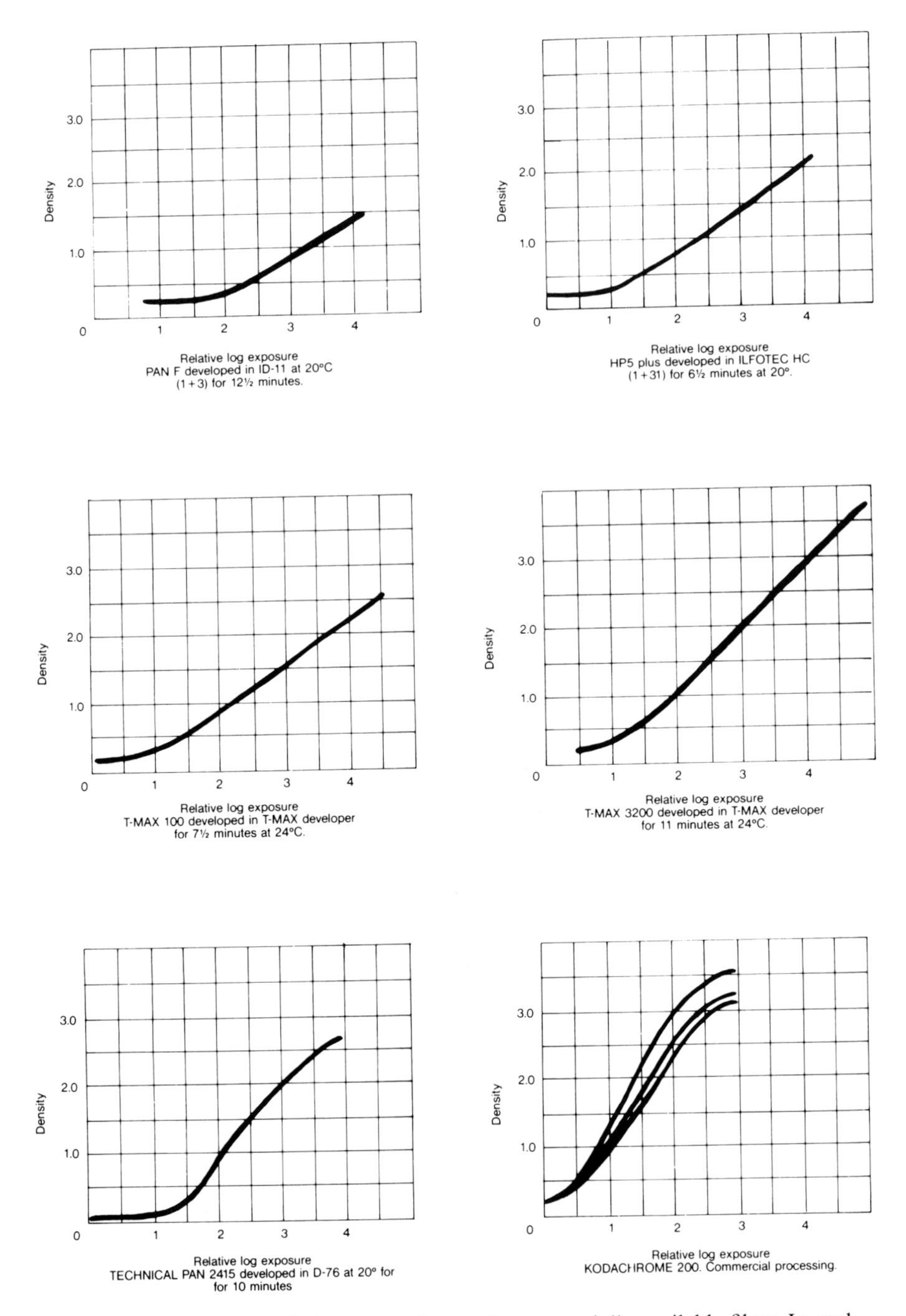

Figure 4.3. Characteristic curves of several commercially available films. In each case these correspond to normal processing in accordance with the manufacturers instructions, based on relevant Kodak and Ilford technical literature, and used with their special permission.

Reciprocity failure and spectral response

For the light levels encountered in everyday photography it is true that if the brightness of the object to be photographed is reduced to a half (maybe the light illuminating the object is reduced by this amount) then the exposure time needed to produce an identical image on the processed film is doubled. However, this relationship breaks down at very low light levels. Halving the brightness of the object then necessitates using more than twice the exposure time to get a correctly exposed image. It is as if the film works at a much lower ISO number when responding to very low light levels. Go below a certain brightness limit and the emulsion fails to respond altogether.

A further complication is that the film, when responding to low light levels, becomes effectively less sensitive with time during the exposure. By this I mean that after the initial period of image formation the image density does not continue to build up at the same rate (even if the density of the image is still well below the saturation limit for the film). As time goes on the rate of further image build up slows down. These effects are known as *reciprocity failure*.

Fast emulsions show the effects of reciprocity failure to a greater extent than medium speed or slow emulsions. In addition to their inherent lack of resolution, reciprocity failure is another reason why high speed films are not always the most suitable for astronomical purposes. With exposures of just a few minutes a high speed film may well image fainter details than a slower film exposed under the same circumstances. However, with very long exposures (perhaps of the order of half an hour or more) the slower film may well overtake the fast film and show details that are totally invisible on the fast film. One is reminded of the parable of the tortoise and the hare!

The human eye is most sensitive to light of wavelength 5.5×10^{-7}m. This corresponds to the yellow/green part of the spectrum. The sensitivity becomes progressively less with wavelengths both higher and lower than that of the maximum response. The earliest photographic emulsions were most sensitive to violet light of wavelength 4.4×10^{-7}m. These could respond to electromagnetic radiation well into the ultra violet part of the spectrum, though the sensitivity gradually decreased with wavelength at wavelengths progressively longer and shorter than 4.4×10^{-7}m respectively. They failed to respond to light of wavelengths longer than about 5×10^{-7}m altogether. Thus the old photographs tended to show red objects much darker and blue objects rather lighter than they appeared to the naked eye.

These old 'blue sensitive' emulsions are termed *orthochromatic*. More modern black and white photographic emulsions have ranges of sensitivity which extend into the red part of the spectrum. These are *panachromatic*. They produce photographs which accord more closely to the eye's response (ie. red objects do not look unnaturally dark).

Colour films have three sensitive layers of emulsion. One layer responds to blue light, another responds to green light and the third responds to red light (though the wavelength responses of the different layers do overlap). These films are designed to produce reasonably faithful colour reproductions of objects photographed in ordinary integrated light (light composed of a mixture of wavelengths), such as daylight. For instance a yellow object will reflect sufficient red and green light to activate the corresponding layers in the film and the final photograph will show the object looking reasonably near to its natural colour.

However, many astronomical objects (particularly emission nebulae) do not emit integrated light. They emit only at specific wavelengths. Thus photographs of a particular nebula taken using different films may well show large variations in colour – none of the reproductions being accurate. Another problem of colour emulsions is that each layer may well suffer from reciprocity failure to a different degree. Thus the colour reproduction becomes less accurate as the length of the exposure increases. The spectral response curves of several commercially available films are shown in Figure 4.4.

Camera lens focal length, image scale and size of field

The same relationship exists between focal length and the resultant image scale for camera lenses as for telescope objectives (see Chapter 1):

$$\text{Image scale} = 206\,265/F,$$

where F is the focal length of the lens in millimetres and the image scale is measured in units of arcseconds per millimetre.

Strictly speaking this figure is only accurate at the centre of the image (where the optical axis of the lens meets the film). At distances further from the centre the image scale progressively differs from this value. Normally the number of arcseconds per millimetre decrease away from the centre of the field (producing pin-cushion distortion), though the amount of distortion present varies with lens design. This distortion is of critical importance in astrometry but it can normally be tolerated for all other astronomical purposes. As far as the image covering a frame of film is concerned, lenses of long focal length produce less distortion than do lenses of short focal length. A single frame of 135 film covers 24 mm×36 mm. The image scale

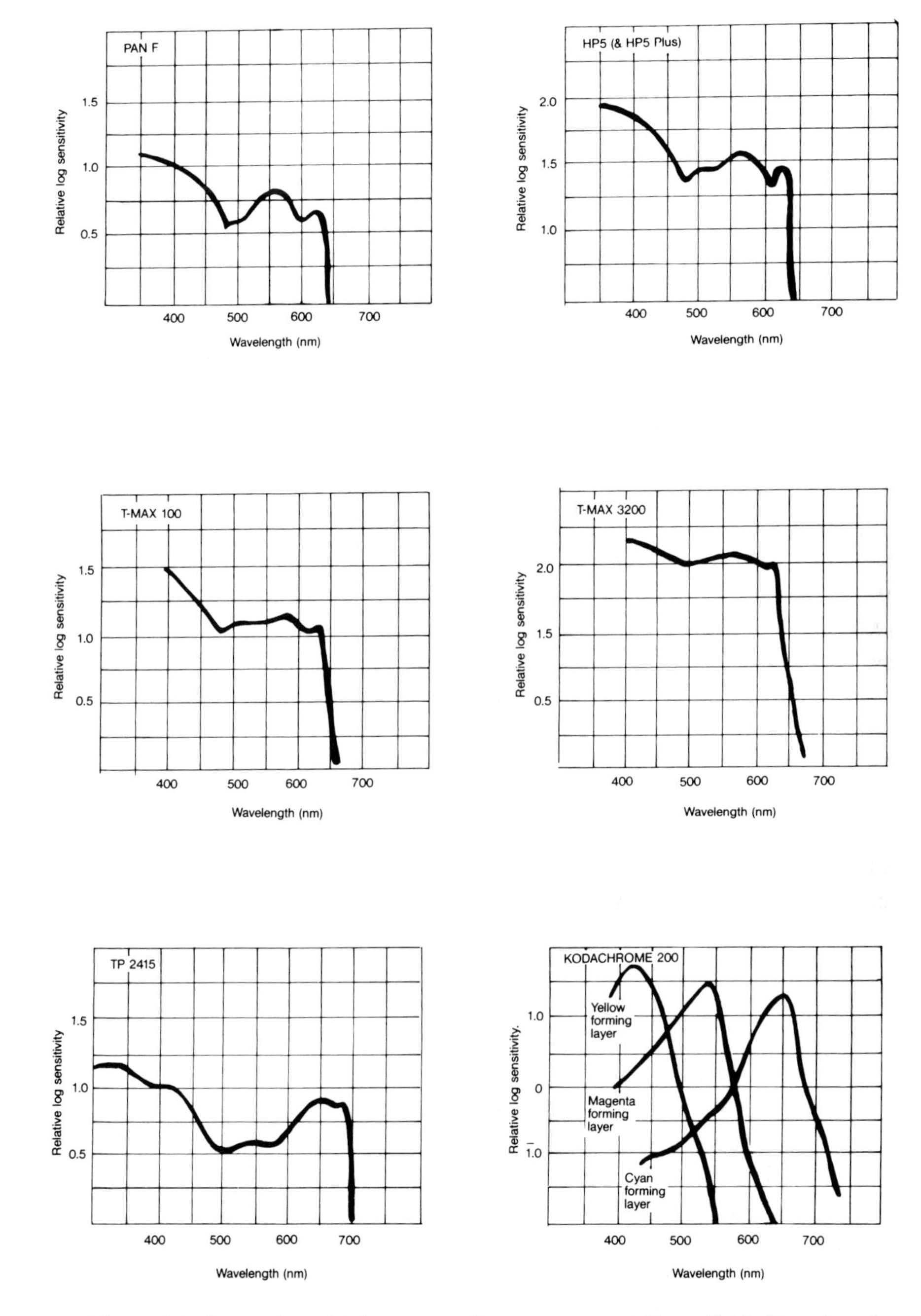

Figure 4.4. Spectral sensitivity curves of some commercially available films, based on relevant Kodak and Ilford technical literature, and used with their special permission.

Table 4.2. *Image scales and corresponding angular sizes of field covered by a 135 format film frame for lenses of various focal lengths*

Focal length (mm)	Image scale (arcseconds mm^{-1})	Angular field covered by a 135 film frame (degrees)
50	4125	27.5×41.3
135	1528	10.2×15.3
200	1031	6.9×10.3
300	688	4.6×6.9
500	413	2.8×4.1

can then be used to calculate the total field of view covered by a frame of this film (see Table 4.2).

Aperture, focal length and focal ratio

Aperture, focal length and focal ratio are defined for the camera lens in exactly the same way as for the telescope objective (see Chapter 1).

Consider two lenses of 25 mm aperture, one having a focal length of 50 mm (and thus a focal ratio of *f*/2) and the other having a focal length of 100 mm (*f*/4). If these lenses are used to produce focused images of a given star on a piece of photographic film, how does the brightness of the star image produced by one lens compare with that produced by the other? In each case the size of the optical image of the star is less than the resolving limit of the film (ignoring any lens aberrations that may be present). We can think of the star images as being 'point images' in each case. The brightness of the image produced by each lens will then only depend upon the aperture of the lens (assuming that the optical efficiency of each lens is the same). The focal ratio of the lens makes no difference to the brightness of the star images formed. Hence **the required exposure time to photograph point objects, using a given film, depends only upon aperture and not focal ratio**.

The situation is a little different when photographing 'extended objects' (those which produce images of larger dimensions than the resolving limit of the film). Imagine using the same lenses to photograph the full Moon. The *f*/2 lens will produce an image of the Moon that is about 0.5 mm across. The *f*/4 lens will produce an image that is about 1 mm across. The amount of moonlight intercepted by each lens is the same. However, the *f*/4 lens produces an image of the Moon that has twice the diameter and four times the surface area as that produced by the *f*/2 lens. Since the same amount of light is spread over four times the surface area, the image brightness produced by the $f/4$ lens is only one quarter that produced by the $f/2$ lens. By the same reasoning an $f/6$ lens of the same aperture would produce an image of the Moon of surface brightness one ninth that produced by the $f/2$ lens. Thus for a given film speed (neglecting any reciprocity failure effects) the length of exposure required to correctly record a given extended object is proportional to the square of the focal ratio:

$$\text{Required exposure} \propto f^2$$
$$\text{(for extended objects)}$$

In fact, the same sort of analysis would show that **when photographing extended objects, using a given film, the exposure required depends only on the focal ratio of the lens irrespective of its aperture.**

The aperture of a camera lens is rendered variable by means of an iris inserted between the lens elements. If the camera operator selects a setting of $f/4$ and the lens has a focal length of 50 mm, then the effective aperture of the lens becomes 12.5 mm.

Astrophotography with a fixed camera

Owing to the Earth's rotation, celestial bodies move across the sky at an angular rate, ω, given by:

$$\omega = 15 \cos \delta,$$

where δ is the declination of the celestial body measured in degrees (ignore the negative sign of the declinations of celestial bodies south of the Celestial Equator) and ω is the apparent rate of drift in arcseconds s^{-1}. Thus an object on the Celestial Equator ($\delta = 0°$) has an apparent drift rate of 15 arcseconds s^{-1}.

I have found that using a camera fitted with a 55 mm focal length lens and a fast (therefore poor resolution) film a total drift of 150 arcseconds can be tolerated before star images become significantly trailed. When photographing stars on the Celestial Equator this is the amount of drift built up over 10 seconds. A lens of twice the focal length would produce an image on the film that is twice as sensitive to position changes (half the number of arcseconds mm^{-1}). The maximum exposure time before significant trailing would then be halved. In general:

$$\text{Maximum exposure time (seconds)} = \frac{550}{F}$$

where F is the focal length of the lens in millimetres.

When photographing celestial bodies away from the Celestial Equator longer exposures can be given. The formula can be refined to:

$$\text{Maximum exposure time (seconds)} = \frac{550}{F\cos\delta}$$

where δ is the declination of the object being photographed. However, do bear in mind that a photograph centred on a particular celestial body will also cover a wide range of declinations north and south of it, where trailing may be greater.

In order to image the faintest objects possible in the few seconds available when using a fixed camera, a fast film is normally advisable. Figure 4.5 is a 10 second exposure of the constellation of Orion, made on 3M's Colour Slide 1000 film, which was commercially processed. The camera lens was set to its lowest *f*/number (*f*/2). The 'belt stars' in Orion are on the Celestial Equator and so show the maximum amount elongation due to trailing for this exposure. The effect of lens aberrations, particularly coma, on star images can be clearly seen in the photograph. Near the edges of the frame the stars appear to be severely distorted, with tails pointing away from the centre of the field of view. A better quality lens would improve matters but all low focal ratio camera lenses show this effect to some extent when set to their lowest *f*/ numbers.

Guided exposures

With long exposures a fixed camera will record star trails (see Figure 4.6). This sets the limit to the faintness of stars that can be recorded. Once the image of a star has moved off from one piece of the photographic emulsion it cannot further build the image density on that piece. The star is then activating the adjacent piece of emulsion. Thus to record fainter stars one has to move the camera so that it accurately follows their apparent motions.

Figure 4.7 shows a 2 minute exposure of the constellation of Orion photographed with the same camera, lens, *f*/number and film type as that shown in Figure 4.5. This time the camera was mounted 'piggy-back' fashion on a clock-driven telescope. The telescope automatically tracked the stars and the result is a photograph that reveals considerately fainter details.

Attaching the camera to the telescope is a purely mechanical problem. A small wooden platform could be made with feet to sit astride the telescope tube. A strong elastic cord then wraps around the tube and, perhaps, attaches to hooks on opposite sides of the platform. The platform has a

Figure 4.5. The constellation of Orion photographed by the author on 1986 December 24^{d} 01^{h} 11^{m} UT. The photograph was taken with a tripod mounted camera fitted with a 58 mm focal length Helios lens set at *f*/2. The length of exposure given was 10 seconds. At the time of the exposure the sky transparency was such that stars less bright than the fourth magnitude could not be seen. The original transparency shows stars of the eigth magnitude in the upper half of the frame.

Figure 4.6. Star trails. This is a 5 minute exposure with a fixed camera fitted with a 58 mm lens, set at *f*/4. It shows the constellation of Cassiopeia circling over the rooftops on 1982 April $15^{d}\ 00^{h}\ 15^{m}$ UT. The exposure was made on HP5 film which was developed in Microphen developer to give an effective speed of about 800 ISO. Photograph by the author.

standard camera ball and socket mount fitted near its centre and the camera is attached to this. You might wish to drill a small hole in a convenient position to take a bolt for attaching the ball and socket mount directly to the telescope tube or part of the mounting. Yet another solution, possible with German equatorial mountings, is to fashion a clamp which wraps round the declination axis and has a bolt, or ball and socket mount, to attach the camera. The important thing is to ensure that there is no relative motion between the telescope and the camera during the exposure.

One pitfall to watch out for is part of the telescope, or its mounting, entering the field of view covered by the camera. Mounting the camera at the 'sky end' of the telescope tube is one way of avoiding this. Another is to have the camera angled so that it and the telescope are looking in somewhat different directions. You should try to ensure that the camera does not seriously upset the balance of the telescope in both R.A. and declination. This normally means re-balancing the telescope. A little trial and error in

Figure 4.7. The constellation of Orion photographed by the author on 1987 Jan $31^d\ 20^h\ 48^m$ UT. The camera, fitted with a 58 mm Helios lens set at *f*/2, was mounted onto a clock-driven telescope for this 2 minute exposure on 3M Colour Slide 1000 film.

Table 4.3. *The accuracy necessary when guiding cameras fitted with various focal length lenses*

Focal length (mm)	Maximum allowable error (arcseconds)	Equivalent time of drift due to diurnal motion (seconds)
50	165	11
135	61	4.0
200	41	2.7
300	27.5	1.8
500	16.5	1.1

Note
The second column shows the maximum error, in arcseconds, that can be allowed to build up during the exposure. The third column shows the corresponding times a star would take to drift the equivalent amount, due to diurnal motion, if it were on the Celestial Equator.

attaching suitable objects as counterweights will soon overcome this problem. Once again elastic straps may well prove useful.

How accurate the guiding has to be depends upon the focal length of the camera lens. Table 4.3 provides some approximate figures. I have taken guided photographs with a makeshift platform strapped to the tube of a 6-inch (152 mm) Newtonian reflector on a German equatorial mount. The telescope had no driving clock, nor any slow motions. Yet I was able to guide a camera with a 50 mm focus lens accurately enough to get good round star images after several minutes exposure by merely pushing the tube westwards in R.A. every few seconds, with the declination axis locked. My method was to keep a suitable guide star on the edge of the the field of view of the eyepiece. The success rate was very high. However, this arrangement would not have sufficed for guiding a camera lens of longer focal length.

On the subject of the guide star, you need not necessarily select one that is close to the object you are photographing. The camera and the telescope do not need to be pointing in the same direction (provided that there is no relative motion between them during the exposure – beware of mechanical flexture!). A reasonably bright star near the Celestial Equator would be a good choice since its apparent drift rate due to the diurnal motion is greatest. This allows for more precise guiding.

The additions of a manual slow motion to the R.A. axis of the telescope and a crosswire eyepiece would probably be sufficient to guide a camera lens of up to 300 mm, or even 500 mm. Of course, a clock drive would make guiding less onerous and more accurate. It is an added convenience if the crosswire eyepiece has some internal illumination that allows the wires to be seen against the sky-background. A magnification of around ×100 will allow guiding that is potentially accurate to a few arcseconds. A higher magnification will make good guiding easier.

One factor that becomes very important when guiding camera lenses of long focal length (especially where long exposures are given) is the accuracy of alignment of the polar axis of the telescope mounting (see Chapter 3 for details). The provision of a declination slow motion would certainly relax this requirement, though even then the error cannot be too large. Declination corrections then generate a rotation of the image about the guide star. The magnitude of these effects on the final photograph depends upon the position of the object in the sky, the latitude of the observation site, the focal length of the camera lens and the length of the exposure. In general, for exposures up to 10 minutes and a camera lens of 135 mm focal length, you could expect the errors to be negligible provided the polar misalignment is less than ½°.

Astrophotography with a barn door

If the telescope has an accurate clock drive, and an accurately aligned polar axis, there is no need to monitor a guide star or to give any corrections during the exposure. Figures 4.8 and 4.9 show two photographs, both made with a 135 mm telephoto lens, taken in this way. Provided the camera can be accurately driven to follow the diurnal motion there is no need for the telescope at all. An arrangement nicknamed the *barn door mounting* is becoming quite popular amongst astrophotographers.

This type of mounting consists of two platforms, one movable and one fixed, joined at one edge by a hinge. The arrangement is orientated so that the hinge is aligned with the Earth's rotation axis. The camera is attached, via a ball and socket mount, to the movable platform. The platform is then driven by a screw, so that the attached camera follows the stars in their diurnal motion. Figures 4.10(a) and (b) shows one variant designed and built by Nick James. In his version the screw is driven by a stepper motor through a gearbox.

With reference to Figure 4.10(a), the required linear drive rate, D, is given by:

$$D = r\tan(15'),$$

where D is the linear drive rate expressed in arcminutes per minute. Of course, the drive rate can only be strictly accurate at one angle, since the drive screw is linear and does not follow an arc centred on the hinge.

Figure 4.8. The 'belt and sword' of Orion photographed by the author on 1988 February 6^{d} 20^{h} 48^{m} UT. The exposure was 7 minutes on 3M Color Slide 1000 film. The camera was fitted with a 135 mm telephoto lens, set to *f*/2.8, and was mounted onto a clock-driven telescope.

Figure 4.9. The Pleiades, a 10 minute exposure, on 1988 February 6^{d} 20^{h} 55^{m} UT. Other details are the same as for Figure 4.8.

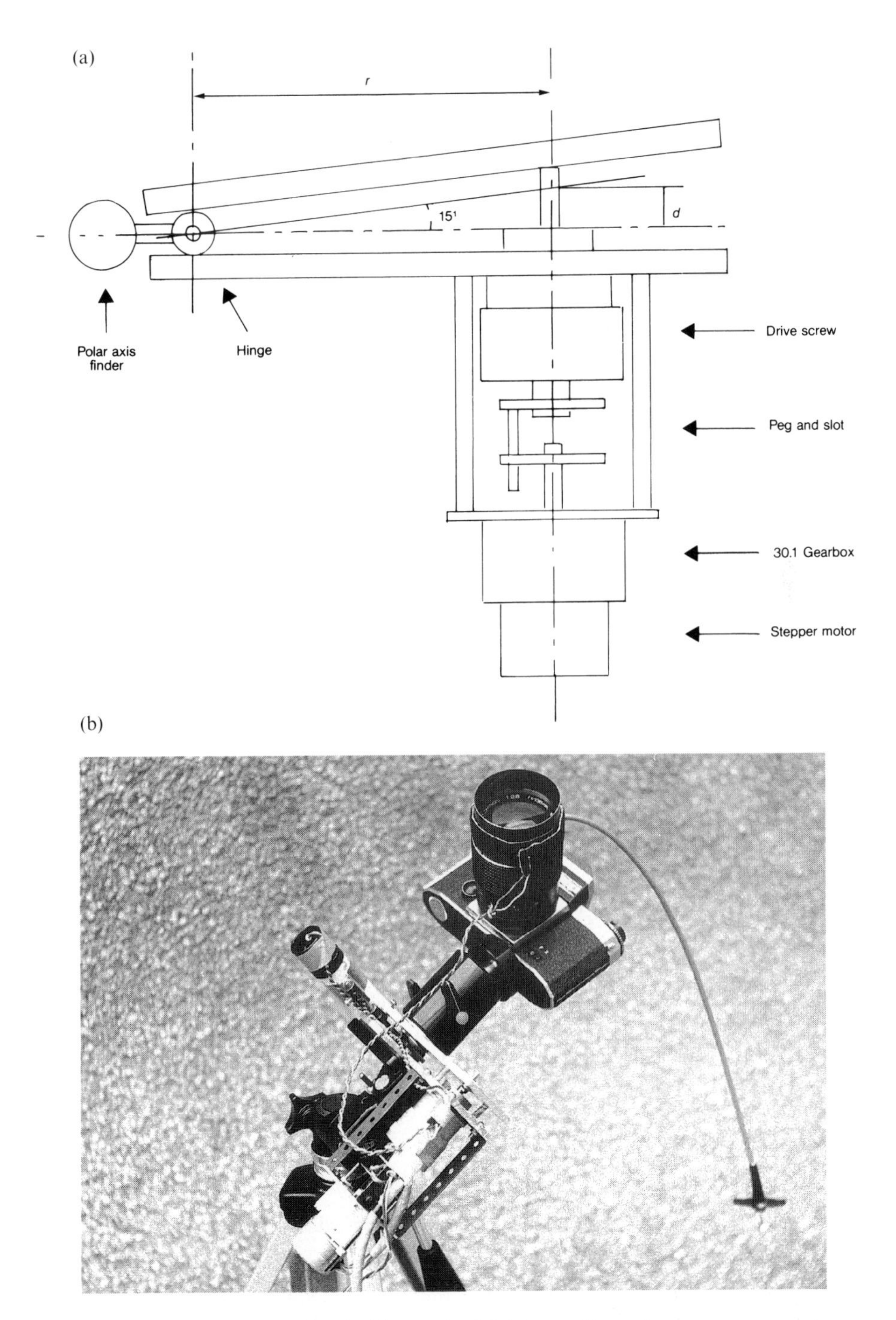

Figure 4.10. (a) The mechanical layout for Nick James' barn door camera mount. See text for details. (b) Nick James' barn door camera mount. The fixed platform is attached to the top of the tripod. The camera is mounted on the movable platform.

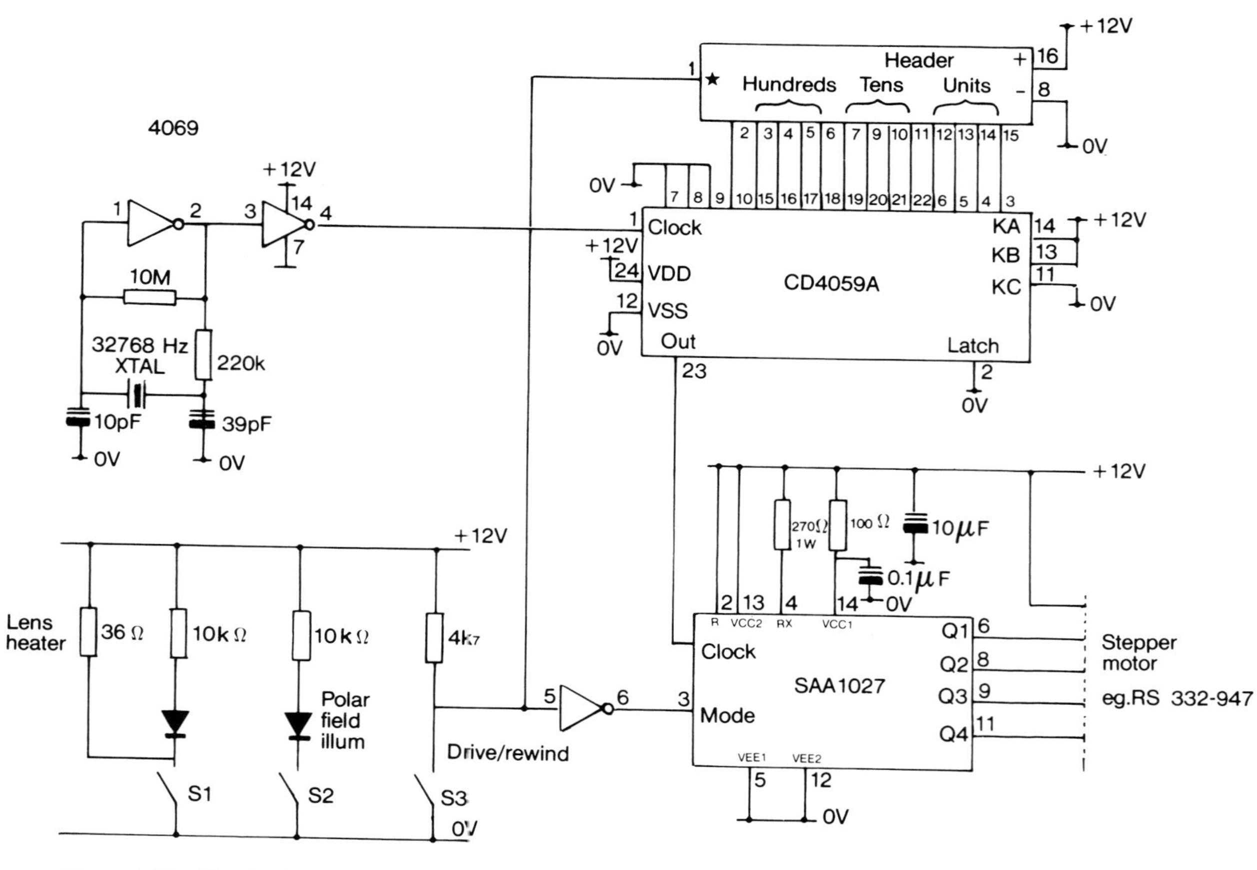

Figure 4.11. Circuit diagram of the oscillator Nick James designed and built for his barn door camera mount.

However, the error is always negligible in practice. The required rotation rate of the screw, R, is the required linear drive rate divided by the pitch of the thread on the screw:

$$R = 360\ D/P,$$

where P is the screw pitch expressed in millimetres and R is the required rotation rate expressed in degrees per minute.

When constructing such a device do ensure that the hinge you select has little side play and is sufficiently strong to support the weight of the camera without twisting or sagging. Nick James' barn door mounting incorporates a small polar axis sighting telescope (made from a 5×22 finder-scope) set parallel to the hinge. A red L.E.D. illuminates the crosswires. Since Polaris lies nearly a degree away from the true celestial pole, he uses a home-constructed star-chart showing the field of view surrounding the celestial pole. He orientates this chart according to the position of Ursa Minor in the sky at the time of observation. Sighting through the telescope then allows him to align it, and hence the barn door's hinge, to within a fraction of a degree of the true pole.

Figure 4.11 shows the circuit diagram of the crystal oscillator Nick built for his barn door mounting. The unit was designed to be portable and works off a 12V lead–acid battery (having a gel electrolyte – much more safe and convenient than a car battery). At the end of several exposures the drive can be rewound by closing switch S3. S2 operates the polar alignment telescope illumination. Another feature of this unit is a camera lens heater, made by sewing a 1.8m length of 20Ω/m resistance wire into a Velcro strip. The strip is wrapped around the top of the lens. Closing the switch S1 causes 4W of heat to be dissipated, enough to keep the lens dry even on very dewy nights.

Figure 4.12 shows one of the very many excellent photographs Nick has taken using this arrangement. Other astrophotographers have built less sophisticated barn door mountings and still got very good results. Some have even dispensed with the stepper motor and gearbox and have cranked the screw by hand, in step to audible clicks from a metronome.

How long the longest exposure?

A ten minute exposure on a fast film, taken using an $f/2.8$ 135 mm telephoto lens, may well show stars of the thirteenth magnitude. What about longer exposures?

Even allowing for tracking errors and the effects of reciprocity failure, there is still one factor that will limit the length of the exposure one can normally give: the brightness of the sky-background. In part the sky

Figure 4.12. The Double Cluster (NGC 869/884) in Perseus photographed by Nick James on 1986 October $12^{d}\ 00^{h}\ 13^{m}$ UT, utilising his barn door camera mount. This is a 3 minute exposure on HP5 film, using a 135 mm telephoto lens set at *f*/2.8.

brightness is caused by natural effects, chiefly auroral emission. However, for most of us the 'light pollution' due to artificial lighting is by far the greatest source of the background sky-glow. An exposure of more than 10 minutes on 1600 ISO emulsion through an *f*/2.8 lens may well fail to record fainter stars, even at moderately dark sites. Prolonging the exposure just increases the 'greyness' of the sky background on the final photograph.

Since the sky-background is effectively an extended body, its effect on the film may be reduced by going to higher focal ratios. Fainter stars could then be recorded, provided the higher focal ratio is achieved without reducing the aperture. Unfortunately with a standard camera lens this is exactly how it is achieved. Also, using a higher focal ratio is of little help if the celestial body to be photographed is itself an extended body, such as a nebula or a comet.

One way of partially overcoming the light pollution problem is to use a filter which selectively blocks out the wavelengths of light emitted by the sources of light pollution (usually mercury and sodium street lighting) but freely transmits all other wavelengths. A number of such filters are on the market (see the listing of manufacturers and suppliers in Chapter 17). Figures 4.13(a) and (b) show the effects of using such a filter to take photographs from a light-polluted site.

Figure 4.13. The effect of using a Lumicon Deep-Sky filter on photographs taken from a light-polluted site. Each photograph is of the area of the sky around Vulpecula and was taken by Nick James using a driven camera fitted with a 55 mm focal length lens. Both photographs were recorded on T-Max 400 film and taken within a few minutes of each other. Photograph (a) is a 1½ minute exposure made with the camera lens set to *f*/2.8 and with no added filtration. Photograph (b) is a 4 minute exposure made with the camera lens set at *f*/2 with the filter in position. Though the filter does cut down the light of all wavengths to some extent, the selective blocking of the chief sources of light pollution allows much longer exposures to be given. The improvement is obvious.

Through-the-camera astrophotography has applications in many of the fields open to the amateur astronomer. Recording meteors and aurorae, monitoring flare stars, variable stars, comets and asteriods, as well as searching for novae and supernovae, easily come to mind. Each field requires its own techniques and these will be covered in the relevant chapters of this book.

5

Astrophotography through the telescope

A natural progression from taking long exposure photographs with a camera mounted piggy-back on a telescope is to take the photographs through the telescope itself. This chapter follows directly on from the last, where the basic principles were covered. Long exposures through the telescope require particularly good guiding and a steady telescope drive. However, it is also possible to obtain some useful results with a telescope not even equipped with a drive! The specialist topics of comet and solar photography will be covered in Chapters 10 and 11, respectively.

Photography at the principal focus

By 'principal focus' I mean the focal plane of a reflector's primary mirror, without the addition of further optics to enlarge the image; and similarly for the focal plane of a refractor. In most cases this means the Newtonian focus of a reflector. The number of amateur reflecting telescopes that have a prime focus position (no secondary mirror) is vanishingly small.

The first problem is how to mount the camera at the focus. Many commercially made telescopes have a focal position which is quite close to the tube wall. Even with an adapter that fits the lensless camera body into the drawtube, the drawtube may not rack far enough in to allow the image to focus on the film. Making or purchasing a 'low-profile' focuser may solve the problem. If not, then moving the secondary mirror and the eyepiece focusing mount down the telescope tube towards the primary mirror may be necessary.

There may be no need to say this, but if you do take that drastic action be sure that all your measurements are very carefully made and proceed with caution. Make sure that all the fresh holes you drill, etc., are correctly placed. Make sure that all the optics are removed and safely stored while you are making the alterations. I would hate to think of any enthusiast ruining his/her expensive telescope! Also, it is as well to check that the size of your telescope's secondary mirror allows for full illumination of at least a small area surrounding the optical axis after the secondary is moved (it will then intercept a larger area of the cone of rays from the primary mirror). The formula for calculating this is given in Chapter 1. You might have to purchase (or make) another secondary mirror for your telescope if the original is not large enough.

Another possibility is to make or buy an achromatic relay lens. If the distances to the foci are equal then the effective focal ratio of the telescope is preserved. Nonetheless, the distance separating the telescope's focal plane from the film at the back of the camera is then four times the principal focal length of the relay lens. Many telescope manufacturers will custom-make a relay lens unit for you, complete with the correct fittings for your camera and the eyepiece focusing mount of your telescope. In a similar vein, many manufacturers supply telecompressors which convert (usually halve) the effective focal ratio of a telescope. These are very useful for use on higher focal ratio telescopes (refractors, Cassegrains and catadioptrics) but the purchaser should be aware that these often severely vignette the field.

On the subject of vignetting, drawtubes can severely limit the size of the fully illuminated area on the film, especially for low focal ratio telescopes (see the equation given in Chapter 3). Really, the camera should be mounted as close to the telescope tube as possible and using the widest possible drawtube. Even the mounting ring of the camera body itself can cause vignetting with telescopes 'faster' than $f/5$.

However your camera is mounted, the next problem is precisely focusing the image on the film. Here a single-lens reflex camera is particularly useful. The photographer, being able to see the field of view that will be imaged onto the film, can compose and focus the shot much more easily than can someone who is 'working blind'.

Of course, there is a problem with faint objects. Finding and then focusing on reasonably bright and conveniently close stars is a help when the object you intend to photograph is virtually invisible on the focusing screen. Even so, an SLR camera fitted with a plain matt focusing screen will not allow for critical focusing, even with bright objects. The best that can be done is to rack the focuser in and out and try to set it half way between the positions which cause identical small amounts of blurring. The situation is rather better if the camera is fitted with a Fresnel screen. The 'fine-focusing dot' in the middle allows a much more critical focus to be achieved and this type of screen has a greater through-put of light, giving brighter images.

Many advanced astrophotographers use a simple (but accurately made) device to improve focusing – the so-called *knife edge focuser*. The device looks like a large eyepiece but it contains no lenses. A knife edge is mounted close to the eye aperture. In use, a star is positioned just off-centre in the field of view and the knife edge is rotated. When the knife edge is positioned exactly at the focal plane of the telescope the star appears to abruptly disappear and reappear. At any other position the star appears to slowly fade out and brighten. Without further disturbing the focus, the focuser is exchanged for the camera.

The focuser is accurately made so that the film plane of the camera coincides with that previously occupied by the knife edge. In practice, photographers can first use their camera to locate the required field and achieve an approximate focus, then exchange the camera for the focusing aid, only to replace the camera when critical focus has been achieved. With care they can be used with non-SLR cameras. Many companies supply the knife edge focuser (see Chapter 17 for a listing of companies) and other types of focusing aid. The manufacturers always include full instructions for their use.

See Chapter 1 for the formula for the image scale at the focal plane of a telescope. It is the same equation as given for calculating the image scale at the focus of a camera lens. A focal length of 2 metres produces an image scale of 103 arcseconds mm^{-1}. A 24 mm×36 mm film frame then covers an angular extent of the sky of 0°.69×1°.03. This is, by a wide margin, easily big enough to capture the full extent of the Moon (Figure 5.1), though not enough to cover the whole of the Pleiades.

How long is the longest exposure that can be given without tracking the telescope at the sidereal rate? Based on the empirical formula presented in the last chapter, the maximum time is given by:

$$\text{Maximum exposure (seconds)} = \frac{550}{F\cos\delta},$$

where F is the focal length in millimetres and δ is the declination of the centre of the field of view. For photographing objects on the Celestial Equator with a telescope of 2 metres focal length, the maximum exposure time that can be given is approximately 0.25 second. For an object at a declination of 60° the maximum exposure time becomes 0.5 second. Star trailing will then be barely noticeable **provided firing the camera shutter does not cause the telescope to vibrate during the exposure**. A combination of a 16-inch (406 mm) aperture telescope of 2 metres focal length and a 1000 ISO film ought to show stars as faint as the tenth magnitude with a

Figure 5.1. Earthshine on the Moon, photographed by Martin Mobberley at the *f*/5 Newtonian focus of his 14-inch (356 mm) reflector. The exposure, made on 1989 August 28^{d} 03^{h} 20^{m} UT, was 5 seconds on T-Max 400 film.

0.25 second exposure. Not bad, when one considers that this is roughly the practical limiting visual magnitude of a good pair of binoculars!

Obviously to record fainter objects one must make the telescope accurately track. A rough guide to the maximum tracking error, in arcseconds, allowable during an exposure can be calculated from the formula:

$$\text{Maximum allowable error} = 7200/F,$$

where F is the focal length of the telescope measured in millimetres. Achieving this required accuracy is easy only when the telescope has been properly designed and built. Too often commercial telescopes are beset by problems which prevent long exposure photography. Does the telescope drive motor really run at a constant rate? Do the meshing gear teeth really impart a smooth motion, or is the result an apparent east–west oscillation of the object to be photographed? Is the motion smooth, or is it jumpy – the gears giving the telescope a shove, then relaxing until they have wound up more tension? Is the driving-rate correct? Is the telescope tube, tube-fittings, and mounting sufficiently rigid or does the optical axis gradually move off the object to be photographed as the exposure continues? Is the polar axis accurately aligned to the celestial pole? If the answer to any of these questions is no then well-tracked long exposure photography will be difficult to achieve, even if not actually impossible.

Taking the points in order, the periodic error (one complete east–west oscillation for every rotation of the worm) due to a poorly mating worm-and-wheel can only be partially cured. The remedy is to dismount the telescope and uncouple the drive mechanism from the worm (leaving the worm in position). Next, plaster the gear wheel with carborundum paste. By attaching the worm to the bit of an electric drill and running the drill, the worm-and-wheel can be ground into a better fit, renewing the carborundum paste as required. Be prepared for several hours work if the worm-and-wheel are poorly matched to start with.

Backlash may be partially overcome by deliberately making the telescope out of balance in R.A. Most authorities suggest adding weights so that the drive motor has to work harder in pushing the telescope round to the west. It is certainly true that the backlash can be taken up in this way but do make sure that you do not overload the motor, and perhaps even burn it out. Personally, I think that it is better to add weights so that load on the motor is reduced (ie. with the drive disengaged the telescope would tend to move round to the west on its own), the drive mechanism then acting as an escapement. Even so, be careful that you do not cause the mechanism to bind up. Backlash and period errors are both avoided if your telescope has a friction drive, like that described in Chapter 3.

Providing the telescope with means of altering the drive-rate (such as an adjustable frequency-stabilised variable frequency oscillator for an a.c. synchronous motor, an adjustable voltage-stabilised supply for a d.c. motor, or a computer (or pulse generator) driven stepper motor, etc) will allow for fine adjustments to the drive-rate.

Flexture can be a very great problem. Apart from using one of the so-called 'off-axis' guiding devices (described shortly), this factor may force the astrophotographer to use the shortest exposures, or else to rebuild the telescope.

Dealing with the last difficulty on the list, the procedure for accurately aligning the polar axis of an equatorial mounting is described in Chapter 3.

Guiding for long exposures

Ron Arbour's 16-inch (406 mm) *f*/5 Newtonian reflector is sufficiently well designed, well built and accurately adjusted to enable him to make exposures of several minutes and still achieve practically perfect tracking

Figure 5.2. The Dumbbell Nebula, M27, in Vulpecula photographed by Ron Arbour with his 16-inch (406 mm) *f*/5 Newtonian reflector. No guiding corrections were needed for this 5 minute exposure on Tri-X film.

without having to make any corrections while the exposure is taking place. Figure 5.2 shows a 5 minute exposure he made of the Dumbbell Nebula. Even so, in order to successfully use longer exposures the telescope's tracking has to be monitored and small adjustments have to be given from time to time. Even if the mechanical errors of the telescope can be made to be vanishingly small, atmospheric refraction (the effect of which varies with altitude) makes its presence felt for long exposures.

Attaching a suitable telescope to the main tube allows one way of monitoring the tracking. The principle was described in the last chapter. However, for taking photographs through the main instrument, the selected guide star must lie within a few degrees of the object to be photographed. If not, the displacements caused by differential atmospheric

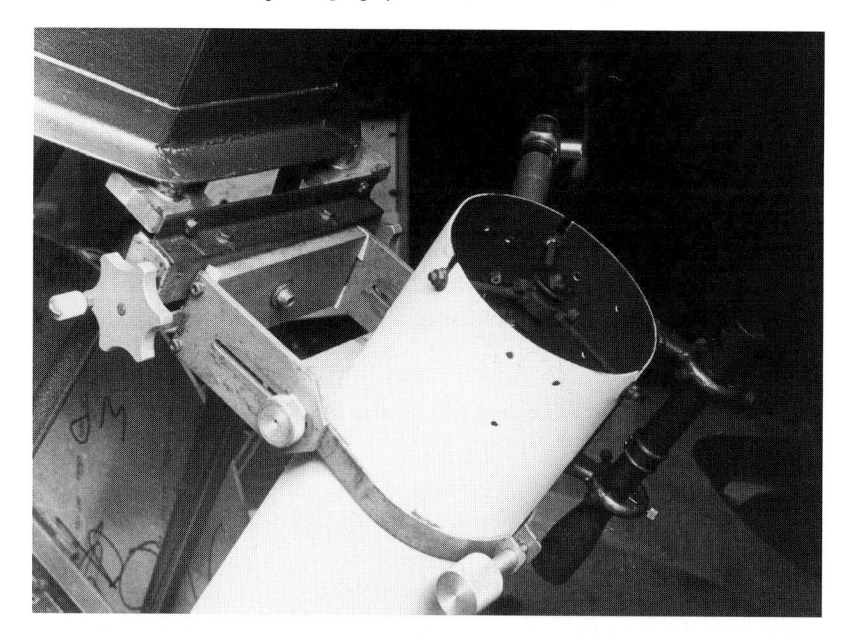

Figure 5.3. Notice the rugged mounting of the guide-scope attached to Ron Arbour's 16-inch reflector. The provisions for adjusting the pointing of the guide-scope are also visible, the tube pivoting at its lower end.

refraction will vary sufficiently during the exposure to cause noticeable trailing (remember, during a long exposure the object and the guide star will both move an appreciable amount across the sky). Any differential mechanical flexture between the main telescope and the guide-scope will also produce a trailed image. Hence the need for rugged construction.

Figure 5.3 shows the 6-inch (152 mm) reflector Ron Arbour uses as a guide-scope on his 16-inch reflector. Notice its rugged mounting and the provision for altering the aim of the guide-scope in order to select a suitable guide star, once the main telescope has been set on its target. After locating and centering a suitable candidate, the wide-field eyepiece is replaced by a Barlow lens and an illuminated crosswire eyepiece for guiding. The magnification used for guiding should be as high as possible (at least $\times 250$ for arcsecond accuracy). Ron has routinely obtained some very fine photographs using his home-built equipment (see Figure 5.10).

'Off-axis' guiding provides a way of monitoring the tracking of a telescope without the need for a separate guide-scope. It also takes care of the problem of flexture. A number of manufacturers produce these units and they can be specially designed for given types of telescope. They usually consist of a very small prism or mirror which intercepts a little of the light from the telescope's focal plane, sending it sideways into a relay lens, then into an eyepiece via a diagonal mirror. The unit is plugged into the telescope before the camera body (sometimes an additional relay lens is needed to transfer the focus to a position further from the telescope tube). The intercepting small prism or mirror is movable, so that a suitable star can be selected (usually near the edge of the field of view) for guiding. As usual, consult the list of manufacturers in Chapter 17. Whatever the system used for monitoring, the fine adjustments are made to the telescope's pointing by the means outlined in Chapter 3.

Films, exposures, filters and focal ratios for deep-sky photography

The details outlined in the last chapter apply equally well here. Fast films allow faint objects to be recorded in relatively short exposure times but suffer from graininess and poor resolution. Low focal ratios are especially useful in allowing faint extended details to be recorded, though a limit is reached when the sky-background is recorded to the extent that its density on the processed emulsion places it on the linear portion of the characteristic curve. For an $f/5$ telescope this *sky fog* limit may well be reached in about three quarters of an hour for a fast (around 1000 ISO) emulsion, even at a dark site. City lights will drastically reduce this time. Also don't forget the effects of reciprocity failure. A fast film is like an athlete trying to sprint in a long distance race. A medium speed 'jogger' is much more likely to complete the run!

What is the faintest stellar magnitude one can expect to record? Assuming the film speed and exposure length are sufficent to reach the limit imposed by sky fog, then the following expression will give a rough guide to the likely limiting magnitude on the very best nights:

$$\text{Limiting magnitude} = 5 + 5\ \text{Log}\ F,$$

where F is the effective focal length of the telescope in millimetres. If you are surprised that the telescope's aperture does not appear in this expression, remember that the magnitude limit is imposed by the recorded sky fog and this behaves as an extended body. Its visibility depends on the film, exposure length and **focal ratio**. The formula can only be a rough guide as many factors will effect the outcome. For instance, atmospheric turbulence plays a more important role with longer focal lengths, since star images are smeared out into discs and their light is effectively diluted.

Of course light-pollution, or other, filters may help, especially if a film is chosen with a spectral response that complements the filter. For instance Kodak's Technical Pan 2415 is especially good when used with a red filter, because of its enhanced red-sensitivity (see the film data given in the last chapter). This film also has superbly fine grain and ultra high resolution. Though its nominal speed rating is quite low (25–200 ISO, depending on processing) it can be increased to something like 1200 ISO by hypersensitisation (discussed later in this chapter).

A red-sensitive film (used with a red filter) will respond well to the Hα light of emission nebulae, while the filter will reduce the effects of the mercury and sodium artificial lighting. TP2415 is also very blue-sensitive (as are most other standard emulsions). A blue filter will partially block the effects of the mercury emissions and will drastically reduce the effects of sodium lights. Reflection nebulae and the spiral arms of galaxies will show up especially well with a blue filter–film combination.

Unlike high quality camera lenses, Newtonian reflectors suffer badly from outfield coma. With a 135 film frame coma will be noticeable at the corners for telescopes of focal ratio *f*/5 and under (see Chapter 1 for a fuller discussion). This can be compensated by the addition of a coma-correcting lens (commercially available). Alternatively, the budding astrophotographer can purchase one of the many forms of telescope suited to wide-field photography, particularly those based on the Schmidt design.

If adequate coma correction can be provided, as well as the absence of vignetting, then the astrophotographer can take advantage of larger film formats: 120 roll film, or even larger sizes of sheet film. Much larger areas of the sky can then be photographed, while maintaining a reasonable image scale and resolution.

High resolution astrophotography

The image sizes at the principal focus are usually far too small to allow for high resolution photographs of the Moon and planets. For instance, at the 2.59 metre focus of my largest telescope the image scale is 80 arcseconds per millimetre. In perfect seeing the telescope should show details as fine as 0.3 arcsecond. This would require an emulsion that is capable of resolving more than 260 lines per millimetre. Even providing this (of the more common types of film only Kodak TP2415 allows this resolution, and then only if suitably processed), one is still left with the problem of focusing the image to that degree of precision. To circumvent both problems one must enlarge the primary image. There are four principal methods used: Barlow projection, eyepiece projection, relay lens projection, and ∞ to ∞ (infinity to infinity) focusing, using the eyepiece and camera lens. The following notes detail each of these methods.

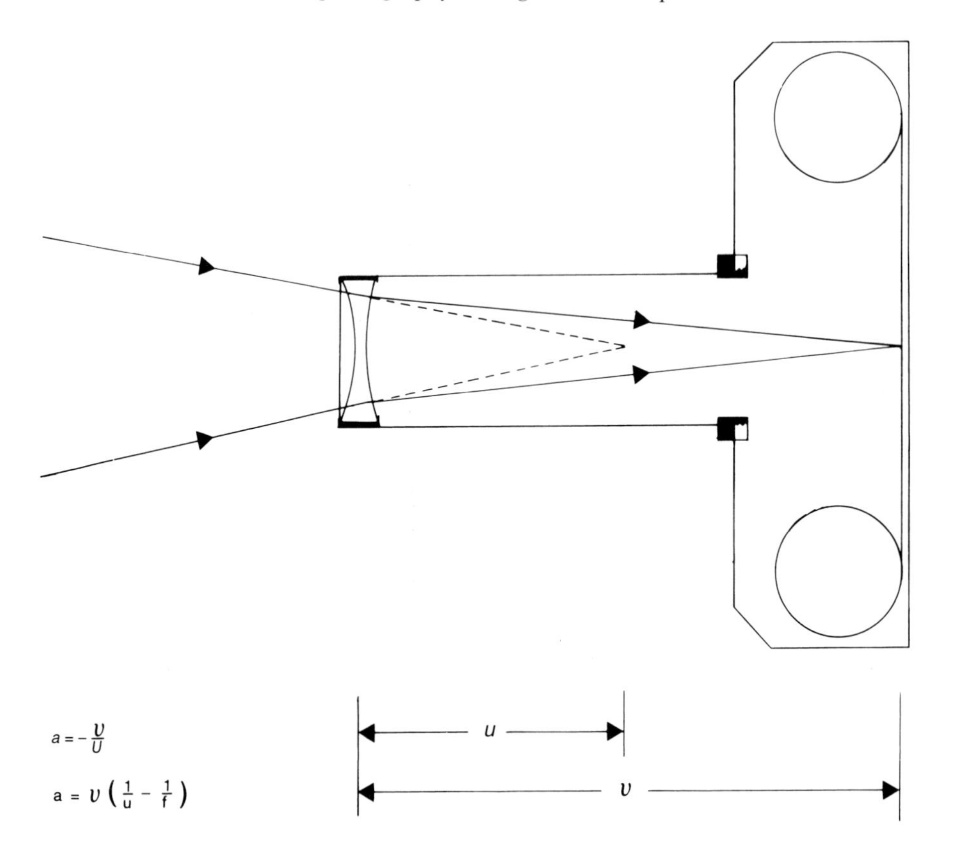

Figure 5.4. The arrangement for enlarging the primary image by projection with a Barlow lens. Alternative formulae for calculating the amplification factor, *a*, are given in the diagram.

Barlow projection

The Barlow lens is described in Chapter 2. The formula for calculating the resultant amplification factor is given there and in Figure 5.4. The lens must be mounted in a tube with the appropriate attachment (screw thread or bayonet fitting) for plugging in to the camera (Figure 5.4). The Barlow's tube then fits into the telescope drawtube. The only real problem with Barlow lenses is that they are usually designed for one specific amplification factor (usually somewhere in the range ×2 to ×3). This amplification may not be enough to match the telescope's potential resolution to that of the film. Using the Barlow lens to provide a higher amplification would involve increasing the aberrations (particularly chromatic) produced by the lens itself.

Eyepiece projection

Figure 5.5(a) shows the general method of attachment used. The equation for calculating the resultant amplification factor is given in the diagram. In general, eyepiece projection can be very successful for photography of the planets, where the object only occupies a small area of the field of view. It is less successful for large extended objects, such as the Moon. With the central zone of the image in focus the outer zones can be very blurred. Figure 5.5(b) shows an early experiment of my own with eyepiece projection.

Relay lens projection

The reason that eyepiece projection produces blurred outfield images is that the eyepiece is designed to produce parallel ray bundles and not the converging sets of rays that are necessary to form an image on the film. Replacing the eyepiece with a more suitable lens system will improve matters dramatically. Enlarger lenses are particularly suitable, as are the eyepieces fitted to projection microscopes.

One note of caution though, do make sure that the focal ratio of the lens system you use is lower than that of your telescope. When you are using the relay lens to produce an amplification factor of unity it should have a focal ratio no more than half that of your telescope. Where it is used to provide a larger amplification its focal ratio can be a little higher. Otherwise the lens will effectively stop your telescope down. The equation used to predict the amplification factor is the same as that for eyepiece projection. Under this heading I include the commercially produced 'tele-extenders' and 'teleconverters', which (if of good quality) can work excellently with telescopes.

∞ to ∞ focusing, using camera lens and eyepiece

Very good high resolution photographs can be obtained by keeping both the camera's lens and the telescope eyepiece in position. In this arrangement the eyepiece is focused in the normal way (and hence to ∞ (infinity), at which condition the rays exiting from the eyepiece are in parallel bundles) and the camera lens is itself set to ∞ focusing. The camera is then positioned so that the camera lens looks squarely, and on-axis, into the eyepiece. In effect, the camera lens replaces the observer's eye. This method can work with a non-SLR camera, though the uncertainty in precise infinity focusing of the eyepiece caused by the accomodation (or long sight, or short sight) of the observer's eye gives the SLR user a distinct advantage. Fine focusing is then

(a)

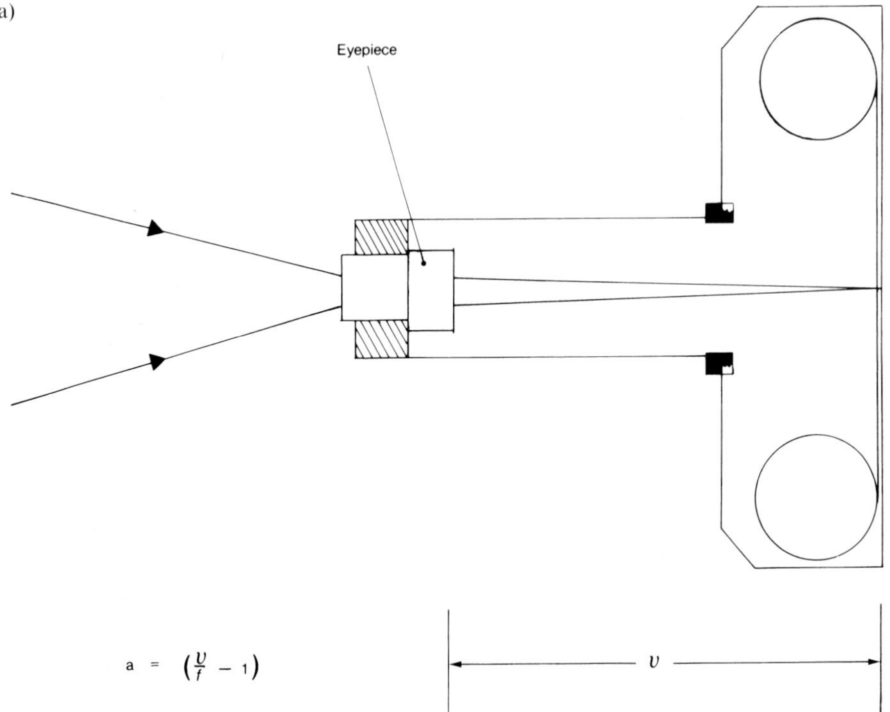

effected by looking through the SLR's viewfinder. The amplification factor is given by the ratio of the camera lens and eyepiece focal lengths:

$$a = f(camera)/f(eyepiece)$$

Figure 5.6 shows a photograph of the Moon I took using this method.

Photographing the Moon and planets

As with many things in life, high resolution photography of the Moon and planets involves a degree of compromise. Even for these relatively bright objects, short exposure times demand that the image is not vastly enlarged from that produced at the telescope's principal focus. However, the limiting resolution of the film and uncertainties in focusing demand that the image **is** enlarged. Using the most sensitive photographic emulsions to reduce the exposure times necessarily also increases the film grain and reduces its resolution, necessitating geater amplification of the primary image, and so a longer exposure! Short exposures are also desirable to

(b)

Figure 5.5. (a) [left] The arrangement for enlarging the primary image by eyepiece projection. The formula for calculating the amplification factor is given on the diagram. (b) [above] Notice the blurring of the outer zones of the photograph. This is a severe disadvantage of eyepiece projection applied to large extended bodies. The photograph was taken by the author with his 18¼-inch *f*/5.6 Newtonian reflector, using an 18 mm orthoscopic eyepiece to enlarge the image to an effective EFL of *f*/17. The exposure was 1/30 of a second, made on FP4 film, on 1977 December 20^{d} 20^{h} 25^{m} UT.

minimise the blurring effects caused by atmospheric turbulence, as well as problems with vibration of the telescope and errors in its drive.

Whatever amplification method is used, the effective focal ratio (EFR) and effective focal length (EFL) of the telescope are simply given by:

$$\mathrm{EFR} = aF,$$

and

$$\mathrm{EFL} = af,$$

where F is the focal ratio and f is the focal length. The user of a catadioptric or a Cassegrain telescope starts off with an advantage over the Newtonian telescope user because of the former's already large (few arcseconds per millimetre) image scale at the principal focus. Not only is less amplification of the primary image needed but the higher focal ratio generates less aberrations in the enlargement lens-system.

Figure 5.6. The southern highlands of the Moon photographed by the author using the ∞ to ∞ method. The camera, fitted with a 55 mm focus lens, was **hand-held** to an 18 mm orthoscopic eyepiece plugged into the author's 18¼-inch Newtonian reflector. The EFR of this arrangement is the same as that for the photograph in Figure 5.5(b) but notice the much more even focus achieved by this method. The photograph was taken on 3M colourslide 1000 film and was commercially processed. The exposure was 1/60 second on 1987 January 7^d 19^h 20^m UT.

Table 5.1. *Exposures for given subjects, calculated for 125 ISO film at an EFR of f/20*

Subject	Exposure (seconds)
Moon (3 days, 25 days)	1/2
Moon (7 days, 21 days)	1/8
Moon (10 days, 19 days)	1/15
Moon (full)	1/60
Mercury	1/8
Venus	1/60
Mars	1/15
Jupiter	1/8
Saturn	1/2
Uranus	1
Neptune	3

Note
These values can be used as the basis for calculating the correct exposures at other EFR's and for other film speeds, as explained in the text.

Table 5.1 shows some exposure times for different objects, based on my own experience. The values in the table are intended as a starting point for calculating the required exposure. In each case the exposure is that for the subject photographed at an EFR of *f*/20 on a film of 125 ISO. The Moon exposures are based on the correct exposures for the terminator regions (where applicable). This data is only provided as a rough guide to get you started. You will undoubtably replace my values with your own as you gain experience. In practice it is always advisable to take a series of exposures. As an example, suppose that you decide to photograph the terminator of the first quarter Moon at an EFR of *f*/20, using 125 ISO film. An exposure of 1/8 second ought to be about right but I recommend taking a series of at least three different exposures – 1/4 second, 1/8 second and 1/15 second. The correct exposure values for other film speeds and focal ratios can be easily calculated by remembering that the exposure time (for short exposures) is inversely proportional to the film speed and is proportional to the square of the focal ratio:

$$\text{Required exposure time} \propto 1/\text{film speed}$$

$$\text{Required exposure time} \propto F^2$$

For instance, to record terminator details in the first quarter Moon with a 250 ISO film at $f/20$ the recommended exposure would be 1/15 second. With the same film, but the telescope working at $f/40$, the exposure should be 1/4 second.

Keeping the exposure as short as possible minimises the deleterious effects of the turbulence of our atmosphere. It also minimises the effects of poor tracking. Since the Moon and planets at least approximately share in the 15 arcsecond per second diurnal motion, keeping the exposures shorter than 1/15 second allows the potential recording of arcsecond-level detail, even if the telescope is not equipped with a drive. 1/30 of a second permits detail as fine as 1/2 arcsecond to be resolved and this is close to the limit of the best photographs attained by amateur astrophotographers. With normal backyard seeing conditions the amateur astrophotographer will be doing well to obtain photographs of arcsecond resolution.

The lunar photographer should beware that the Moon moves about four per cent more slowly than the diurnal motion. It also changes its declination particularly rapidly when at a declination of 0° (the rate of change then being 0.3 arcsecond per second). The total differential motion may amount to nearly an arcsecond per second.

With 1/60 second exposures the amateur can get some fair results just by **hand-holding** a camera up to the eyepiece of the telescope (using the ∞ to ∞ method). This is a method I have often used when I wanted to take a few quick photographs without going to the trouble of setting up the equipment for attaching the camera to the telescope. Of course, exposures that short demand a fast film in the camera. Even then only fairly low amplification factors can be used. For instance terminator details in the first quarter Moon can be photographed with a 1/60 second exposure on 1000 ISO film provided the EFR does not exceed about $f/20$.

A 300 mm aperture telescope working at $f/20$ has an image-scale of 34 arcseconds per millimetre. A 1000 ISO film has a resolution of about 40 lines mm^{-1} and so might just record arcsecond-level details if the errors in focusing and atmospheric turbulence are not too severe. To prove that this 'low tech' approach can work, the lunar photographs, taken by myself, which I have included in this chapter were obtained this way. Of course, the images of the planets are all very small at low focal ratios. Even the mighty Jupiter's image spans only just over a millimetre on the film photographed with a 300 mm $f/20$. Nonetheless arsecond-level detail can still be recorded. Obviously, higher EFR's can be used for the gibbous and full Moon and for the brighter planets Venus and Mars.

However, I must say that the very best photographs I have seen of the

Moon and planets were all made using relatively slow films, with telescopes working at high EFRs and with consequently long (circa 1/2 second) exposure times. I would especially recommend Kodak's Technical Pan 2415 because of its high contrast, high definition and high-resolution characteristics.

In the forgoing I have made one tacit assumption: that is, that vibration does not spoil the photograph while the exposure is being made. For exposures longer than 1/30 second this may not be the case. In many instances vibration can be a bigger problem than any inaccuracy in the tracking. The shutter mechanisms of most modern cameras (particular the motion of the mirror in SLR cameras) may cause considerable problems. One solution is to install a second shutter in the light path before the camera. This can take the form of a rotating blade, or an iris shutter salvaged from an old camera.

Just before the exposure is made the second shutter is closed. Then the camera shutter is locked open on a brief-time setting. Allowing a few seconds for vibrations to die down, the second shutter is then opened to make the exposure, after which the camera shutter is, of course, closed. Some people have even used a large sheet of black painted cardboard, shaped like a large paddle, which is hand-held in front of the telescope tube to act as the second shutter. A little ingenuity will often overcome any practical problems.

Enhancement of the image on the film

Obviously there is little point in enhancing the sensitivity of a film for deep-sky photography if the combination of focal ratio, exposure length and its normal speed mean that the limit imposed by sky fog is already reached. On the other hand, where exposures must be kept short and where slow (and hence higher resolution and less grainy) films are to be used, then there would be some benefit in up-rating the film to a higher effective speed. The following notes outline some methods which the experienced astrophotographer might try. In all cases the precise details vary enormously depending on film, subject of photograph, etc., and so only a rough guide can be given. A little experimentation will prove invaluable. Obviously, practical experience of basic darkroom techniques is assumed.

Pre-exposure

Introducing a small amount of fogging on the film by exposing it to a flash of dim light before the exposure may seem an odd thing to do but it does help by off-setting the reciprocity failure of the film to extremely dim illuminations. The dimmest images then effectively add to the level of greying caused by the pre-fogging, instead of being lost at the toe of the characteristic curve and not being recorded at all. To have the greatest effect the level of fogging should **just** be at the bottom of the linear portion of the characteristic curve. This occurs at a density value of around 0.2, depending on the film.

The brief flash of light can be provided by a lamp in a light-tight box with a small aperture covered by a shutter mechanism. A strong diverging lens in front of the shutter then gives the beam a fairly even spread. The film is unrolled in the dark and a brief exposure is given. The ideal length of the exposure can only be determined by prior experiment with the arrangement and film used.

Force-processing

In general increasing the length of development time, during processing, increases the effective ISO number of the film. However, it also increases image contrast (which is sometimes desirable, sometimes not) and film granularity (which is never desirable). The speed of the film can easily be more than doubled by this technique. The effective speed of the film also depends on the particular developer used to process it, as well as the temperature of the solution. A practical limit is reached when the film significantly greys due to chemical (dichroic) fogging. What is possible, or desirable, with given combinations of film and developer can only be found by experiment.

Pre-treating the film

Few amateurs go this far, as the equipment is expensive and the technique is a little involved. The most used method is to bake the film in forming gas (a mixture of 8 per cent hydrogen and 92 per cent nitrogen) for several hours at a temperature of about 50°C. This technique is known as *gas hypersensitisation*. The film is loaded onto a spiral and is then set in position in a pressure vessel (Figure 5.7). The tank is then well flushed with forming gas and finally the valves are closed. The thermostatically controlled heater is then switched on and the film is left to 'cook' for several hours.

A number of companies supply gas hypersensitisation kits, or rolls of ready treated film. Full instructions are always provided with these. The

Figure 5.7. Ron Arbour's hypersensitisation tank. Notice the wrap-around heater elements.

February 1981 issue of *Sky & Telescope* magazine carries three articles concerned with this technique. However, once the film has been treated it should be used fairly quickly (on the same night if possible) if the maximum sensitivity gain is to be preserved. According to one Kodak official I have contacted, their TP2415 film can have an effective speed rating of about 1200 ISO if properly hypersensitised. Do beware of a shift in the colour balance of colour films, when they are treated this way.

Cold camera

In this technique the film is cooled to dry ice (frozen carbon dioxide) temperatures while the exposure is being made. Cold camera designs vary and they can be made or purchased. Figure 5.8 shows one made by Ron Arbour. A small chamber is provided to hold a few granuals of dry ice. This chamber is in contact with the pressure-backing plate behind the film. The film chamber is closed by a thick perspex window. If this perspex is not in contact with the front of the film then any air in the camera has to be displaced by dry nitrogen gas, to stop any condensation or frosting problems. Colour films preserve their colour balance well with this technique and all

Figure 5.8. Ron Arbour's cold-camera, which he has mounted onto its own focusing plate for direct attachment to his telescope. The small canister at the top takes the dry ice.

films show an apparent increase of sensitivity, sometimes amounting to several times their original rated values.

Copying

Making new negatives or transparencies from existing ones often allows barely perceptible details in the original to be emphasised. The copies can be made using a standard slide copier. Manipulations of the colour balance can be effected using filters. Sandwiching copies together, to make further copies, can further bring out details as well as increasing the general contrast of the image. I refer interested readers to an article entitled 'Enhanced-color astrophotography' in the August 1989 issue of *Sky & Telescope* (see also Figure 14.6, in Chapter 14 of this book). There the authors present some stunning results obtained using this technique. One other advantage of this method is that the random grain effects in each of the copies tend to cancel each other out, producing a much smoother looking final photograph.

Techniques to bring out detail in printing

As in the foregoing section, only rough notes can be given as a guide to getting started. A little practical experience will, as always, work wonders!

High contrast versus low contrast and 'dodging'

The total range of brightnesses (dynamic range) recorded on the film may be of the order of 1000:1. However, photographic paper can only represent a brightness range of about 50:1 when printing with normal contrast. A very low contrast paper (and/or developer) would have to be used to compress the brightness values on the film and show a greater range in the final print. However, individual details on the image would then be less well seen (see Figures 5.9(a) and (b)).

A compromise solution is to use high contrast paper and expose different parts of the image by different amounts, depending on their photographic density on the film. Figure 5.9(c) is an example of this. A piece of cardboard was used as a mask to cover all of the photographic paper as the enlarger was switched on. Then the cardboard mask was slowly removed, first uncovering the darkest portion of the negative image. The mask was continually withdrawn over several seconds until all the photographic paper was exposed to the image. After several further seconds the enlarger

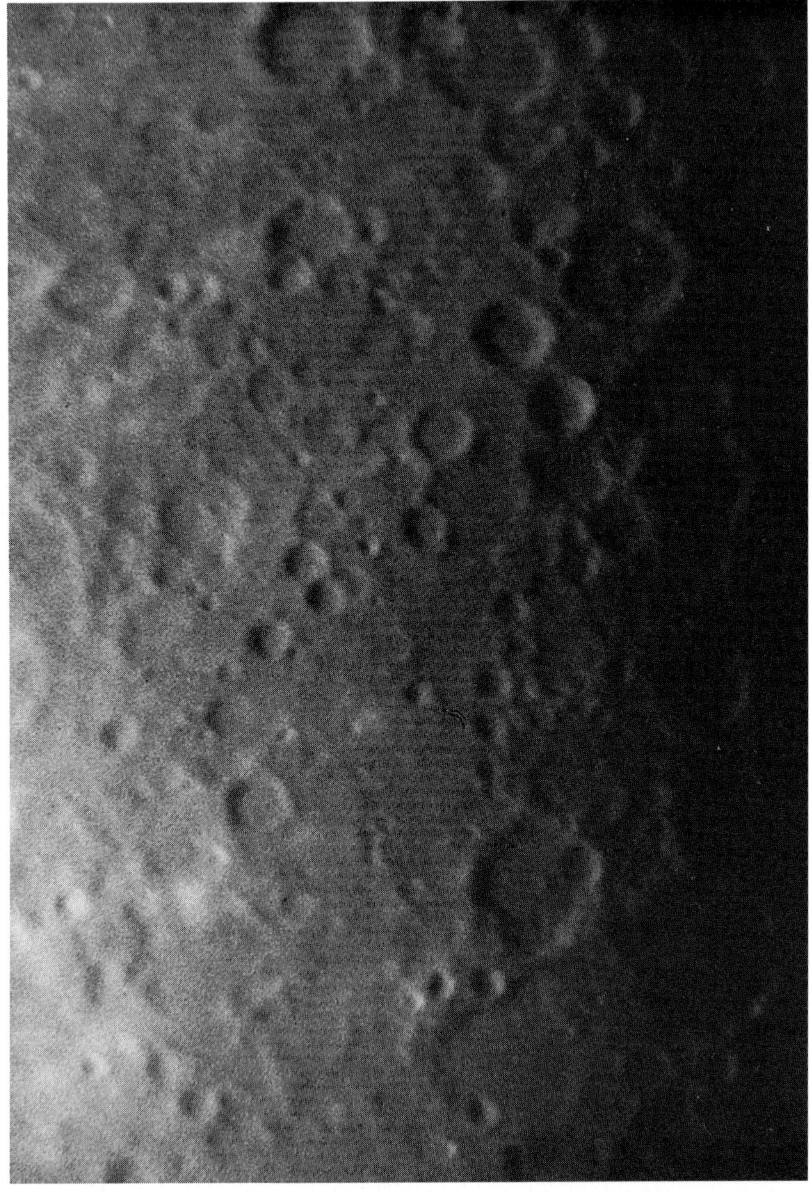

Figure 5.9. The correct techniques in printing can be vital in bringing out the best in an amateur's astronomical photographs. Each of these are prints from the same negative (taken on the same date, with the same film, and in the same way as the photograph in Figure 5.6). (a) [above] Is a normal print onto grade 2 (normal contrast) photographic paper. (b) [right] Is a print onto grade 4 (hard contrast) photographic paper. (c) [overleaf] Is the same as for (b) but using a moving mask during the exposure, held under the enlarger, to even out the image density.

(b)

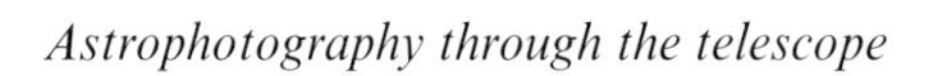

(c)

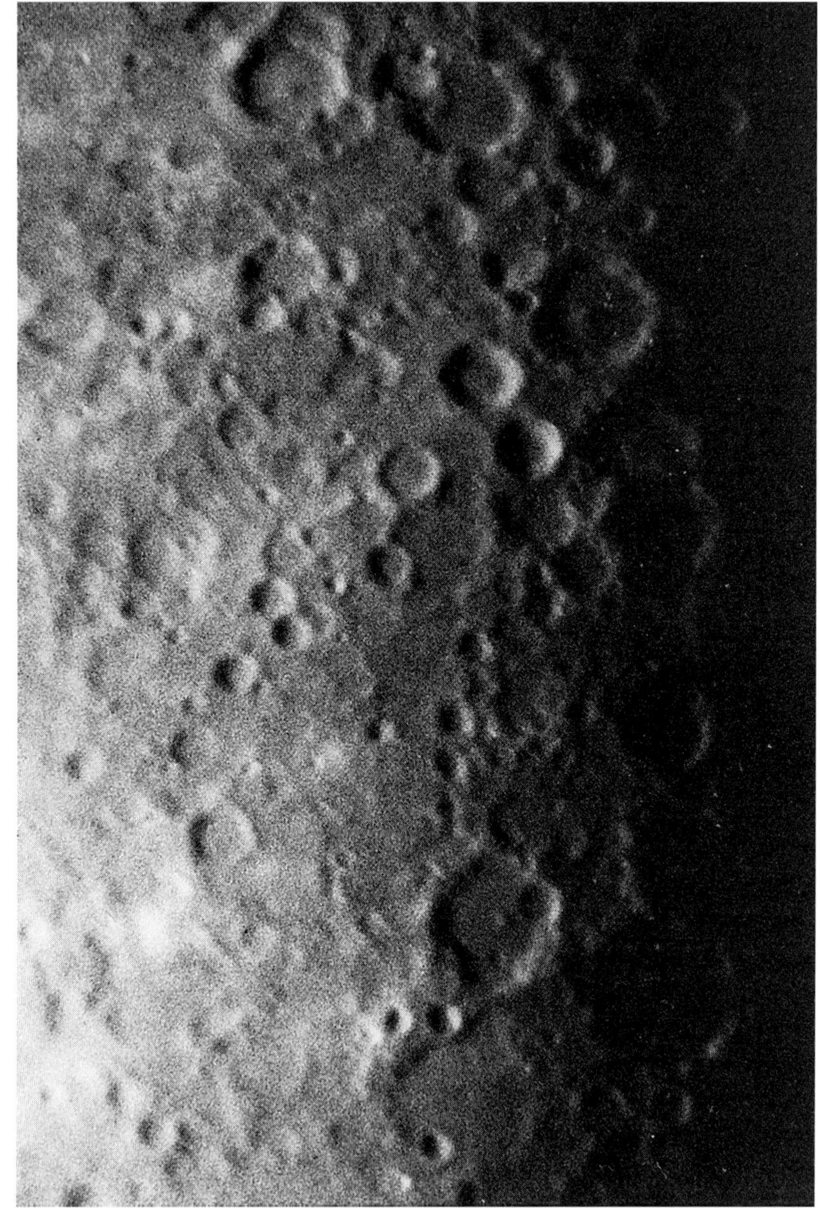

was switched off and the photographic paper developed. In this way the photographic paper was exposed to the darkest portions of the negative image for longer than than the lighter portions and so the range of brightness values was compressed. Nonetheless, full definition of the details is maintained.

Keeping the mask well above the photographic paper ensures that its edge is not in sharp focus and so no boundaries of different brightnesses show up on the final print. Masks can be made any size and shape to suit the image on the film. For instance, a small nebula at the centre of a star field can be effectively 'dodged' by means of a small mask held on a thin wire handle. Keeping the mask in motion avoids creating artificial-looking boundaries in the final print.

Unsharp masking

This is really just a refinement of the previously described method. The principle is just the same: compressing the range of brightness values on the film by counteracting the exposures of the regions of widely differing density. A thin (underexposed) and deliberately out-of-focus copy is made of the original image on negative film. The copy is then sandwiched together with the original and the print is made through both (sharply focused on the original). Figure 5.10(a) and (b) (overleaf) shows the improvement in the final image that can result using this technique.

Examples of the results obtainable by the methods covered in this chapter can be found spread throughout the pages of this book.

(a)

Figure 5.10. Unsharp masking. (a) is a normal direct print of the negative. (b) is a print made through the same negative, sandwiched together with a thin and out-of-focus positive (a copy of the original negative on negative film). Notice how the burnt out regions of the normal print are shown in much more detail in (b), yet (b) shows the faint and tenuous regions just as well as the normal print. The original negative is a 20 minute exposure taken by Ron Arbour with his 16-inch *f*/5 Newtonian reflector on hypersensitised TP2415 film, on 1988 January $12^{d}\ 21^{h}\ 55^{m}$ UT.

(b)

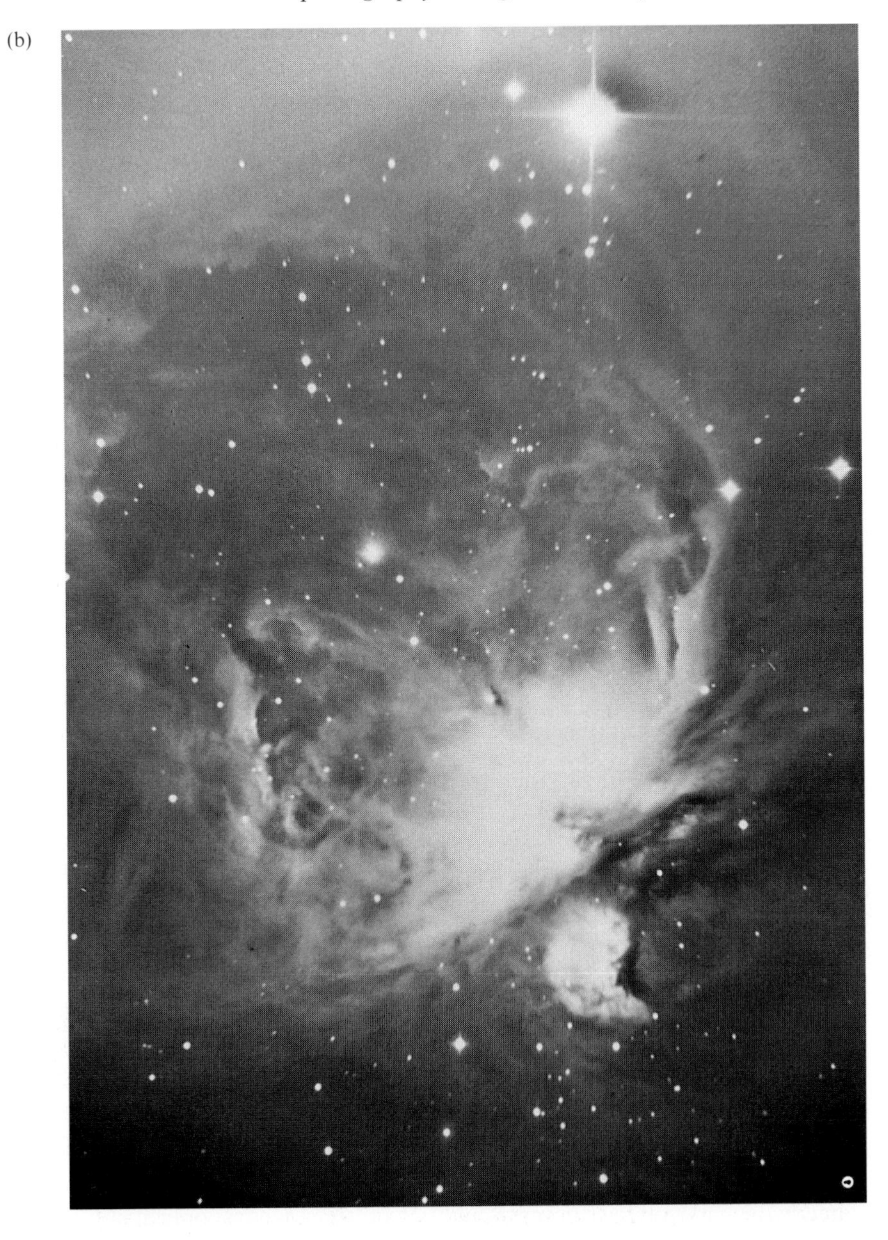

6
Electronic imaging

When I wrote the first edition of this book barely a handful of amateur astronomers were using electronic means to capture and store astronomical images. How things have changed in just a few years! While it is true that the practitioners of electronic imaging are still in the minority of amateur astronomers, certainly the rapidly growing numbers demand that the former brief notes be expanded into a full chapter for this edition.

If it is really true that 'the difference between men and boys is merely the price of their toys', then I think this distinction is equally applicable between amateur and professional astronomers. Amateurs had to wait until the required equipment became available at affordable prices before they also could begin to exploit the powerful advantages of electronic techniques newly enjoyed by the professionals.

Even so, the equipment is still relatively expensive and the techniques necessary to do what the professionals can do are somewhat involved. If you think 'why bother?', then the answer is given by the results achievable. Your backyard telescope can be transformed into a powerful research tool capable of producing results to rival those obtained with the large mountaintop observatory telescopes, when they used to record on the photographic plates of yesteryear.

I have just said that the equipment is expensive and the techniques a little involved. Throughout this book I try to provide simple and cheaper alternative ways of doing things wherever possible. Happily, I can do so in this field, too, and so before we consider 'proper' electronic imaging and all it entails, let us make a start by utilising a piece of equipment you might already own and yet have never thought of putting to astronomical use . . .

Astrophotography with the home video camera

Take a look at Figure 6.1. It shows a small part of the Moon's Mare Imbrium, near the crater Archimedes. This was taken by me using an

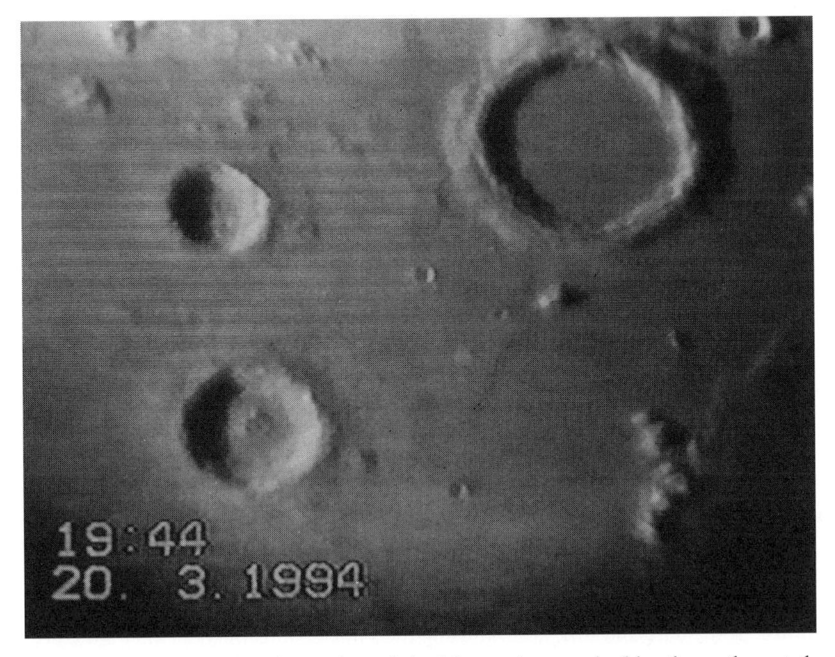

Figure 6.1. The Archimedes region of the Moon photgraphed by the author at the date and time shown, with his 18¼-inch Newtonian reflector and a video camera as described in the text.

ordinary domestic video 'palmcorder' attached to my 18¼-inch telescope, with its drive not working, and under very ordinary seeing conditions. Yet the detail shown approaches the 1 arcsecond level that is a rarity for conventional astrophotography from my observing site. A fluke? Then have a look at Figures 6.2 and 6.3, which were also recorded that night. There is a further example in the next chapter. You could easily do at least as well, even with a telescope of half the aperture of the one I used. If you have good seeing you will do better still. The following notes detail how.

Selecting a video camera

If you already have a video camera, then use that one. You will be very pleased with the results. If you are setting out to buy one with astrophotgraphy in mind, then there is scope for making your choice carefully.

After the money you can afford to pay for it, one of the first considerations must be its weight and bulk. Remember that you will have to attach it to your telescope – and add additional weights to the telescope for counterbalancing. If your telescope quivers under the weight of your largest

Figure 6.2. The lunar crater Regiomontanus and environs photographed by the author, at the date and time shown, using the same arrangement as for Figure 6.1.

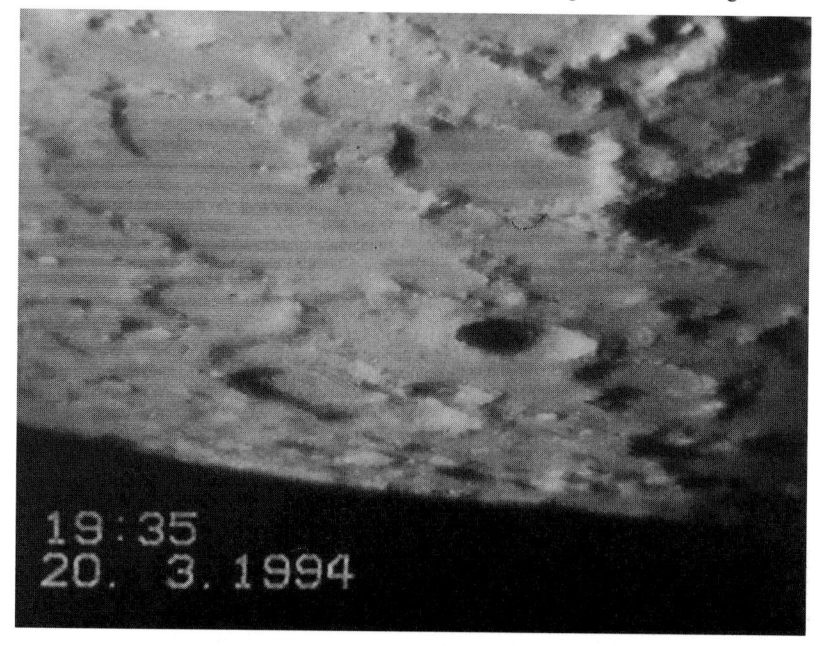

Figure 6.3. The north polar region of the Moon is well shown because of an unusually favourable libration in this video image obtained by the author at the date and time shown, using the same arrangement as for Figures 6.1 and 6.2.

eyepiece, it is clearly futile to attempt to attach a large and heavy camera to it! I think that a 'palmcorder' would be more than heavy enough for most amateur telescopes. Even these will have a mass of around a kilogram and so will demand a similar mass for the counterweight. Don't forget, also, the weight of the contrivance you will need to physically attach the camera to the telescope. The heavier the camera, the more rigid (and usually this means heavier) the contrivance will have to be.

The resolution of the camera is also important and, of course, should be as high as possible. Older cameras had 'vidicon' tubes set at the focus of the lens in order to detect the image but all the common modern domestic cameras use *CCD*s (*charge-couple devices*, explained later). A medium quality camera will have a '1/3-inch CCD Image sensor' as it is usually called on the specifications sheet (usually to be found at the back of the instruction manual). The horizontal resolution of these cameras is usually poorer than that in the vertical direction, and ought to be more than 200 lines across the full width of a TV frame for a camera with a '1/3-inch chip'. Again, this will be stated in the specifications sheet.

You may be surprised that I haven't put sensitivity uppermost in the list of requirements. Frankly, as long as the camera can operate in a light level of about 30 lux or less it will be more than able to cope with the images of the Moon and planets delivered by your telescope. For an old 'tubed' camera this wasn't always attainable but almost any modern CCD camera will do at least as well as this. As far as imaging fainter objects goes, even the 1 lux ratings of most medium and good quality video cameras isn't good enough. So, when using a domestic video camera we are dealing with an instrument which we can use to get spectacular images of the Moon and the planets (and perhaps the Sun, with proper filtering!), plus a selection of brighter stars.

The camera I chose was a National Panasonic NV-S20B. It cost just under £600 (about $1000) at the beginning of 1994. It is a fairly typical 'palmcorder' having a mass of just under 1Kg without its batteries and a so-called '1/3-inch' CCD with a stated horizontal resolution of 'more than 230 lines' (the vertical resolution was not stated but is judged by me to be superior to its horizontal resolution).

Among the various nice pieces of 'gadetry' on this camera, including the facility of on-screen recording of date and time, is a selection of 'shutter speeds'. Though these make no difference to the rate of recording the images (there are 25 complete frames every second for a British camera, 30 every second for an American camera), allowing one to choose the shutter speed to suit the subject is still useful. The slowest 'shutter speed' (purely

Figure 6.4. The arrangement the author uses to attach his video recorder to his telescope. It can be set up in a matter of minutes. Note how adjustments are provided for 'squaring on' the camera to the eyepiece – essential if the best quality images are to be obtained. Note also how the camera can be racked back and forth, which is a convenience when using different eyepieces and in initially setting up. Counterweights (not seen in this photo) attach to the lower end of the telescope tube to afford balance.

electronic, there is no mechanical shutter) setting allows one to shoot in the lowest light conditions and the fastest (usually 'sports') setting allows one to photograph moving objects with the minimum of smear. I will elaborate on the other advantages of this facility later.

One thing to be avoided at all costs is a camera with no manual override for the focusing. Don't even bother to try and use a camera set on automatic focusing with your telescope. You will find that the electronics will take a distinct dislike to the images of the Moon and the planets and will cause the camera to restlessly zoom and jitter about the proper focusing position.

Using the video camera on the telescope

As always, attaching the camera is purely a matter of mechanics, as is providing the counterweighting. Figure 6.4 shows the arrangement I use. It is described in the caption to that figure, but note that it does allow the camera to be set squarely looking into the eyepiece and to be racked back and forth from the eyepiece. The former requirement is essential if good images are to be obtained. The latter is a great convenience when dealing with the focusing of different eyepieces.

Unless you are either very brave (even if technically adept) or very foolhardy (if not) you will not wish to attack your newly purchased video camera with a toolkit and remove its lens. So, all your images must be obtained using the 'infinity to infinity' technique described in the last chapter.

In use, the telescope eyepiece is set roughly to a normal focus by eye and the camera focusing is set to 'infinity'. The lens on my camera has an 8:1 zoom range, the corresponding focal lengths being 40 mm down to 5 mm. For use with your telescope the lens should be set on its fullest zoom setting (this corresponds to an effective focal length of 40 mm with my camera). The camera is then brought up so that its lens is close to the telescope eyepiece. Looking through the camera viewfinder you can make any fine adjustments necessary to bring the image into sharp focus **but be very careful not to run the eyepiece into the camera lens!**

Assuming that you are trying this out for the first time with the Moon, you should now be seeing some impressive images of the lunar mountains and craters through the camera viewfinder and so can now set the camera recording these marvels onto its own tape.

If you wonder why the camera lens has to be set on full zoom, try altering the setting to wide angle. You will see that the image, while of lower magnification, has become a small patch in the middle of the field of view with a sea of darkness surrounding. The reason is that on a wide-angle setting the complete set of light rays emerging from the eyepiece does not find its way through the lens system of the camera and so fully cover the chip. To get a picture which fills the TV screen, set the camera lens on full zoom.

When I tried this for the first time I was surprised to find that my 18¼-inch telescope gathered **too much** moonlight for the camera to cope with. The camera was clearly overloaded (producing a black curtain effect which descended over the image) on all but the 'sports' exposure setting.

I found that on the 'sports' setting all was fine. This rapid exposure speed also has the advantage of producing the sharpest images because the effects of atmospheric turbulence are virtually frozen on individual frames, as are any tremors and movements of the telescope.

An equatorialy mounted telescope with a drive is very much a luxury item for this work. You will get superb results even if your telescope is on an undriven altazimuth mounting.

As well as having the facility to record commentary at the time, using a domestic video camera has another bonus that comes along with it – the images are in full colour!

The field of view and magnification obtained

The effective focal ratio and effective focal lengths are calculated in exactly the same way as for a conventional camera/telescope combination in the 'infinity to infinity' configuration (see the last Chapter). The images presented in Figures 6.1, 6.2 and 6.3 were all taken using an 18 mm Orthoscopic eyepiece plugged into my 18¼-inch (464 mm) telescope of 2.59m focal length. Hence the effective focal amplification factor is 40/18, or 2.22 and the effective focal length of the arrangement becomes 5.76 m and the effective focal ratio is *f*/12.4.

Of course, knowing the effective focal length one can calculate the image scale on the CCD. In my case it is 206265/5760, or 35.8 arcseconds mm^{-1}.

This can help you make a prediction as to the size of the field of view you can expect to image. The picture area of the chip is only roughly known, for instance the quoted '1/3-inch' is merely a category rather than a precise figure. It also refers to the length of the **diagonal** across the CCD; the same deceit commercially perpetuated with TV screen sizes. A '1/3-inch' CCD will probably have a picture area of something like 4.0 mm×5.3 mm.

With the arrangement I have used and discussed here, the actual size of the field imaged and displayed on my TV screen during playback is approximately 190 arcseconds×140 arcseconds and the lunar photographs shown are virtually full-frame representations from my TV screen.

Getting the maximum resolution in your images

Here the key is knowing the minimum resolution of your camera, together with the size of the field imaged. The resolution should be stated in the 'specifications' sheet. For my camera it is 230 lines spanning the full frame in the horizontal direction and so with my arrangement, the horizontal resolution of the image in arcseconds is then simply 190/230, or 0.8 arcsecond.

On most nights from my observing site the seeing allows only glimpses of details at the arcsecond level of resolution and so this limit imposed by the telescope/eyepiece/camera combination is about right. Further amplification would serve only to reduce the field of view and produce a larger but blurrier image. However, on nights of better seeing the amplification factor must be increased if the best possible image resolution is to be recorded. This is done simply by changing to a shorter focal length eyepiece.

Exchanging my 18 mm eyepiece for one of 6 mm focal length would produce a field size of 63 arcseconds×48 arcseconds and a potential resolution of 0.28 arcseconds, a close match to the theoretical limit of my telescope.

Playback!

Of course the completed tape of your observation can be played from the camera or a copy can be made onto a standard tape in your VCR and this played. You will be delighted and your friends will be stunned by the images that you should be able to show of the Moon and bright planets on your television set. It really is bringing the Universe into your living room! More than that, since these images are actual hard-copy they have a proportionately increased scientific value. If you have a VCR with a good and steady 'pause' facility, or better still a 'frame-by-frame advance', then you can select out those images where for a split second the absence of turbulent distortion allows particularly fine details to be glimpsed. How many still photographs would you have to take to get a reasonable chance of just one showing such fine detail? If your observing site is anything like mine, the answer is more than a hundred. You can simply play your recording and search out that one detailed view at your leisure.

Photographing from a TV screen

Many practitioners of electronic imaging say that the most difficult part is getting hard-copy from the system. No problem if you have a high quality (and therefore expensive) graphics printer. If not, then you are forced into photographing the results from the TV screen or computer monitor.

Even this need not be a problem if you go about it the right way. Obviously the camera should be set up in front of the TV screen and the screen properly framed and focused through the viewfinder. The room should be heavily darkened to avoid reflections appearing in the glass of the TV screen.

The film in the camera ought to be of medium speed (perhaps 125 ISO) and an exposure meter used, while the TV displays the image to be photographed, in order to set the camera for taking the exposure.

All very standard so far – but wait, there is one snag, and that is caused by the way the TV generates the viewed picture. The TV picture is built up from a spot of varying brightness that scans from left to right and then rapidly flies back to start the next line, leaving a gap between it and the last. Having started in the top left of the screen, the spot sweeps out a complete set of horizontal lines covering it in 1/50 second (1/60 second in the U.S.). In the next 1/50 second the spot goes back and fills in the gaps, as it were. In this way a complete new picture is created on the TV screen every 1/25 second (1/30 second for US televisions). It is only persistence of vision that allows us to see the illusion of a flicker-free and complete TV picture.

The snag arises from how the photographic camera's shutter speed interacts with the moving spot. The exposure time you should select must obviously be longer than 1/25 second in order to avoid either a partial picture or spurious bands of lighter and darker details on the photographed image. I use an exposure of 1/8 second when photographing from the TV and this proves to be just about sufficient, though some banding is still evident and 1/4 second might be better. Obviously the freeze-frame on the VCR must work well. If it doesn't, then try adjusting the 'tracking' control. This often works wonders.

Normal darkroom procedures will be followed in making the final prints but you might like to try slightly defocusing the image in order to soften the appearance of the TV's phosphors and so produce an aesthetically more pleasing result.

The dedicated CCD astrocamera: basic principles

Does obtaining images of nineteenth and twentieth magnitude stars and detailed views of spiral galaxies through a telescope of less than half metre aperture, with an exposure of no more than 20 minutes, and from a light-polluted site seem a mere fantasy? Well, this, and more, is possible if you have a telescope with a good guiding system, and can afford to buy a mid-range personal computer (capable of handling high quality graphics and with a high resolution monitor), plus at least another £1000 ($1600), or so, expenditure on one of the CCD camera systems now available.

At the time of writing these words, the market for camera systems is still virtually newborn and fresh developments are coming into the arena almost on a monthly basis. Consequently any advice or information I give here can only reflect the state of things at the time of writing (the beginning of 1996). Therefore I strongly recommend that all prospective purchasers of CCD equipment subscribe to a new quarterly magazine issued by Sky Publishing called *CCD Astronomy*. Also, gather as much data and information from the manufacturers of the various systems about their products (see Chapter 17 for the listing) as you can before making the final choice.

If one can define a 'typical' CCD suitable for amateur astronomical work, then it will have an array of something around 500×360 *pixels*, each pixel being about 15 μm square and the total covering an area of about 7.5 mm×5.4 mm, though there are large variations from these figures even with the CCDs currently available. A pixel is a picture element; the equivalent of an individual tile in a mosaic. The array is mounted on a 'silicon chip' type base with about 20 separate electrical connections. It works by

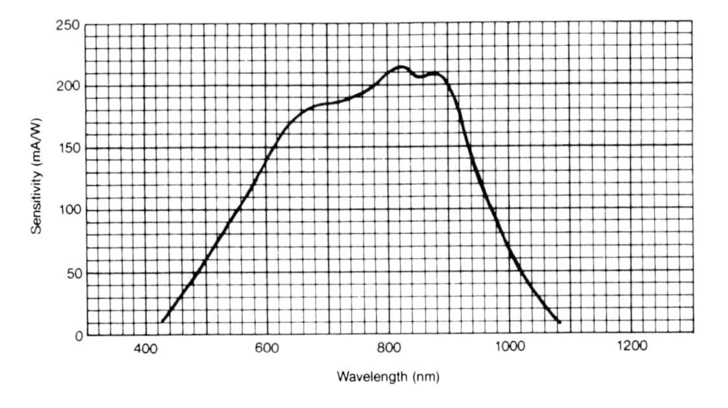

Figure 6.5. The spectral response of EEV's P8602 CCD – typical of monochrome CCDs. Reproduced by kind permission of EEV Ltd.

photons of light creating charges (electron – hole pairs) in each of the pixels. The more light that falls on a given pixel, the more charge is created in it. After a given interval, or *integration time*, the array of charges is then read off the CCD into a 'metal box' of supporting electronics. The electronics then deliver the 'image' (in the form of an electronic signal) either directly to a television monitor or, more usually, to a computer.

A 'typical' CCD will respond in a somewhat different way to differing wavelengths of light than do our eyes, or photographic emulsions. CCDs tend to be red-sensitive things, though some of the newer examples have added coatings which render them more responsive to blue light. In particular, the Philips FT12 CCD (used in the 'Starlight Xpress SXL8 camera) has a peak spectral response, at 530 nm and is half as sensitive at wavelengths of roughly 400nm and 700nm. Figure 6.5 shows the spectral response curve of a more typical monochrome CCD manufactured by EEV Ltd.

CCDs generally have values of *detector quantum efficiency* (DQE) greater than 50 per cent, at least for a limited range of wavelength. A 100 per cent DQE is as sensitive as one could possibly have any detector: all the incoming photons would then be detected. By comparison, a photographic emulsion has a DQE of something around 1 or 2 per cent. The advantage of the CCD is clear. The more 'visual' response of the Philips FT12 CCD is achieved at some expense in sensitivity, the peak DQE being 30 per cent.

A typical CCD can respond to extremely faint levels of light and fully saturates (gives the maximum possible output signal) when light of brightness about 3 lux (30 foot-candles in old fashioned, though more descriptive, units) shines on it. In between these limits it has a potential dynamic range (explained in Chapter 5) of several thousand.

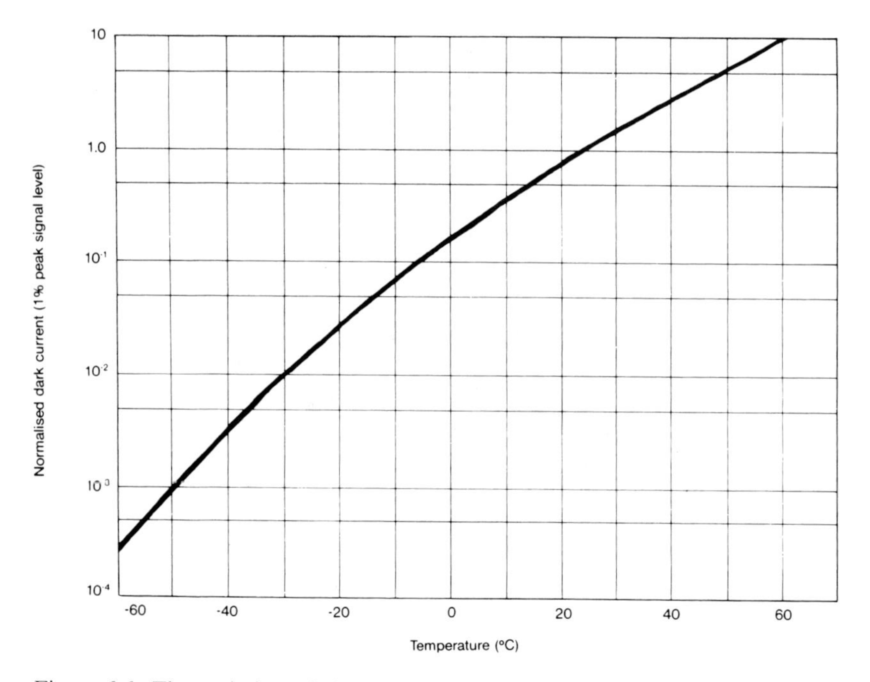

Figure 6.6. The variation of dark current with temperature, for a typical monochrome CCD. Reproduced by kind permission of EEV Ltd.

The practical CCD camera

There is one fundamental problem which bedevils CCDs – and that is something figuratively called *dark current*. During the integration time thermally generated charges build up in each of the pixels. At room temperature this dark current can fully saturate a CCD after just a few seconds integration. Even before that stage, it is obviously reducing the total dynamic range recordable.

Fortunately this problem is surmountable. Cooling the CCD dramatically reduces the dark current and allows much longer integration times, as Figure 6.6 shows. Consequently practical CCD astrocameras that are intended for use with integration times of longer than a few seconds have built-in thermoelectric coolers. Figure 6.7(a) shows one of the 'Starlight Xpress' commercial units: the camera head SXL8. Notice the fins projecting from the back of the camera head, these are to radiate the heat pumped away from the CCD itself. In operation, you could expect a typical camera to take about 10–15 minutes to stabilise (to a pre-set level) at a temperature of about −25°C after switching on. The CCD temperature is monitored and displayed by the supporting electronics unit as shown in the

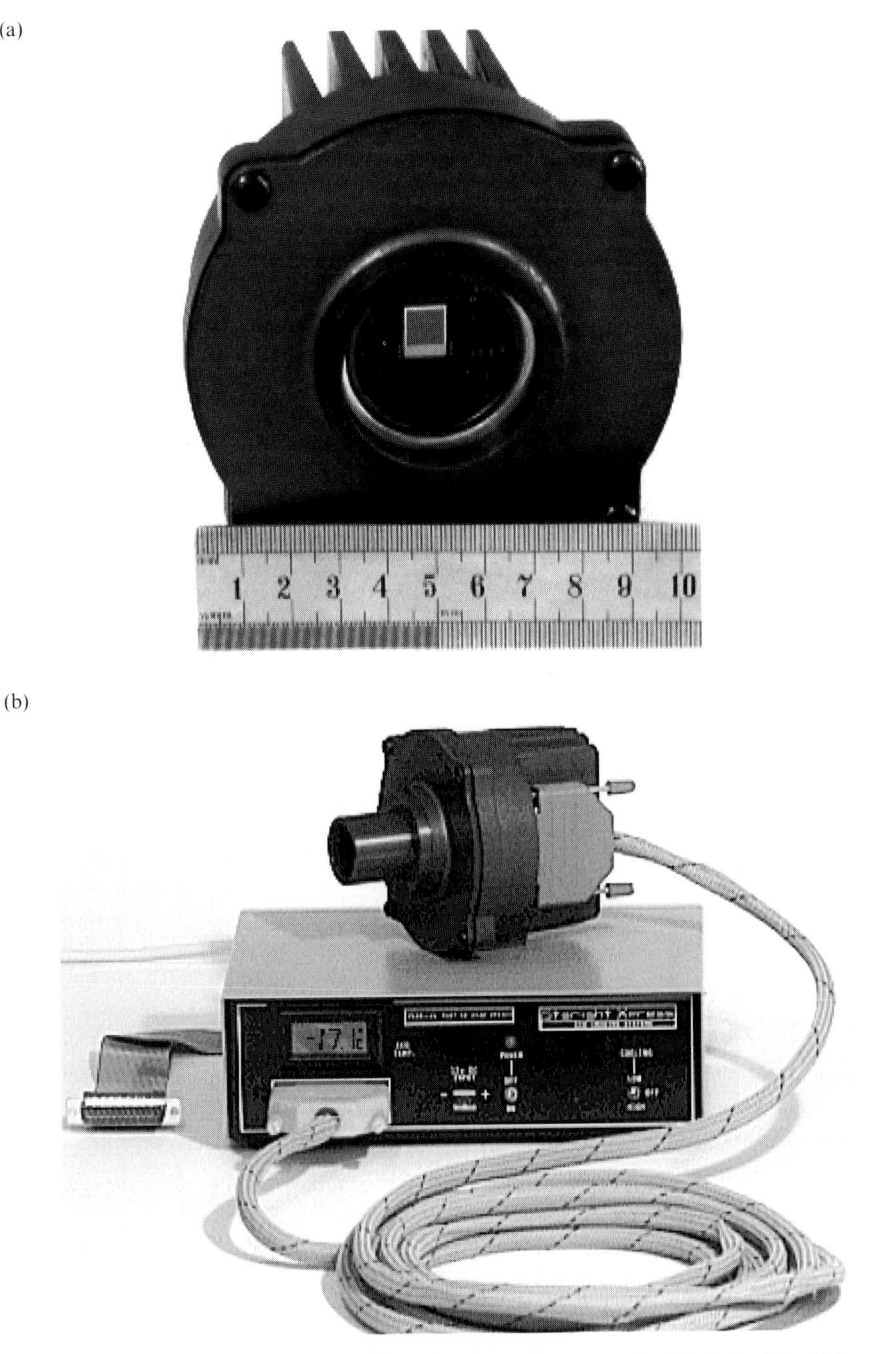

Figure 6.7. (a) The CCD can be seen within the camera head of the 'Starlight Xpress' SXL8 unit. (b) The 'Starlight Xpress' SX system, typical of commercial units.

Figure 6.7(b). Some units operate at lower temperatures and will take a little longer to be ready, though they normally have more powerful cooling units fitted.

Referring again to Figure 6.7(b), the open end of the tube screwed into the front of the camera head plugs into the telescope drawtube in the same way as does an eyepiece, the chip being centrally mounted in a position that corresponds to that of the field stop. A typical CCD camera head might have a mass of about a kilogram, a little more for some models, a little less for others, obviously necessitating re-balancing of the telescope.

CCDs, themselves, can be either of two types: *interline transfer*, or *frame transfer*. These refer to the manner in which the electrical charges are 'read' from the chip at the end of an exposure. Much is made in the current literature about the various pros and cons associated with either of these types. At this time of rapid technical advances in CCD manufacture, the seeming advantages of one type become overtaken by those of the other as better quality CCDs of either type are brought out. The proof of the pudding might be in the eating but we will first have to wait for it to finish cooking!

The system shown in Figure 6.8 is unusual amongst those currently available because it is a 'stand-alone' system capable of capturing and displaying images (on a monitor) without the need for a computer. However, a computer is still needed to save and to process images, though it does not have to be particularly fast. This might be ideal for the practitioner who does not mind using an old computer in the dampness of night by the telescope and would then take the results indoors to his/her latest expensive aquisition to do the actual image processing. The system shown in Figure 6.7(b) is more typical, the electronics being merely an interface between the CCD and the computer. The downside is that a reasonably fast computer (33MHz, or better) is needed.

The digital nature of images from CCDs make them particularly conducive to computer manipulation. The possibilities are manifold: contrast enhancement, the selective enhancement of small-scale or of large-scale features, the manipulation of brightness representations (for instance, colour-coding to bring out subtle details), etc. Indeed, some basic manipulation is most often required if the image is to be at least cosmetically acceptable.

Along with the hardware needed, all the manufacturers and suppliers of CCD equipment provide full instructions and the computer software necessary to operate their units.

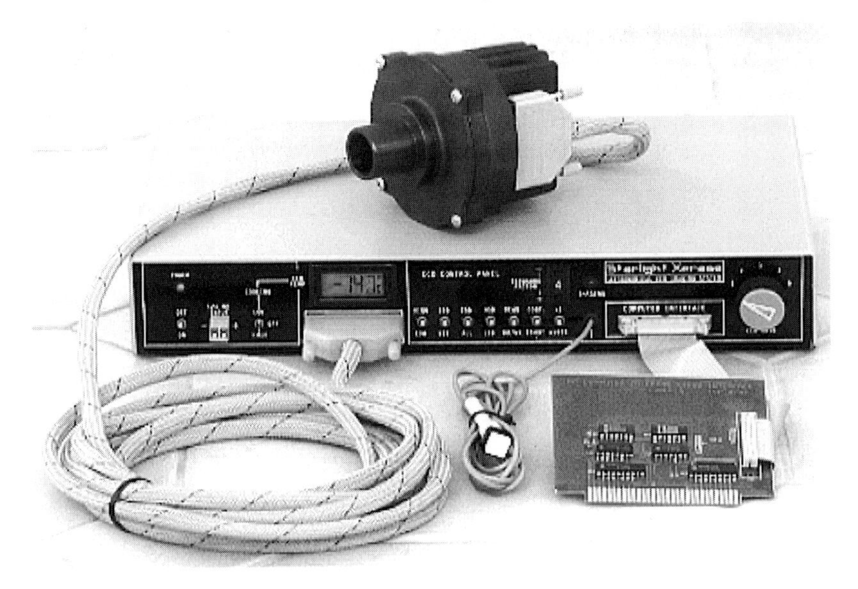

Figure 6.8. The 'Starlight Xpress' SXF camera system is unusual in that it is a 'frame grabber' system that doesn't need the intervention of a computer to capture and display images. However, a computer is needed if the images are to be saved and subsequently computer processed.

High resolution CCD imaging

Here the main requirement is the matching of the smallest resolved details in the image to the limiting resolution of the CCD. How this is done is described in the last chapter (for imaging onto photographic emulsions) and earlier in this chapter (imaging with a video camera) but as a summary: if the CCD has a pixel size of 10 μm an EFR of *f*/15 will only just allow the recording of details at the theoretical resolving limit of a given telescope. This increases to *f*/22.5 for a CCD with 15 μm pixels and to *f*/30 for one with 20 μm pixels. A large aperture telescope might be prevented from attaining its potential resolution by the ambient seeing conditions but some of the lost resolution could be recovered by careful processing. Certainly for a smaller aperture these *f*/numbers might be increased somewhat in order that the 'blocky' appearance of the pixels do not assume undue prominence.

Figures 6.9 and 6.10 show early results obtained by Terry Platt with a system he built himself based on a Sony ICX 021CL monochrome CCD chip. It has an array of 500×576 pixels. The CCD was driven by a 'frame grabber' which was provided with 256K×8 bits of dynamic memory and an 8 bit analogue to digital converter to digitise the video from the camera. He used a spectrum 128 computer to store the images and he monitored the images on a television screen, rejecting all but the best images obtained in moments of steady seeing. These best images were 'grabbed' and then subjected to computer processing. Finally hard-copies were obtained by photographing the TV screen.

Figure 6.9(a) shows a 'raw' image and (b) shows the same image after enhancement filtering, and boosting the contrast with a square law characteristic (three times the intensity difference is transformed into nine times the intensity difference, etc.). He built the entire system himself, including the 12½-inch *f*/24 tri-schiefspiegler telescope he used to secure the images! Since those early experiments, Terry Platt has gone on to found the 'Starlight Xpress' range of astrocameras and accessories now sold worldwide.

Figure 6.11 shows some processing software in action on an image of part of the Moon.

CCD images of 'deep-sky' objects

The methods of capturing images of deep-sky objects on the CCD follow exactly those described for doing the same on photographic emulsions in the last chapter. Focusing is a little more tricky but can be carried out using very short integrations on a bright star, before setting on the chosen object. Some manufacturers are now supplying focusing aids that plug into the telescope before the camera. Even with these some fine adjustment will often be necessary.

Even with short exposures on a properly cooled CCD the 'raw' images obtained may be lacking in contrast and showing various blemishes and variations of sensitivity across the field due to imperfections in the CCD itself. The problems are compounded when long exposures of faint objects are required. Hence the first piece of image processing that is necessary is the procedure of *flat-fielding*. Figure 6.12 shows this procedure in action for an image of the variable star HT Cas obtained by Nick James, using a 'Starlight Xpress SX' camera on his 12-inch (305 mm) *f*/5.25 Newtonian reflector. The step-by-step procedure is described in the captions accompanying the images but notice the dramatic improvement in the final image (Figure 6.12(d)) over the original (Figure 6.12(a)).

As ably demonstrated by Figure 6.12, the real power of CCD astrophotography lies in the computer processing that is done after the capturing of the image. Obviously software packages vary from manufacturer to manufacturer. A few 'computer-wizards' like Nick James, write their own

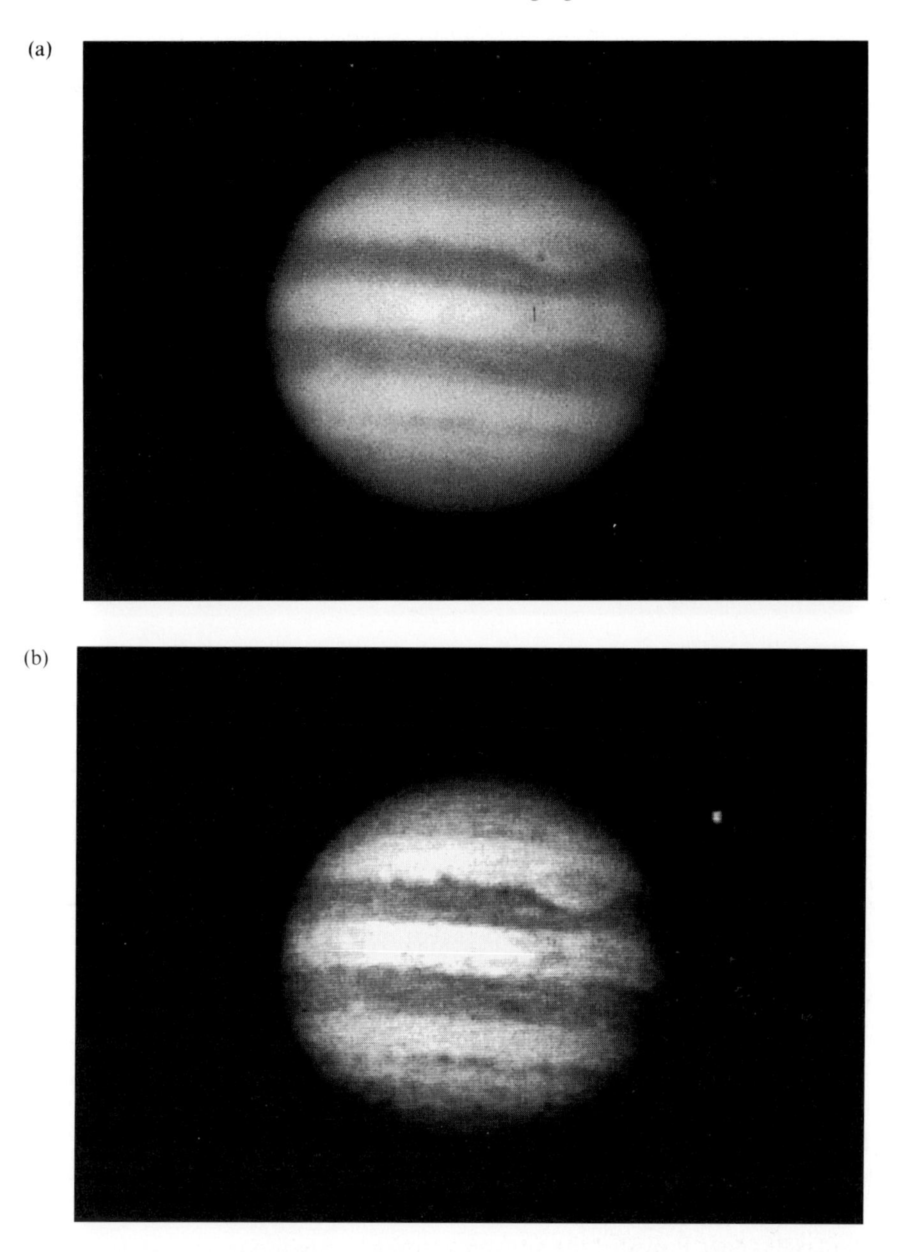

Figure 6.9. (a) A 'raw' image of the planet Jupiter obtained by Terry Platt on 1988 December 28^{d} 22^{h} 10^{m} UT, using his 12½-inch (318 mm) tri-schiefspiegler reflector and prototype CCD imaging system. (b) The same image after computer processing (high-pass filtering and square law contrast boost). Notice the image of Ganymede to the upper right.

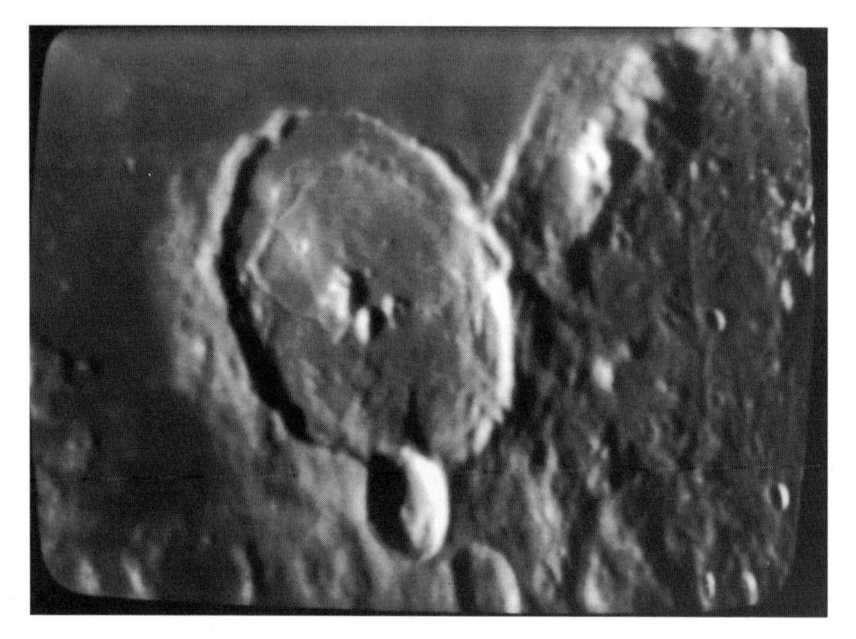

Figure 6.10. Another of the outstanding results from Terry Platt's CCD prototype imaging system, this time of the lunar crater Gasssendi on 1989 January 18^{d}. The EFR for this photograph and that shown in Figure 6.9 is approximately *f*/30 and the integration times in each case were 0.5 second.

software but the majority of amateur astronomers will use either the software supplied by the manufacturer of the astrocamera units or perhaps one or more of the multivarious image processing software packages now commercially available.

The possibilities are almost limitless. For instance, Figure 6.13(a) shows a single raw image of Comet Szczepanski that Nick James took using his telescope. It was made using an integration time of just 60 seconds – enough to freeze the differential motion of the comet amongst the starfield. Figure 6.13(b) shows the result of co-adding 10 of these 60 second exposures but aligning each so that all the comet images are superimposed. Hence the result is the same as of a 10 minute exposure on the comet but without the need to track the telescope on it.

The same principle can be used if your telescope does not track on the stars accurately. Provided the apparent motion is not too great, a series of short integration times can be given and the final result stacked together in such a way that all the star images coincide.

Figure 6.14(a) and (b) shows image processing in action for another type of subject. This time the target is the barred spiral galaxy NGC7479 in

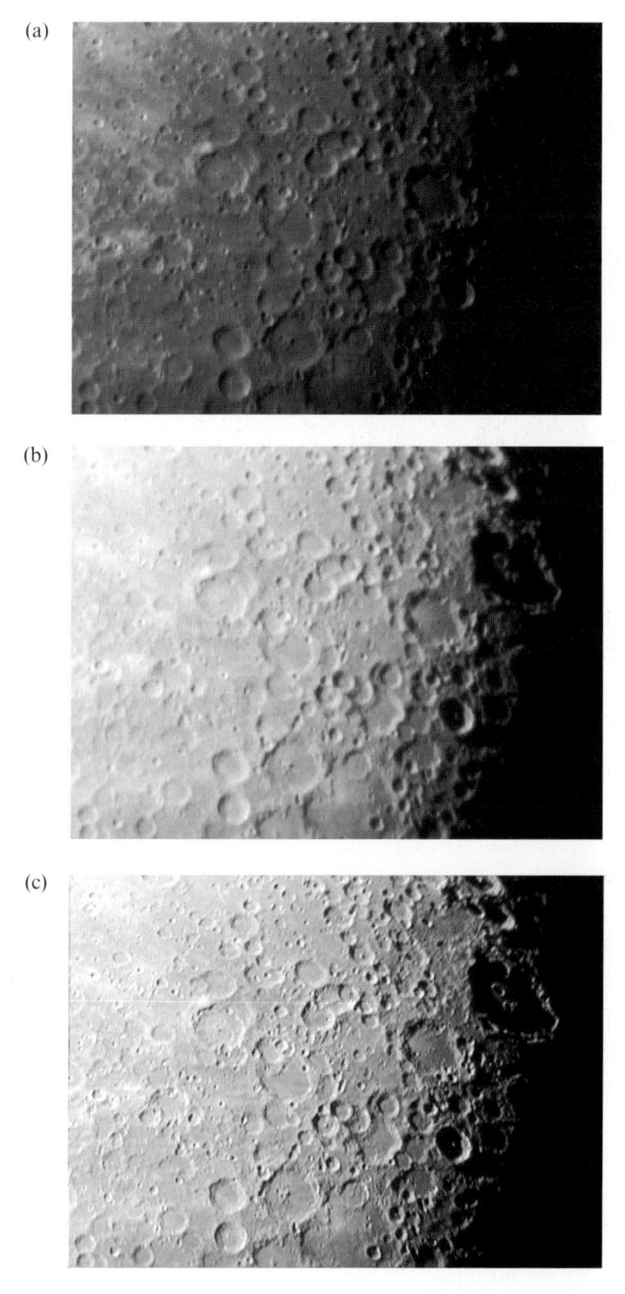

Figure 6.11. This sequence, created by Terry Platt illustrates steps in the processing of an image of the Moon. These were taken at the *f*/5 Newtonian focus of an 8-inch reflector **stopped to 3-inches (76 mm) off- axis**, with the 'Starlight Xpress SFX camera. (a) shows the original image and (b) shows the result of a non-linear contrast stretch in order to improve details in the dark terminator region. (c) shows the final result after the sharpness has been boosted by the use of a high-pass filter.

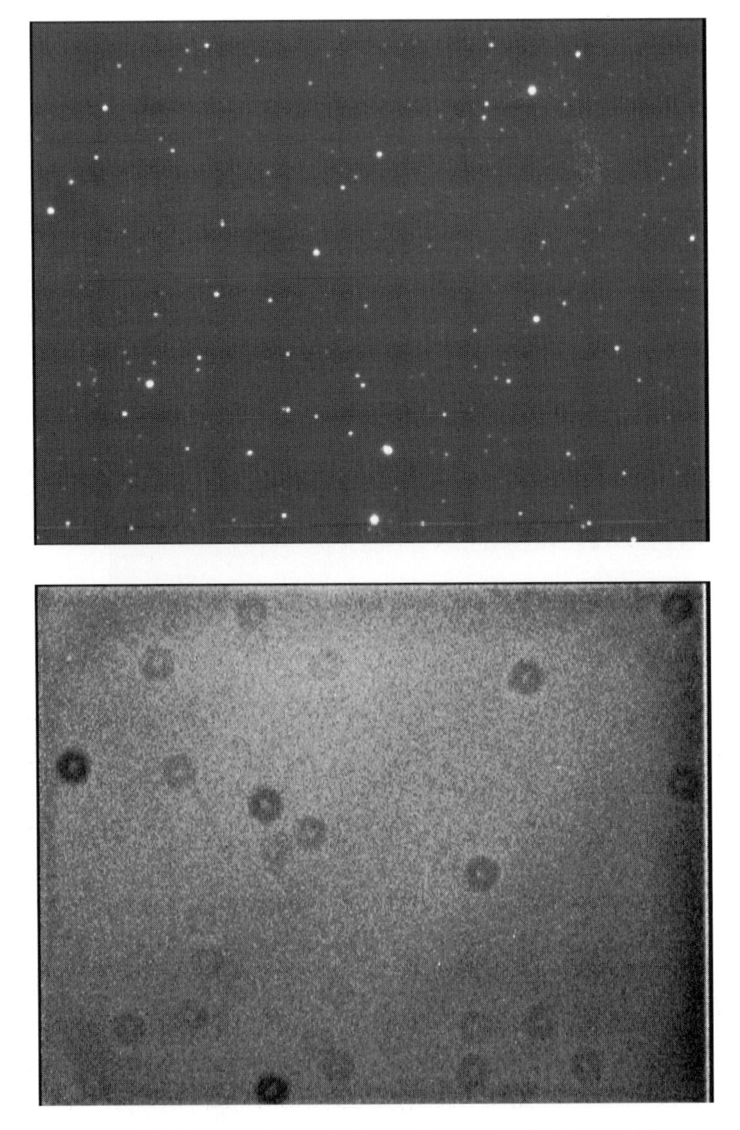
(a)
(b)

Figure 6.12. (a) Nick James obtained this raw image of HT Cas on 1995 December 9^d, using a 'Starlight Xpress' SX CCD camera for a 3 minute exposure through his 12-inch *f*/5.25 Newtonian reflector. On this unprocessed image the sky background level is around 20 000 counts on a scale of 0–65 535 (determined by the software used). It can be seen that the illumination of the field is non-uniform and some dust shadows are particularly evident. (b) This frame is called the flat field. It is the result of an average of six short exposures of the twilight sky with the CCD camera mounted on the telescope. This image has been stretched so that black to white represents a 5 per cent variation of illumination. Dust shadows are evident, as is a general variation of sensitivity across the chip. The flat field can be used to correct the raw image for non-uniform sensitivity.

(c) Even though the CCD chip is cooled it still suffers from thermal noise. This is the dark frame associated with the raw image. It is also a 3 minute exposure but was made with the cap over the end of the telescope. This picture is the result of contrast stretching the dark frame so that black to white corresponds to a 5 per cent change in brightness. (d) The raw image is processed by first subtracting the dark frame. This removes the signal caused by thermal charge. The resulting image is then divided by the flat field in order that the sensitivity of each pixel is normalised to the average across the whole chip. The sky background level is then determined by the software. This is subtracted from the image and the resulting frame is contrast stretched so that white corresponds to a brightness of 10 per cent above the sky. The final image has a limiting magnitude of about 19^{m}. Not bad for a 3 minute exposure from a light-polluted site!

Figure 6.13. (a) A single, uncalibrated raw frame of the Comet Szczepanski taken by Nick James using his 12-inch reflector on 1996 February 1^{d}. The exposure was 60 seconds. The contrast has been stretched for this illustration: note the dust shadows. (b) shows the final processed frame, the result of co-adding ten 60 second exposures and correcting for flat field and dark frame.

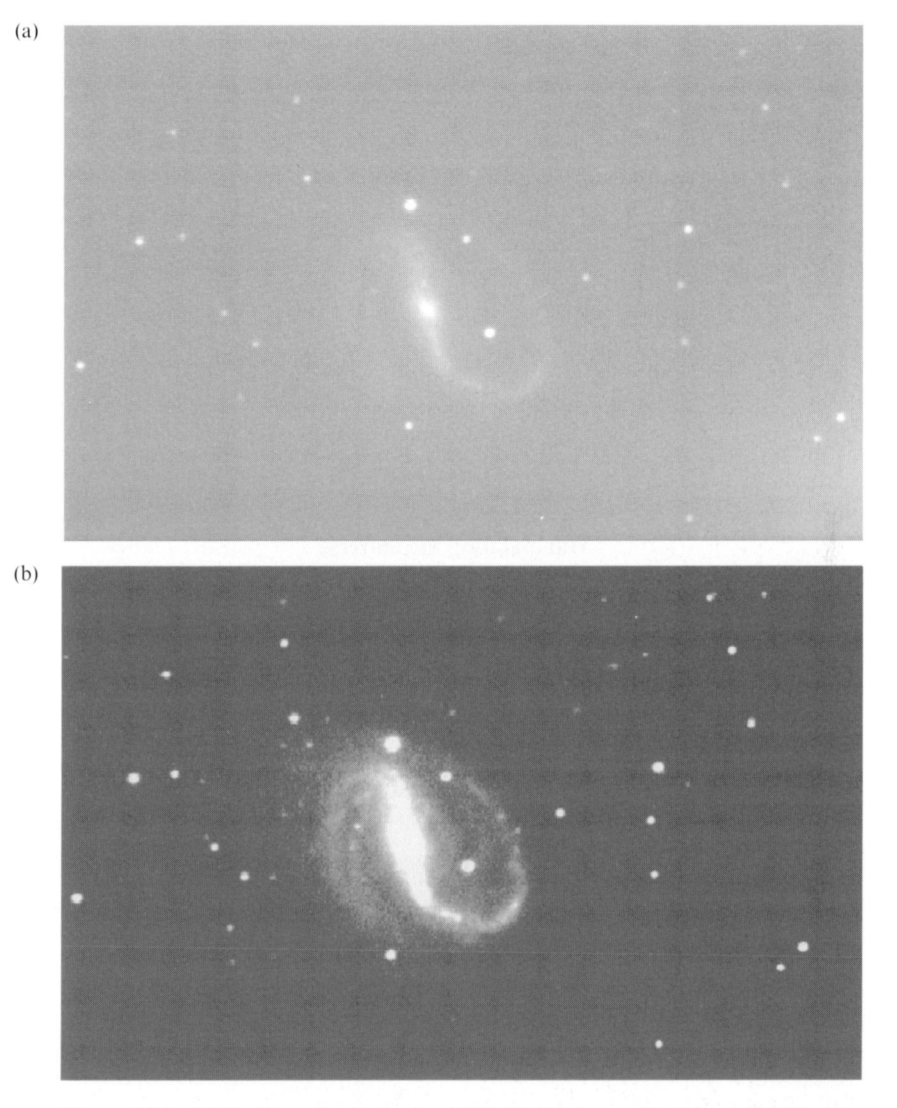

Figure 6.14. (a) The barred spiral galaxy NGC 7479, imaged by Martin Mobberley using the 'Starlight Xpress' SXF camera system with his 19-inch $f/4.5$ Newtonian reflector on 1992 July 28^d. This is the raw image, the result of a single 10 minute exposure. (b) shows the result of image processing by subtracting the dark frame and contrast stretching.

Pegasus, imaged by Martin Mobberley using a 'Starlight Xpress' SXF camera with his 19-inch (490 mm) *f*/4.5 Newtonian reflector. Note the delicate details shown in the spiral arms of this eleventh magnitude object, the result of just a 10 minute integration under heavily light-polluted skies!

If the current CCDs have one major disadvantage when used with telescopes it must be that the field of view covered by one frame is necessarily small. However, this can be increased by use of a telecompressor and if rather larger fields are needed then one can mate the camera to a camera lens, rather than to the telescope. Figure 6.15(a) shows a 'raw' image of our neighbouring Local Group galaxy M31 taken by Terry Platt using the 'Starlight Xpress' SXL8 camera used with a 200 mm telephoto lens. The field spans about 2°.2 and Figures 6.15(b) and (c) show the steps in the subsequent processing needed to bring out the best overall detail.

Manipulating the universe

Take a look through magazines such as *Sky & Telescope* and you will see the sorts of things that professional astronomers get up to with their image processing. Given adequate software there is nothing to stop **you** using the same techniques. For instance, does it interest you to know how the distributions of oxygen and hydrogen compare in planetary nebulae? If so, then why not obtain a wide-band Hα filter and an OIII filter from a company like Lumicon and then use your telescope to obtain CCD images of those nebulae that interest you, through each of the filters. You could then colour-code each pair of images and combine them to produce a spectacular and informative image of each nebula. No need to stop there.

Why not generate luminosity profiles of these or other objects, or perhaps colour-code a particular small range of brightness levels to bring out detailed, but otherwise not obvious, structures? What galaxies have jets emanating from their nuclei? What peculiar structures and distributions of particular elements can you find in various emission and reflection nebulae? This is real science and the list of possible investigations which are open to you is limited only by your imagination.

If those spectacular true-colour images (or, **nearly** true-colour images!) of planets and deep-sky objects seem evocative then you can create those, too. Several manufacturers such as 'SBig' provide filter wheels that plug into the telescope before the camera and allow one to take exposures through standard red, green, and blue filters. The three separate images can be combined during processing to make full-colour images. Also, at the time of writing at least one manufacturer ('Starlight Xpress') produces a

Figure 6.15. A sequence created by Terry Platt showing steps in the processing of an image of the galaxy M31 he obtained using a 'Starlight Xpress' SXL8 camera and an *f*/5.6 telephoto lens of 200 mm focal length. (a) This raw image is the result of an exposure of 160 seconds, (b) shows the effect of a logarithmic contrast stretch and (c) shows the result of subsequently re-defining the black level in the frame. North is approximately uppermost in these views.

'one shot' chip that can register a full-colour image without the need for any filters.

The principal image processing techniques are demonstrated by the examples presented in this chapter. However, these only scratch the surface of what is possible with contemporary software. For instance, the image can be processed with any of several different types of brightness scaling in order to bring out specific types of details. As an example, a logarithmic contrast stretch produces a similar result to unsharp masking in ordinary film photography, whereas gamma scaling allows one to enhance the mid-tones of an image without burning out the details in the brightest portions, nor losing the faintest and darkest parts.

The image can also be subjected to enhancements by means of techniques such as Fourier processing or maximum entropy deconvolution to emphasise small spatial details.

CCD images, because of their nature, contain both brightness and positional information and these can also be exploited. The techniques of photometry and astrometry are described later in this book.

A CCD camera can monitor the tracking of a telescope and the appropriate software and electronics can allow it to operate a telescope's slow motion controls. The 'Meade Instruments Corporation' were the first to provide commercial products that would do this and they currently market a number of high quality astrocamera systems as do other companies listed in Chapter 17.

Examples of electronic imaging are included in the later chapters of this book. It is probably true to say that this particular area of endeavour is one of the most important to have opened up for amateur astronomers in a very long time. With the CCD and computer attached to a telescope it really is a case of 'the sky's the limit!'

7
The Moon

In the post-Apollo years there was a growing feeling among amateur astronomers that lunar observation was no longer worthwhile. Moon mapping had been all but completed by the Orbiter probes of the 1960s and men had been to the Moon's desolate surface and brought back samples of lunar rock and soil. Even so, there is still much we do not know. Even to this day there are still one or two areas of research that can profitably be undertaken by the careful amateur.

Selenographic longitude and latitude

Old and new maps of the lunar surface differ in their definition of east and west. On the classical scheme the Mare Crisium was to the west and the Oceanus Procellarum to the east. The modern scheme, due to the International Astronomical Union (IAU), is that the **Mare Crisium** lies to the **east**. Coordinates measured with respect to the surface of the Moon are termed *selenographic*. Libration affects the apparent positions of features on the Moon's surface but the mean position of the meridian corresponds to a selenographic longitude of 0°. Selenographic longitude increases eastwards (towards the Mare Crisium) and is 90° at the mean east limb. It further increases, through 180° at the mean position opposite the Earth, to 270° at the mean west limb. From the mean west limb the selenographic longitude increases further to 360° (equivalent to 0°) at the meridian.

Selenographic latitude is measured from the Moon's equator and is positive going northwards (towards the crater Plato) and negative going southwards (towards the crater Tycho). Figure 7.1 shows the IAU scheme for selenographic longitude and latitude.

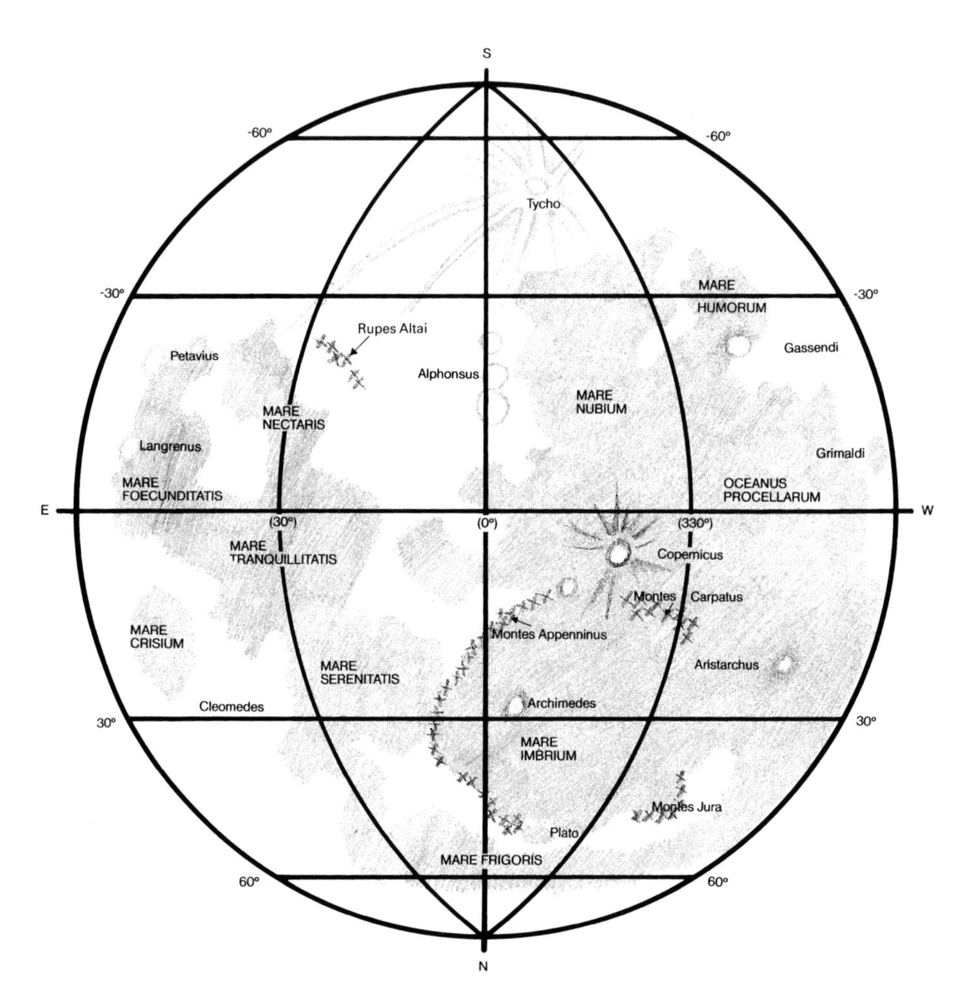

Figure 7.1. Outline map of the Moon.

Topographic study: drawings

Space probes, in particular the 'Orbiter' series of the 1960s and the recent 'Clementine', have photographed most of the Moon's surface to a much higher resolution than is possible by telescopic observation from the Earth. However, any one photograph can only show surface relief under the prevailing angle of sunlight. In fact, the majority of the Moon has still only been covered at very high resolution under one or two lighting angles. Our knowledge of the surface still lacks the completeness afforded by high-resolution study at all angles of illumination. This is an area worthy of amateur effort. In particular low-relief features, such as domes and wrinkle ridges, only show up well under very low angles of sunlight (when the features are close to the terminator). Yet, most of the space probe photographs were taken under relatively high angles of illumination. The amateur can wait until just the right time to observe a feature when it is close to the terminator.

On the negative side, the Moon **has** been well photographed from the Earth. Moreover, some of these photographs have been taken with large telescopes at sites of very good seeing. Also, the Moon's surface has been subjected to high-resolution radar-ranging telemetry by probes such as 'Clementine'. Consequently, the opportunities for genuinely useful contributions are extremely limited.

Still, there is a lot to be said for spending time doing something to increase one's own knowledge. Certainly if you spend some time studying the Moon, and perhaps drawing its ever changing vistas, you will get to know it a whole lot better.

An observer can record what he/she sees by making a drawing. This could take the form of a simple line diagram. At the other extreme it could be a 'photographic quality' sketch showing all the half-tones. What is possible depends largely on the observer's artistic ability. This, like most things, improves with practice. The important requirement is that the drawing is **accurate**. The simplest representation that is accurate is vastly preferable to the most picturesque representation that is not.

The production of line diagrams

The diagram could be generated entirely from the view at the telescope straight onto a blank sheet of paper. Alternatively, a pre-prepared outline of the major features could be used, the work at the telescope then consisting of filling in the shadows and the minor details. The outline could be made by tracing over a suitable photograph. This method should produce greater positional accuracy in the finished drawing. However, the effects of libration should be taken into account. This is particularly important for any region near the limb. The area of the Moon covered by the drawing should not be too great, certainly no more than 200 km × 200 km and preferably smaller. Also, the scale of the drawing should be such that 50 km on the Moon covers **at least** 25 mm on the paper.

Figure 7.2 shows a line drawing made by Rob Moseley in order to show albedo estimates in the crater Aristarchus. Notice how he has used both solid and broken lines to represent outlines. The broken lines indicate

ARISTARCHUS 1984 JAN. 16 UT. 01.10 - 01.25

COLONG. 62°.3 6" SPEC. X 120 SEEING II TRANS. V.GOOD.

ALBEDO ESTIMATES. R. MOSELEY.

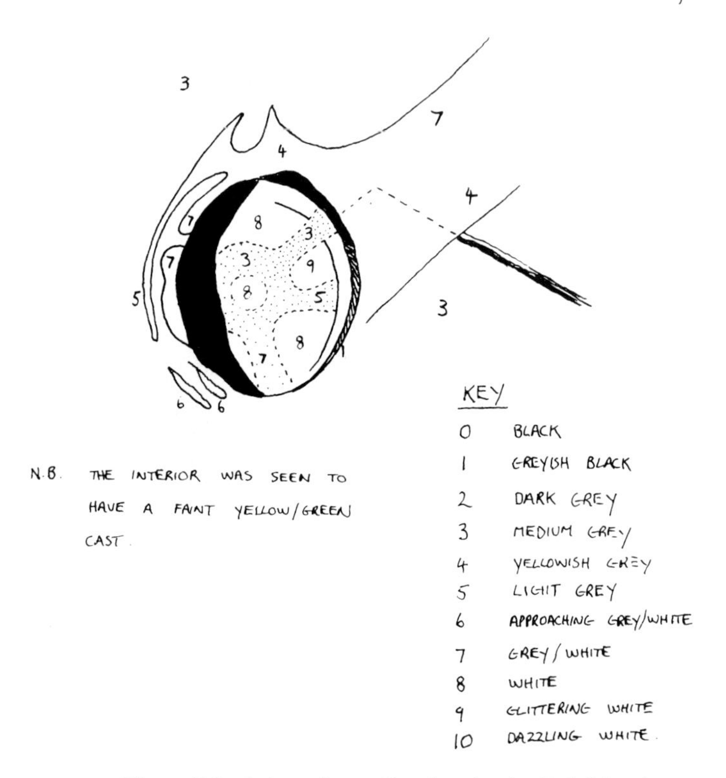

Figure 7.2. Aristarchus, a line drawing by Rob Moseley.

boundaries that are not as sharply delineated. Figure 7.3 shows another line diagram by the same observer, this time of the area around the crater Yerkes.

Normally line diagrams should not be 'worked up' after the observation. It is true that black ink can be laid over the pencil lines and the shadows filled in, though the temptation to make alterations should be resisted. The observer should critically compare the completed pencil drawing at the telescope with the view through the eyepiece. Any anomalies that defy correction can always be noted near the drawing (eg. 'crater pit should be about 10 per cent longer in east–west direction', etc.). The time of completion of the drawing should be recorded, together with the usual details of telescope, magnification and seeing conditions.

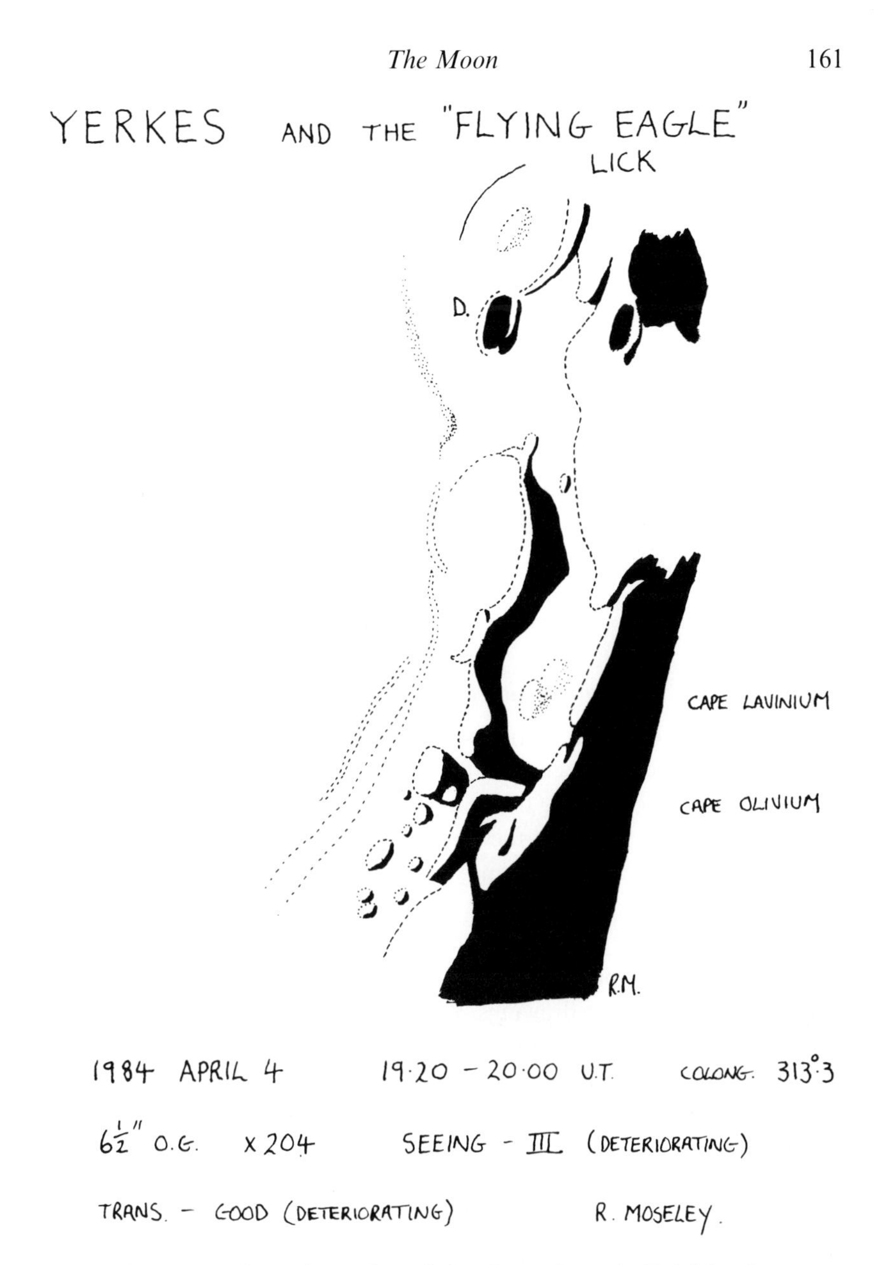

Figure 7.3. The Yerkes region of the Moon, drawn by Rob Moseley.

You will notice that Rob Moseley has also included values for the *Sun's selenographic colongitude* at the time of the observation on each of his drawings. This is a figure (obtained from an ephemeris) which represents the position of the morning terminator on the Moon. It is the selenographic longitude of the morning terminator and has a value of approximately 270° at new Moon, 0° at first quarter, 90° at full Moon and 180° at last quarter.

Though great care should always be taken when making a drawing at the telescope, it is also important that the time taken to complete it is not too long. The appearance of a feature close to the terminator changes rapidly. Ideally the pencil sketch should be finished in under half an hour. Where a feature is observed for a longer period the drawing can carry information on the times that particular features first emerged into sunlight, etc. (see Figure 7.4). In connection with this, sketches made in a sequence showing sunrise or sunset occurring for a particular lunar feature are especially useful.

The production of half-tone drawings

These take longer to produce than line diagrams and so a strategy that allows the maximum efficiency of the use of telescope time is called for. I recommend the prior tracing of the major outlines of the area under study from a photograph. Greying the area of the paper that the drawing will occupy will also save precious time. This can be achieved by shading with a soft pencil or sprinkling on a little charcoal powder and rubbing with the finger until a uniform greyness is achieved.

When at the telescope, proceed adding fine details in the form of outlines, as for the making of line drawings. When this process has been completed then add in the darker tones and make sure that the areas to be blacked are either filled in with dark pencil shading, or else marked for filling in afterwards. A rubber sharpened to a point can be used for removing the grey shading in order to show highlights and the brightest features. As soon as this is finished check the details with the view through the telescope and record all the necessary data, as for the line drawings.

The drawing can then be 'worked up' by filling in the black shadows and the edge of the terminator with black ink if desired (I have found that felt pen works very well). Some observers copy their original drawing and then spend time on this copy making it more artistically presentable. Many use materials other than pencil and ink for the final drawing, as well as a variety of different techniques. All are valid, provided that the final result is an accurate representation of the lunar surface as it appeared at the time of

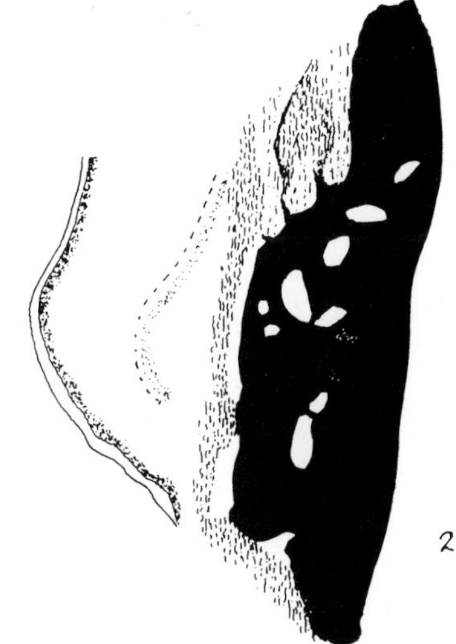

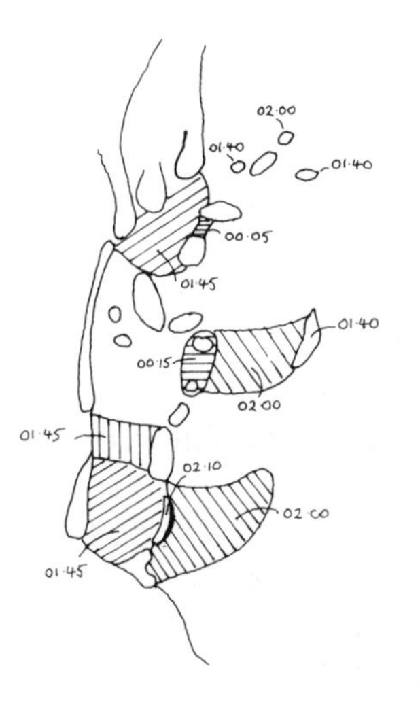

Figure 7.4. The Harbinger Mountains, drawn by Rob Moseley.

the observation. There is no room for artistic licence in science! Figure 7.5 shows another of Rob Moseley's excellent drawings. Notice how he handles the effects of highlights and shadow. A few particularly gifted observers and draughtsmen can produce stunningly beautiful drawings. One such is Andrew Johnson. Figures 7.6, 7.7 and 7.8 show some examples of Andrew's work.

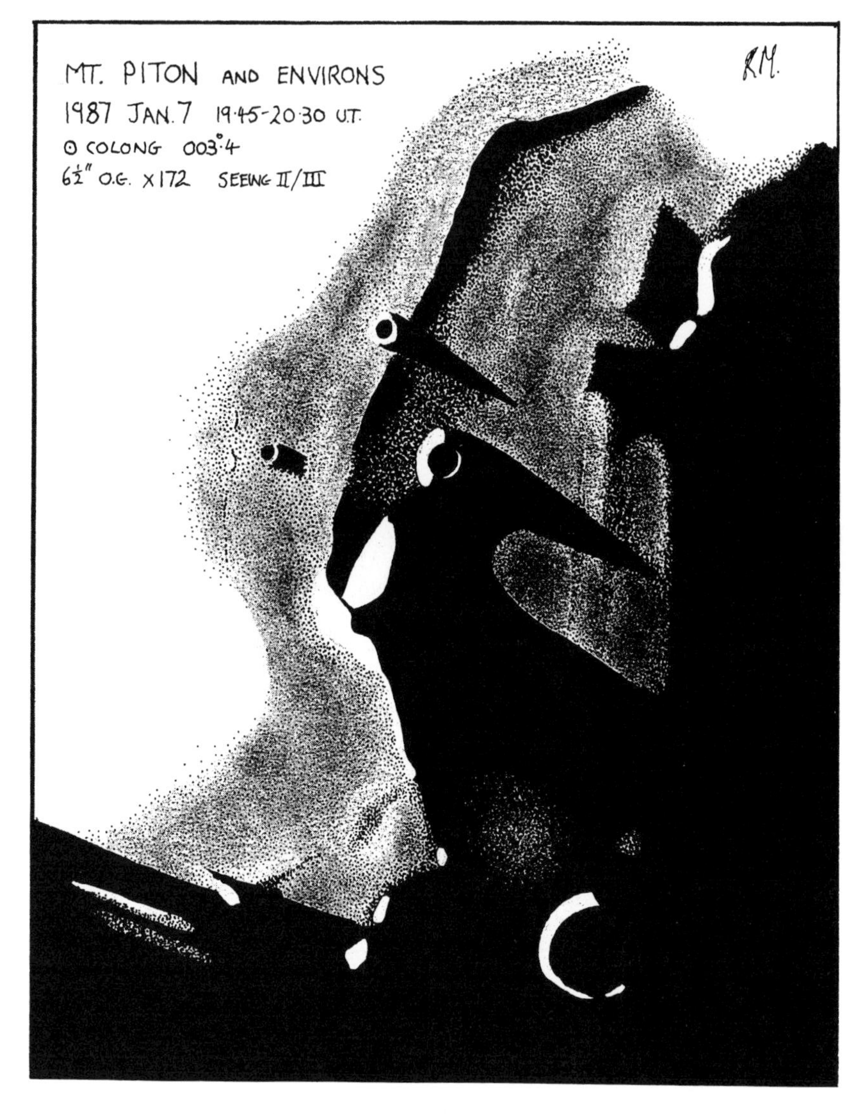

Figure 7.5. Mt Piton and environs, drawn by Rob Moseley.

Topographic study: photography and electronic imaging

High-resolution photography of the Moon is still worthwhile (see Chapter 5 for details of the methods used), though the seeing in backyard observatories will mean that even 1 arcsecond resolution (1 arcsecond corresponds to a linear span of about 1.6 km at the Moon's distance) is only rarely attained. A few skilful photographers in favoured sites can do better,

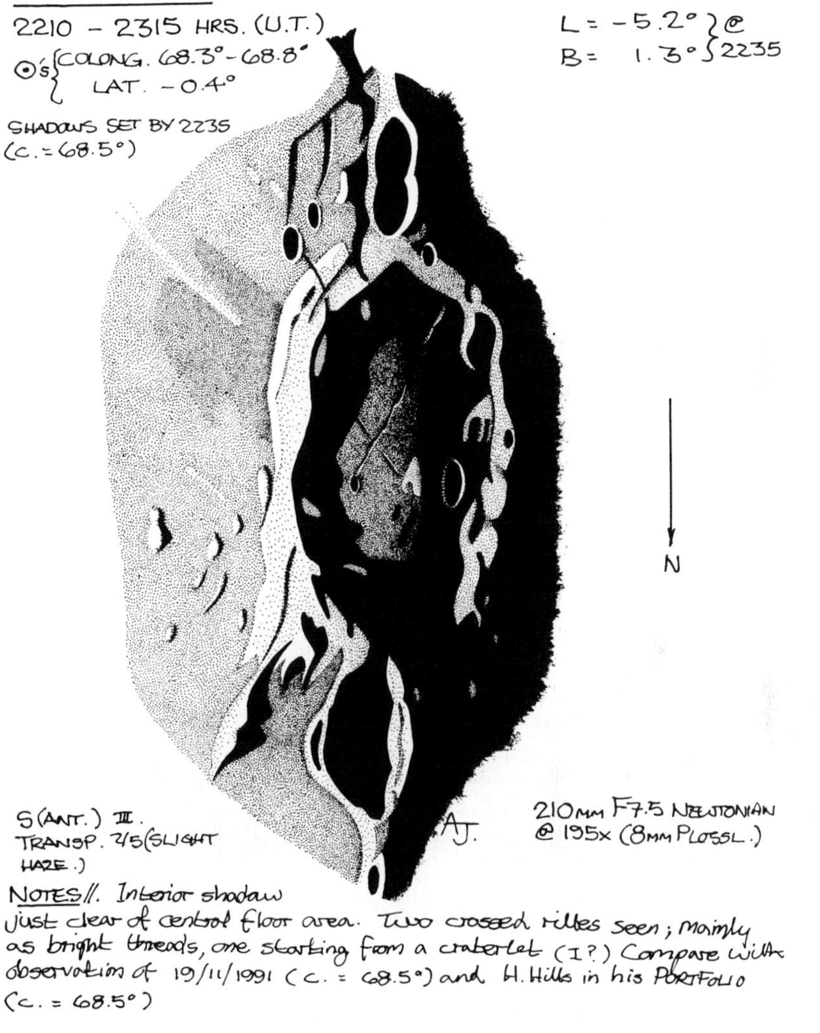

Figure 7.6. Hevelius at sunrise, drawn by Andrew Johnson.

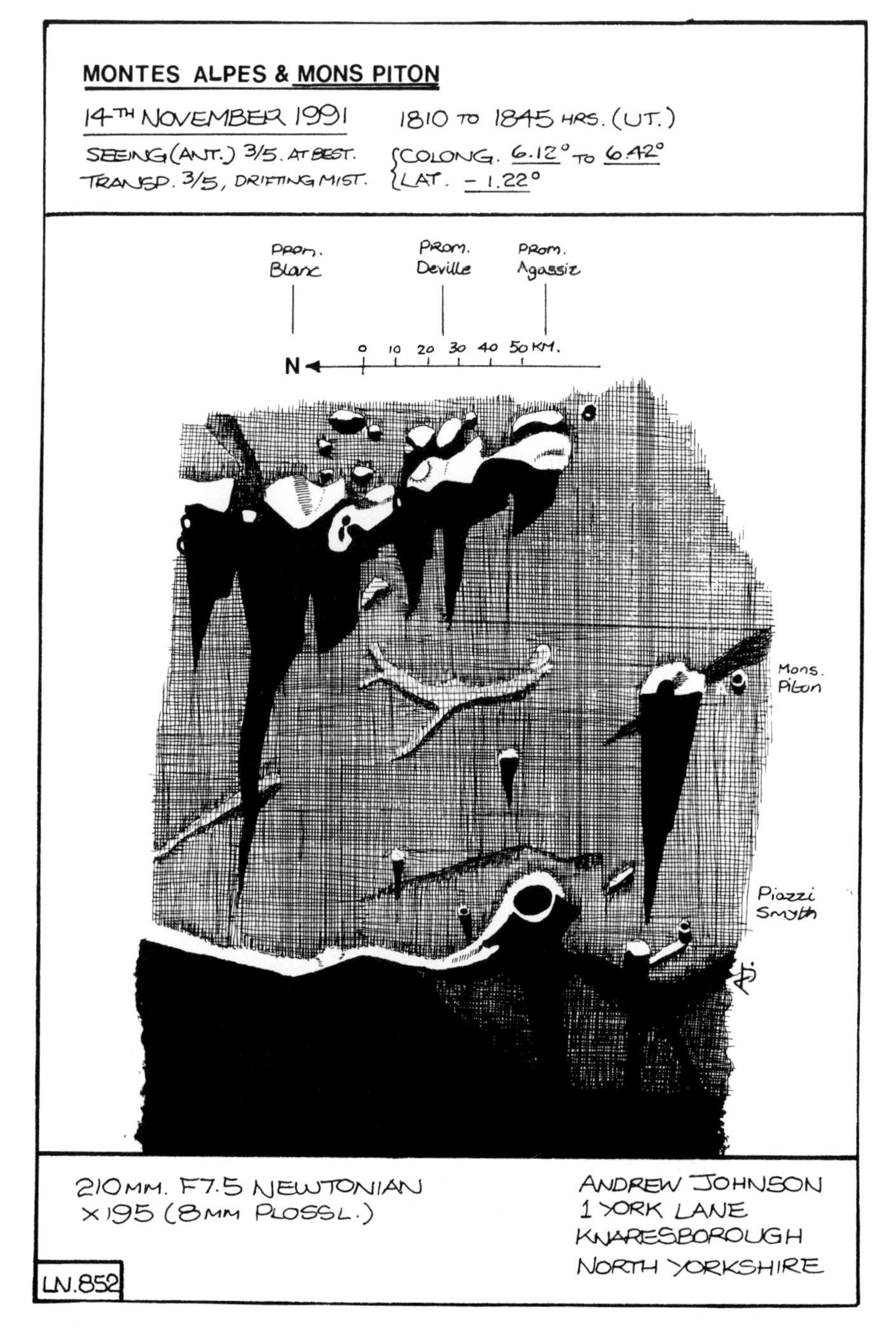

Figure 7.7. Another splendid drawing by Andrew Johnson, this time of the Montes Alpes region of the Moon.

LN.894

SIRSALIS & RILLE

1995 APRIL 12

2000 – 2110 HRS. (U.T.)

⊙'s { COLONG. 61·1° – 61·7°
LAT. 0·4°

GEOC. LIBR. @ 2000 HRS. { L = −6.3°, B = 5.2°

210MM F7.5 NEWT. @ × 195

SEEING. (ANT.) II

TRANSP. 2/5

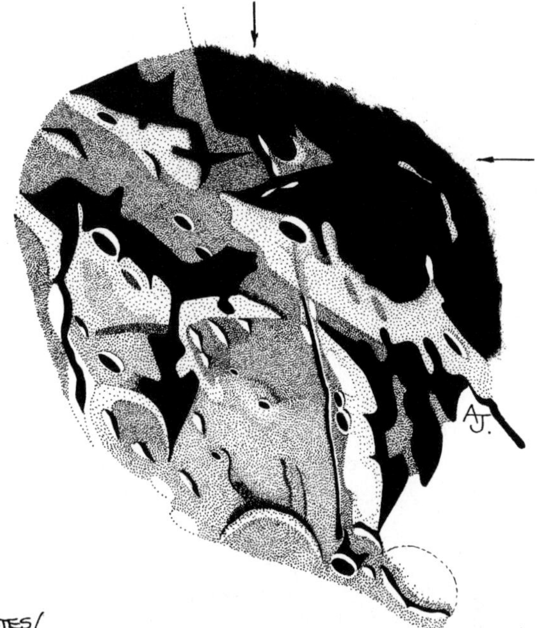

NOTES/
Further observation in search of confirmation that the Sirsalis rille does reach Sirsalis J, climbing the surrounding slope of the crater. This observation was made under a lower angle of illumination, and as the rille was indeed seen, and that it lacked any real shadows, perhaps these conditions indicate that the rille may be quite shallow over this section of its path? Arrows indicate possible extension of rille beyond the terminator.
(Shadows shown as at 2025 HRS.)

ANDREW JOHNSON, KNARESBOROUGH, NORTH YORKSHIRE.

Figure 7.8. Sirsalis, drawn by Andrew Johnson.

Figure 7.9. Plato and part of the Vallis Alpes, imaged at the date and time shown by the author with his video camera and 18¼-inch (464 mm) reflector.

notably G. Viscardy in France, who can produce photographs of resolution close to the theoretical limit of his half-metre reflector. As described in the last chapter, the rest of us can do better by using a video camera with a telescope (see Figure 7.9) or, and best of all, by using a dedicated CCD camera and the appropriate software to extract the very finest possible details.

Topographic study: shadow measurements

The relative heights of features on the lunar surface can be found by measurements of the lengths of the shadows they cast. Even more so than is the case when making drawings, an observer undertaking this study is largely doing it for his/her own personal pleasure and desire to know the Moon. Formerly this sort of work would be carried out by an observer with a good 8-inch (203 mm), or larger, telescope equipped with a clock-drive and an eyepiece micrometer. Now it would be better to use high quality photographs or, best of all, CCD images.

The use of the eyepiece micrometer is discussed in detail in Chapter 14. As for double star work, use the highest magnification the seeing allows when making the measurement. The transverse fixed wire is aligned along the shadow and the movable wires are positioned to intersect the end of the shadow and the peak that is causing it. The drum reading(s) and the known value of the micrometer constant are then used to calculate the apparent length of the shadow in arcseconds. The measurement can be repeated a number of times in order to increase accuracy as well as to provide a figure for the likely uncertainty in the calculated value (from the spread of readings). However, do remember that the shadow length varies rapidly, particularly for an object near the terminator.

If using photographs or CCD images, the shadow lengths can be measured directly but of course it is important to know the scale of the image (see the last two chapters).

The apparent length of the shadow is expressed as a fraction of the Moon's semi-diameter (given in an ephemeris). This quantity is denoted by s. For example, if at the time of observation the apparent semi-diameter of the Moon was 1000″ and the measured length of the shadow was 2″, then the value of $s = 0.002$.

The subsequent reduction can be carried out by the use of four equations. The terms used in the equations are listed together here for convenience. All coordinates are selenographic and may be found from an ephemeris for the time of the observation:

L_p = longitude of peak casting shadow
b_p = latitude of peak casting shadow
$colong.$ = Sun's colongitude
b_s = Sun's latitude
L_E = longitude of Earth
b_E = latitude of Earth

X, the distance of the apparent centre of the lunar disc to the sub-solar point, may be found from:

$$\cos X = \sin b_s \cdot \sin b_E + \cos b_s \cdot \cos b_E \cdot \sin(colong. - L_E).$$

The apparent length of the shadow can then be corrected for the angle the Sun makes to the east–west plane:

$$S = s/\sin X,$$

where S is the corrected shadow length expressed as a fraction of the Moon's semi-diameter. The altitude, A, of the Sun from the peak casting the shadow is found from:

$$\sin A = \sin b_s \cdot \sin b_p - \cos b_s \cdot \cos b_p \cdot \sin(colong. - L_p).$$

Finally, the height of the peak, H, can be found from:

$$H = S sin A - S^2 cos^2 A - \tfrac{1}{8} S^4 cos^4 A.$$

Strictly, a correction should be applied for the effects of parallax since the figures given in the ephemeris are *geocentric* (as would be observed from the centre of the Earth) rather than *topocentric* (as seen from the Earth's surface). However, the error generated by ignoring this correction is always less than the likely instrumental and human errors. Consequently the need for correction can be ignored.

The instrumental and human errors will be largest when measuring short shadow lengths. On the other hand, long shadows will also give rise to a large uncertainty in the calculated height of a peak above its surrounds because the shadow may well fall across terrain of varying height. Ideally a feature's shadow should be measured for a number of lighting angles. This will allow a profile of the area to be generated.

Transient Lunar Phenomena

Certain areas of the Moon's surface seem prone to occasional and short-lived glows (sometimes coloured) and misty obscurations. These events have been termed *Transient Lunar Phenomena*, or TLP (Americans use the name Lunar Transient Phenomena, LTP). Occurrences of TLP have been reported going back centuries but they were widely regarded as tricks of the eye until the spectroscopic confirmation of one event in 1958.

Dr Nikolai Kozyrev was using the 50-inch (1.27 m) Cassegrain reflector of the Crimean Astrophysical Observatory on November 3 of that year when he noticed that the central peak of the crater Alphonsus (Figure 7.10 shows a normal view) became obscured by a reddish haze just after 01^h U.T. Over the next couple of hours he saw the central peak of the crater become very bright and white. Between $03^h\ 00^m$ and $03^h\ 30^m$ the appearance of the crater returned to normal.

He took a series of spectra during these times and they showed that a gaseous emission had taken place. The spectrum taken when the central peak appeared bright displayed strong emission bands. Particularly prominent were the Swan bands of carbon, C_2. The last spectrum showed that the region had returned to normal, showing nothing but the usual solar spectrum reflected from the surface of the Moon.

Dr Kozyrev was not the first to provide hard evidence that TLP were something real and not illusory. However, his observation aroused world-wide interest and after that date the study of this phenomenon became

Figure 7.10. Arzachel (top), Alphonsus (centre) and Ptolemaeus (bottom) photographed by Martin Mobberley using his 14-inch (356 mm) Cassegrain reflector. The exposure was ½ second, made on XP1 by eyepiece projection (EFR = $f/65$), on 1981 November 19^{d} 05^{h} 55^{m} UT.

somewhat more respectable. Over the years a considerable number of research papers have appeared on this topic and much more evidence has been gathered. Some events have been recorded photographically, others photometrically.

TLP sites are not distributed randomly, they are primarily concentrated near the edges of the maria. One feature, the crater Aristarchus (Figure 7.2), is responsible for about a third of all TLP reports. Significantly, the results obtained from the Apollo Alpha Particle Spectrometer during the Apollo 15 and Apollo 16 missions show that radon gas seeps to the Moon's surface. Moreover, the sites of maximum emission coincide with the previously recorded sites of TLP (the data are summarised in a paper by Paul Gorenstein, Leon Golub and Paul Bjorkholm 'Radon emanation from the Moon, spatial and temporal variability', *The Moon* volume 9 (1974)). The radon emission from Aristarchus was particularly high. On one occasion the three astronauts aboard Apollo 11 observed a brightening in the north-west corner of Aristarchus. Earth-bound observers also witnessed this event.

Visual hunting for TLP

Just once in a while a TLP occurs that is striking enough to be seen in a small telescope. More usually the observed effects are too delicate and require larger apertures. Consequently it is not possible to put a definite limit on the smallest size of telescope that can be useful for TLP work, though at least 6-inch (152 mm) aperture is desirable. It is crucial that the observer should be very familiar with the visual appearance of the Moon.

The appearance of any surface formation changes considerably during a lunation. As an example consider the crater Eratosthenes. It stands out in rugged relief at the times of first quarter and last quarter Moon. However, at full Moon it almost blends into its surroundings and its interior looks decidedly 'washed out', almost as if the crater was enveloped in some sort of haze. The effect is caused purely by the change in lighting angle. An observer can only be sure if what he/she is observing really is anomalous by being very familiar with the normal appearance of the feature under the given lighting conditions. This can only come from practice.

The following types of visual anomaly have been reported. Each type of phenomenon usually effects only a very small area of the lunar surface – a small crater, or a mountain peak, etc.

Short-term albedo (brightness) changes

These could be increases or decreases in apparent brilliance. Sometimes these last for hours, at other times just minutes. Sometimes the albedo of a feature remains fairly steady during the period it differs from the norm. At other times pulsations in brightness occur – sometimes on timescales as short as a second.

Obscurations of surface features

Small areas might show blurring or lack of contrast, while the immediate surrounds remain perfectly sharp and clear cut. Sometimes the region affected is initially very small but gradually spreads out. The obscuration often dissipates after an hour or so and the region then returns to normal.

Coloured effects

Sometimes brightness changes and obscurations are accompanied by colourations. Occasionally coloured effects appear without these changes. It now seems that most TLP do not show significant coloured effects. However, with some events the colours seen, ranging through the entire rainbow, can be very vivid. Regions undergoing albedo variations tend to show a bluish colour, if any colour is seen at all.

Brief flashes of light

These are the rarest of observed TLP and manifest as bright twinkles and flashes from the lunar surface, usually lasting less than a second. Despite their rarity at least two examples of this phenomenon have been photographed, though not everyone accepts that the apparent flashes in these two cases really are of lunar origin. When flashes occur they often accompany other manifestations of TLP.

The TLP hunter should adopt a definite strategy. On the occasions I use my own telescope for this work a typical observing session will be split up into two main activities. Initially I 'raster-scan' the whole of the visible surface, taking about 15 minutes to complete the task. This 'raster-scanning' consists of overlapping east–west sweeps across the lunar surface, advancing the telescope a little in declination for each new sweep. While features are moving through the field of view I examine them for any signs of abnormality. With my 18¼-inch telescope I usually use a magnification of ×144 for the first sweep.

I then carefully scrutinise any features about which I might be suspicious. I might well leave the telescope for a moment to check charts or photographs of the area in question (hopefully made under similar lighting conditions). If the area does seem to exhibit an unusual appearance I will keep it under scrutiny. Assuming all is normal I then spend some time examining several other chosen targets. My list includes such features as: Aristarchus, Torricelli B, Plato, Proclus, Alphonsus, Messier and Messier A – all sites of previous TLP. Of course, not all of these objects will be in sunlight at any one time, apart from near full Moon. Assuming all is normal, I then proceed to re-scan the Moon with progressively higher magnifications, at the end of each scan period spending some time individually scrutinising the selected formations; and so the observation period continues.

A special watch should be kept for TLP during lunar eclipses – there is speculation that some events could be triggered by rapid changes of temperature on the Moon's surface. On the vast majority of nights nothing unusual will be seen but occasionally suspicions are aroused. This is where a coordinated team of observers is useful. A telephone call to a central coordinator will then alert other observers to monitor the area in question. It is important that the coordinator passes on only the location of the suspected event and not any details of the type of the abnormality suspected. Observational bias is then avoided. The individual reports can afterwards be examined for corroboration, or otherwise.

A coordinated team of observers also allows various team members to bring different techniques to bear: visual, visual with filters, photometry, photography, video or CCD recording, spectroscopy, etc. No one observer is likely to have the time or equipment to carry out all these techniques.

TLP hunting: photometry and colorimetry

The way that the brightness of a given area of the Moon's surface varies during the lunar cycle depends on the nature of the surface in that area (the local *microrelief*). Colour measurements provide data on the composition of the lunar soil. This has now been done so effectively with the 'Clementine' probe that there is no longer any reason for an amateur to do it – except as part of the TLP hunting programme, where temporary differences from the norm are looked for. The use of the photoelectric photometer for measurements of the brightness and colour of stars is covered in Chapter 13. The lunar observer uses the same equipment and techniques.

The Moon's brightness permits accurate results for short integration times, even when the photometer is attached to a very small telescope. In fact, short integration times should always be used because of the Moon's constantly changing right ascension and declination. The size of the photometer's focal plane diaphragm should be smaller than that normally selected for stellar work. A size equivalent to a few arcseconds of image should be chosen. Nearby and bright stars can then be used as comparisons to allow absolute brightness measurements to be made.

Other techniques are available to observers who do not possess photoelectric equipment. For instance, it is also possible to obtain relative brightness measurements from a photographic image. The processed film could be mounted into a slide projector and the brightnesses of different areas on the projected image could be measured with a sensitive light meter. Ideally, a fine-grained film with a large linear range (see Chapter 4) should be chosen. Kodak's Technical Pan 2415 would be a good choice. This technique allows the recording of whole disc images. The subsequent measurements and analysis can then be carried out at leisure. The light meter can either be purchased or constructed (an LDR, a photodiode, or a phototransistor would make a good detector).

CCDs also allow photometry of the whole lunar disc (or at least a large part of it) to be performed. In this case the output from the CCD is read into a computer and the brightness levels of individual pixels can be read off using the appropriate software.

The normal brightness behaviour of a particular lunar feature (with respect to that of the standard features) over a lunation can be established from several months data. The feature can then be monitored for anomalous brightness changes.

TLP or not TLP

As with any observation, the reports of both positive and negative sightings of TLP should carry such information as: size and type of telescope, seeing conditions (turbulence and transparency), magnifications used, date and precise times. Another detail to be included is the presence, or not, of *spurious colour*. This is the slight colour fringing produced in the image by differential refraction of the light as it passes through the Earth's atmosphere. The amount of spurious colour seen in the image depends heavily on the altitude of the celestial body being observed and is greatest when the object is low over the horizon. It also varies somewhat with the ambient atmospheric conditions.

The complex interplay of light and dark on the Moon's surface can generate a variety of coloured effects. Sometimes these are visible through a small telescope. At other times the image will look colour free even through a large aperture. You should always report the extent of spurious colour in the image. It might be that the 'red glow' along a section of the rim of a certain crater is in fact generated by the Earth's atmosphere. However, if colour is seen and the surrounds are completely free of spurious colour then there are grounds for suspecting that the effect originates at or near to the lunar surface.

Other coloured effects can be caused by the telescope. For this reason reflectors are preferable to refractors for TLP observation. Also beware that even the more complex types of eyepiece show some colour fringing, especially near the edge of the field of view. It is also useful to compare the appearance of a formation with its surrounds using different coloured filters. For instance, does any apparent blurring show up most in a filter of a particular colour? Colour filters can be mounted to screw into the eyepiece barrel, or perhaps in a filter wheel that plugs into the telescope before the eyepiece. Rapid switching between red and blue filters might show a 'blink' reaction but do make sure that the surrounding features are not similarly affected.

Wherever possible try to quantify your observations. For instance if you suspect that the albedo of a particular feature is variable, try to time the variations. Also try to do the same with the variations in the image quality due to the seeing. As with coloured effects, in many cases the apparent albedo variations will be entirely due to the Earth's atmosphere. Where the appearance of a feature is slowly changing, record its appearance at given times.

It is certainly true that this area of research is controversial. Many do not accept that transient lunar phenomena could ever occur. However, those people tend not to be lunar observers. There is plenty of evidence to the effect that something very minor does occasionally happen at or near to the surface of our nearest celestial neighbour. It is tempting to speculate that build-ups of sub-surface radon and/or other gases might be occasionally released, perhaps raising colloidal sized particles in the process and so causing the misty obscurations. Maybe coincidental extra strong gusts of the solar wind could interact with the released gas, causing it to fluoresce if these interact near the lunar surface? Maybe charge separation causes an electrical potential to build up which breaks down in the gas to cause the rarely seen flashes? Of course, all this is mere conjecture. Nobody yet has the definitive answer as to what really happens. TLP research really is 'cutting edge' science and yet it is something that the patient amateur, free from the limitations on telescope time endured by the professionals, can successfully pursue.

Occultations

The timing of the occultations of stars by the Moon may be less glamorous than other fields of lunar study but it still provides useful data. Moreover it requires only very modest equipment. The long-term nature of the data generated by occultation timings is especially useful in studies of the dynamical slowing of the Moon in its orbit, due to its tidal interaction with the Earth. The observation of occultations can also reveal the binary nature of some stars: instead of sharply 'snapping out' as they pass behind the lunar limb, some stars take a brief moment to fade. Yet these binaries may be too close to be individually resolved by conventional observing techniques. Measures of the profile of the lunar limb and stellar proper motion studies are among the other spin-offs generated by occultation timings.

For the observer to work as part of a group is highly desirable in most fields. In occultation work it is essential. Many provincial observing groups coordinate their results and then send them to either the ILOC (International Lunar Occultation Centre) in Tokyo, or the IOTA (International Occultation Timing Association) in St. Charles, Illinois. The prediction data can be obtained from these sources, usually via the offices of the provincial group.

The observer's first step is to obtain predictions for the forthcoming occultations. These give the designation of the star, its magnitude, its coordinates, the date and approximate time of the occultation, as well as whether the event is a disappearance or a reappearance. Another piece of included information, particularly useful for observing reappearances, is the position angle that the star makes with the limb at the time of the event (see Figure 7.11).

One should set up to observe at least half an hour before the occultation is due. This will allow plenty of time to locate the star (if the event is a disappearance) and to check that everything is functional. If the telescope is equipped with a clock-drive this will make the procedure relatively easy. Just set the star in the centre of the field of view and watch for the approach of the Moon's limb. Then record the exact time of disappearance.

If the telescope does not have a clock-drive, then the aim is to move the telescope manually a little before the event, such that the occultation happens with the star close to the centre of the field of view. If you know the field of view of the eyepiece and the declination of the star (and hence

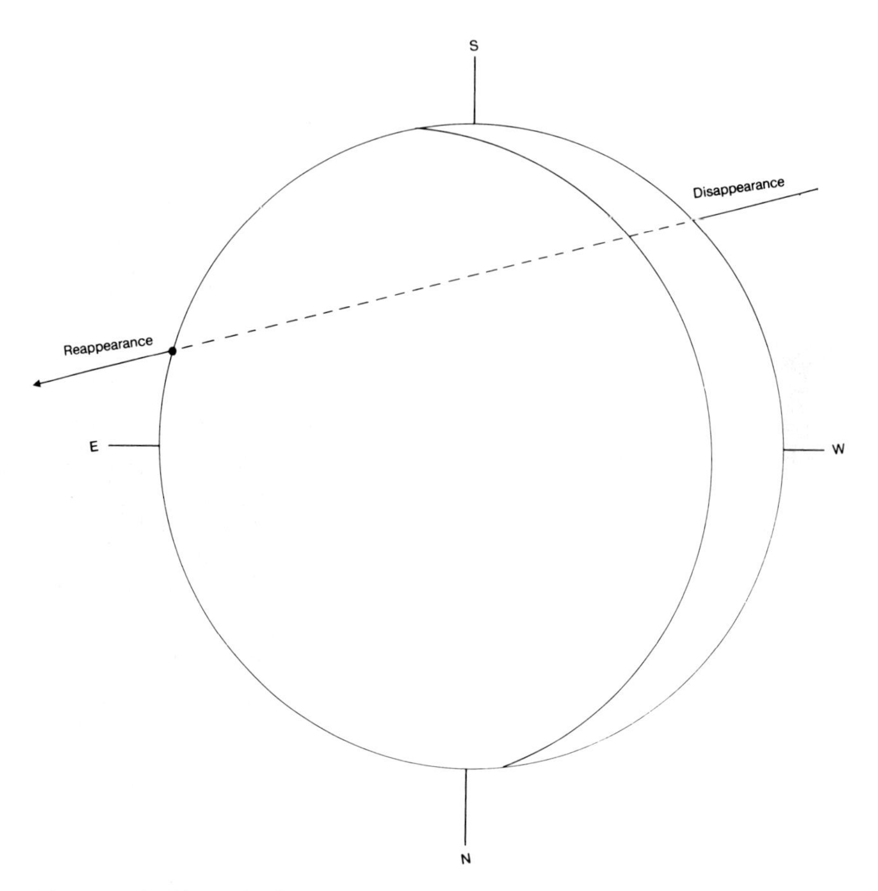

Figure 7.11. The path of a star behind the Moon during an occultation (exaggerated in its inclination for this illustration). Of course, an occultation could happen at any lunar phase so the disappearance could take place at either a dark limb or a bright limb. The same is true for a reappearance. The position angle (always measured from the north point, increasing in an anticlockwise direction) of the disappearance event shown is approximately 135°. That of the reappearance is approximately 250°.

its rate of drift due to diurnal motion), this is not too difficult to arrange. However, do try to avoid moving the telescope at the critical moment and so losing the timing.

Observing reappearances is more difficult and here a clock-drive is especially useful. Ideally the telescope is set on the star before it first disappears behind the Moon's limb and both the disappearance and reappearance events are timed. Otherwise, a telescope with accurate setting circles can be set on the position of the star, as given in the prediction, and a lookout kept for its reappearance. Failing that, keep a watch on the lunar limb at the estimated position angle of the emersion.

With regard to the method of timing, the object is to time the event to an accuracy of 0.1 second. One way of doing this is to start a stopwatch going a minute or two before the event (using radio or telephone signals to record the start-time accurately). Then stop the stopwatch at the instant of the event. If the time of starting the stopwatch is A and the time recorded on the stopwatch after the event is B, then the time of the event is simply A+B. Obviously, the stopwatch must be reliable and accurate. Some observers have constructed their own electronic chronometers, which automatically record the time when a button is pushed. Whatever method is used, the crucial requirement is that the timing is reliable and accurate.

Grazing occultations, where the star appears and disappears several times behind irregularities in the lunar limb, are particularly interesting. Here the most valuable work can be performed by a coordinated team of observers set up across the graze track (the predicted path along which the graze event would be observed). The individual timings then allow a very accurate profile of the limb to be constructed as well as a particularly accurate fix for the position of the Moon at the time of the event.

A video recording of the occultation could be made using a sensitive video camera or a CCD astrocamera. If the video has provision for the display of on-screen timings then the event can be timed to much better accuracy (1/25 second for UK video machines, 1/30 second for US recorders) than is possible by eye. The purely visual observer should aim for a timing accuracy of 0.1 second.

A computer-operated telescope, used with a CCD/video system, allows for totally automated occultation timings. In that case the on-screen timer should be set at the beginning of the session and checked at the end to make sure that it remains accurately synchronised to UT.

When submitting results, it is the observer's responsibility to report the exact *geodetic* coordinates (latitude, longitude and altitude above mean sea level) from which the observations were made. Reference to a large-scale ordinance survey map will provide this information. The observer's reported timings should normally contain no correction for a *personal equation* (the estimate time delay in seeing an event and pressing the button on the stopwatch). The only exception is when the observer is very experienced and has accurate and reliable figures for his/her delay in responding to disappearances and reappearances (the latter will be greater). In either case, the observer's report should state whether a personal equation has been applied or not (and if it has, the magnitude of the correction).

8

The terrestrial planets

Long-term observations of the Solar System bodies have traditionally been the province of the amateur. Space probe missions have come and gone and we have learned much as a result. The terrestrial planets have been mapped to a far higher resolution than is possible from the Earth. The planet Mercury is, frankly, not a very satifactory target for continued Earth-based observation. However, the same is certainly not true of Venus and Mars. These planets are much easier targets and diplay interesting changes and phenomena that can only be appreciated by long-term study. We may have learned much from the space probes but there is still plenty more we don't know!

Telescopes for planetary observation

The first three chapters of this book deal in detail with telescopic equipment. Sheer light grasp is not an essential for planetary observation. Much more important is that the image is as sharp and contrasty as possible. My advice to any planetary observer is to invest in quality optics, rather than the largest affordable aperture .

A large focal ratio is more conducive to high quality imaging (eyepieces, in particular, deliver better quality images when working with higher focal ratios) but is not an essential. Refractors can deliver very fine images provided their secondary spectra are not too prominent. However, if value for money is sought then I would recommend the Newtonian reflector every time. This old warhorse is capable of delivering very fine images, **provided its optics are of first class quality and are in good collimation**. Refractors and off-axis reflectors may produce better images but they will cost **very** much more money. Cassegrain and catadiopric telescopes can also produce images of roughly equivalent quality to the Newtonian of the same aperture but these, too, cost much more than the same size Newtonian telescope.

The central obstruction present in most compound telescopes is the reason why they produce lower contrast images than do unobstructed-aperture instruments, such as refractors and tri-schiefspiegler reflectors. If the central obstruction is as big as 30 per cent of the aperture diameter, the contrast of planetary details is cut roughly in half. However, the resolution of high contrast details is practically unaffected. So, a telescope for planetary observation should have the smallest central obstruction possible. After all, only the central part of the field of view needs to be fully illuminated by the secondary mirror for this type of observation.

Even if you have a large 'light bucket' of a reflecting telescope with a low focal ratio, a large secondary mirror, and mediocre quality optics, you can still convert it to a high quality planetary instrument. Make a mask that covers most of the aperture but has an off-set circular hole in it (see Figure 8.1). Make this hole as large as possible without the shadow of the secondary mirror or its support vanes causing any obstruction. A 16-inch or 18-inch Newtonian reflector can then effectively be converted to a 5-inch or 6-inch unobstructed-aperture reflector. The mask can be removed when you require your telescope to function as a 'light bucket' once more.

Observing Mercury

Mercury is a particularly difficult target for the Earth-based observer. Orbiting close to the Sun's fiery heat, Mercury never strays far from it in our skies. The planet's mean distance from the Sun is 58 million km but this can vary from 47 million km at perihelion to an aphelic value of 69 million km. Mercury's eccentric orbit causes its apparent angular distance at maximimum elongation from the Sun to vary between 18° and 27°. When at *eastern elongation* it can be seen low down in a dusk sky, shining as a 'star' of brightness $+0^{m}.7$. At *western elongation* the planet rises a little before the Sun in a dawn sky. Of course, the Earth's atmosphere will severely degrade the image of any celestial body that is so close to the horizon. A boiling, blurred, blob, fringed with spurious colour is all that one is likely to see of Mercury at such times.

If your telescope is equipped with setting circles then setting on Mercury in full daylight, with the planet high in the sky, should not be too difficult. A much steadier view will then be afforded. However, **do be careful about sweeping around with the Sun close by**. If the Sun should accidently enter the field of view your eyesight will be damaged before you have time to react. I make no apology for giving this warning in a book for advanced amateur astronomers. Familiarity tends to breed carelessness in the best of us.

When close to maximum elongation Mercury, as seen through the telescope, will appear with a phase rather like that of the first or last quarter Moon. The apparent angular diameter of the planet's full disc will then be about 8 arcseconds. This decreases to about 4½ arcseconds when the planet is at *superior conjunction*, on the opposite side of the Sun to us. At that time the planet appears full-phase. At inferior conjuction Mercury is on the same side of the Sun as we are and so appears at its largest (about 13 arcseconds), though the planet then has its night-time hemisphere pointed towards us. The planet is virtually unobservable when it is anywhere near conjuction, then appearing so very close to the Sun. Mercury's *synodic period* (the time between successive conjunctions) averages 116 Earth-days.

Under exceptionally good conditions a telescope larger than about 5-inches (127 mm) aperture will show some of the elusive shadings on Mercury's pinkish coloured disc. At elongations a magnifying power of about ×240 is needed to make the planet appear the same size as the Moon, when the latter is seen with the naked eye. It is sobering to think that the best pre-space-probe map of the planet (by E. M. Antoniadi from his observations using the 33-inch refractor at the Meudon Observatory in France) was found to be hopelessly inaccurate. Astronomers had even confidently ascribed the wrong rotation rate to the planet until as recently as 1965! Even then it was a radar technique, and not visual observations, that provided the true value of 58.65 days.

I would certainly stop short of trying to dissuade anyone from observing Mercury. We can never be absolutely sure that the planet won't ever throw us some unusual appearance, as remote as that possibility might be.

Observing Venus

The second closest planet to the Sun is much more easily observed than the elusive Mercury. Having a nearly circular orbit of radius 108 million km, Venus can appear separated by over 47° from the Sun. It's phases are linked with its apparent diameter in the same way as for Mercury. At superior conjunction it appears as a little 'full Moon' type disc 9½ arcseconds in diameter. At *dichotomy* (50 per cent illuminated) the planet subtends an apparent diameter of 25 arcseconds. A magnifying power of about ×80 will

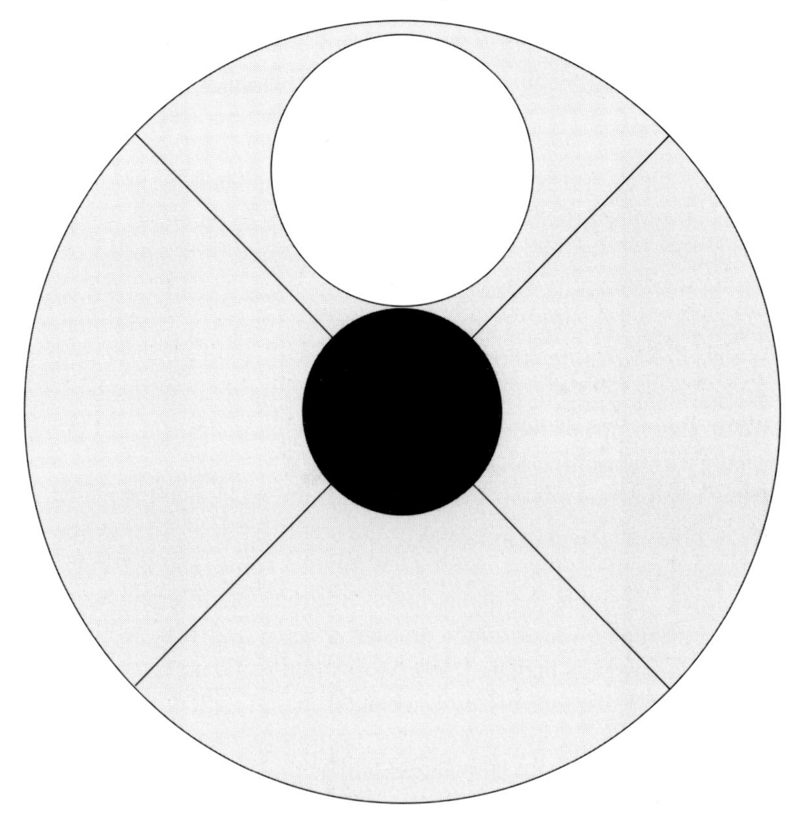

Figure 8.1. This schematically shows the desirable position of an off-axis stop for a telescope with a central obstruction. The stop can be placed over the primary mirror or at the end of the telescope tube.

then render a view of the planet of equivalent size to the Moon as seen with the naked eye. Venus's apparent size further increases to 65 arcseconds at inferior conjuction. When seen against a dark sky Venus is always impressive. At its greatest brilliance the planet shines with an apparent magnitude of $-4^{m}.4$. Venus's synodic period is 584 Earth-days.

Since Venus can set several hours after the Sun or rise several hours before it, the opportunities for observing this planet in good seeing against a dark sky are increased. Even so, there is value in observing the planet in full daylight – especially when it is near conjuction or when its declination causes its altitude to be rather low. Moreover, seen against a dark sky Venus's glaring brilliance may swamp out any of the tenuous shadings sometimes present. However, let me repeat my warning about observing with the Sun close by the planet in the sky.

Drawing Venus

The observer should begin with a pre-prepared circular outline. A size of 50 mm to the planet's full diameter has become the accepted standard amongst the various organisations coordinating amateur work. The drawing board should be light-weight and convenient to hold at the telescope. It should be fitted with a lamp with a shield to block the direct light of the bulb from your eye. The lamp can be powered by a battery, or via a lead from the low-voltage power supply at the telescope. Including a small rheostat is a convenience as this enables the brightness of the illumination to be varied. I recommend covering the bulb with a red filter for deep-sky observing, since red light disturbs one's dark adaption less than white light. However, this is not really necessary for observing the Moon and planets, especially if the brightness of the illumination is variable.

Before committing anything to paper, take time to look at the planet and allow your eye to become adjusted to the scene. At first little will be perceived, apart from the phase. After a few minutes some faint markings may be discerned. Examine the planet under a variety of different magnifications. Occasionally Venus might present a sharp and reasonably steady image with a power as high as ×300. More often ×200. or less, is the limit. How does the appearance of any marking(s) change with different powers?

Begin your drawing by marking the position of the terminator. Don't bother about filling in the sky background at this point. Use the valuable telescope time for drawing details on the planet itself. Any shadings on Venus are usually very difficult to see. The following number scale is generally used for the intensity of the shadings:

0 = Brilliant white
1 = The overall tint of the planet's disc
2 = Very faint shadings, hardly discernable
3 = Definite, though still faint, shadings
4 = Somewhat darker shadings
5 = Still darker shadings (very rare)

As with any similar scale, the observer's estimate can only be very largely subjective. Much also depends upon the size and quality of his/her telescope.

One way of representing any shadings present is by means of dotted lines, the intensity numbers then being used to label the appropriate areas of the drawing (see Figure 8.2). Alternatively, one can make a more artistic rendition (Figure 8.3) using pencil shadings suitably smudged by finger. Figure 8.4 shows another technique, where density of stippling has been used to represent various levels of shading. As always, the important factor

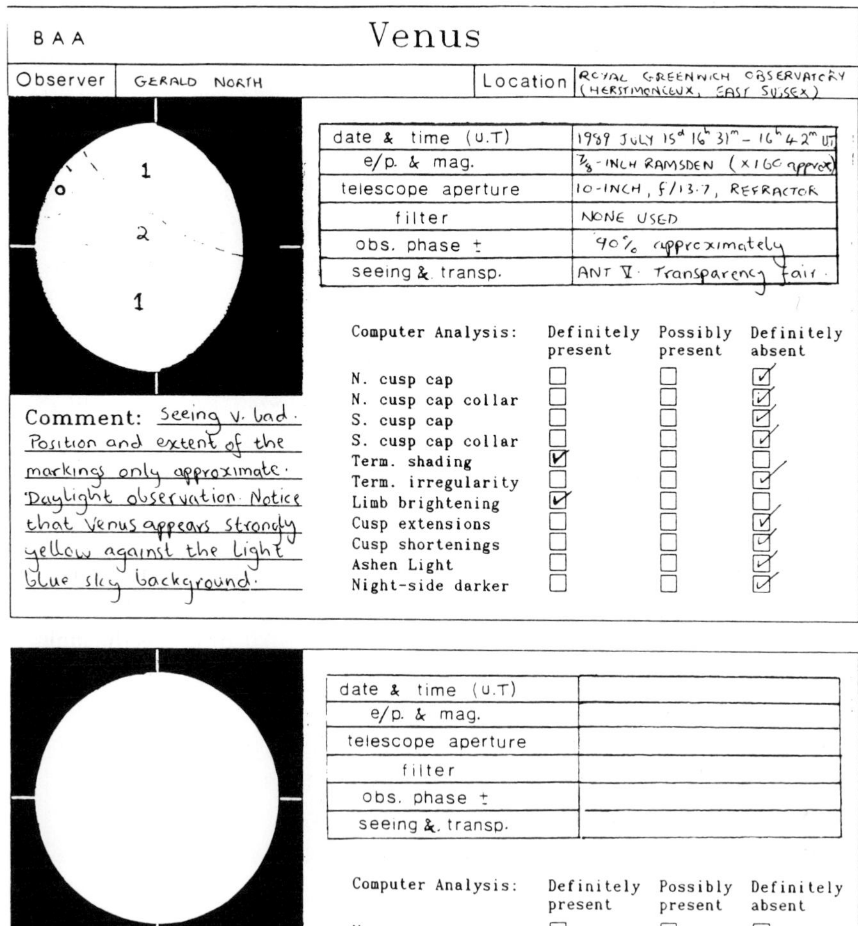

BAA **Venus**

Observer	GERALD NORTH	Location	ROYAL GREENWICH OBSERVATORY (HERSTMONCEUX, EAST SUSSEX)

date & time (U.T)	1989 JULY 15^d 16^h 31^m – 16^h 42^m UT
e/p. & mag.	7/8-INCH RAMSDEN (×160 approx)
telescope aperture	10-INCH, f/13.7, REFRACTOR
filter	NONE USED
obs. phase ±	90% approximately
seeing & transp.	ANT V. Transparency fair.

Computer Analysis:	Definitely present	Possibly present	Definitely absent
N. cusp cap			✓
N. cusp cap collar			✓
S. cusp cap			✓
S. cusp cap collar			✓
Term. shading	✓		
Term. irregularity			✓
Limb brightening	✓		
Cusp extensions			✓
Cusp shortenings			✓
Ashen Light			✓
Night-side darker			✓

Comment: Seeing v. bad. Position and extent of the markings only approximate. Daylight observation. Notice that Venus appears strongly yellow against the light blue sky background.

date & time (U.T)	
e/p. & mag.	
telescope aperture	
filter	
obs. phase ±	
seeing & transp.	

Computer Analysis:	Definitely present	Possibly present	Definitely absent
N. cusp cap			
N. cusp cap collar			
S. cusp cap			
S. cusp cap collar			
Term. shading			
Term. irregularity			
Limb brightening			
Cusp extensions			
Cusp shortenings			
Ashen Light			
Night-side darker			

Comment:

Computer analysis: Please tick the appropriate box. – Intensity estimates: 0 = brightest to 5 = darkest. – Please use a separate form for each day of observation.

Figure 8.2. An observation filled out on a standard BAA (Terrestrial Planets Section, Venus sub-section) issue observation report form. The layout of this form was designed by John McCue and Detlev Niechoy who are the former Venus coordinator of the BAA (in England) and coordinator of the VdS (in Germany), respectively. Note the intensity estimates given on the drawing.

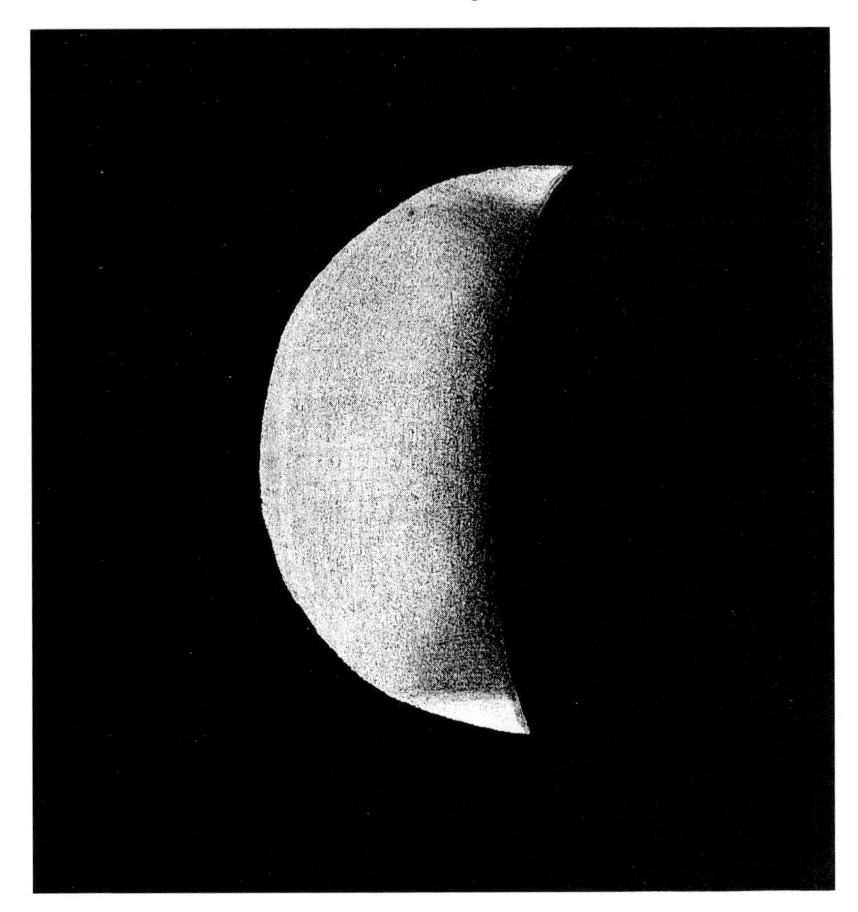

Figure 8.3. A pencil-shaded drawing of Venus, from an observation by the author on 1988 April 23^{d} 19^{h} 19^{m} UT, made with the 36-inch (0.9 m) Cassegrain reflector of the Royal Greenwich Observatory, Herstmonceux. Magnification ×312 (this was the lowest magnification available). Seeing ANT. IV.

is scientific accuracy. If you adopt the second method you will have to draw in the shadings rather darker than you observed them for the benefit of others analysing your observation. You should clearly state this and it is as well to include numerical intensity estimates, either in the acompanying notes or in a separate drawing.

Your report should also carry the date and time(s) of your observation and drawing, the aperture and type of telescope used, the magnification(s) used for the observation and for your drawing, an assessment of the seeing conditions (atmospheric turbulence and transparency) as well as any other information pertaining to the observation (filters or other accessories used

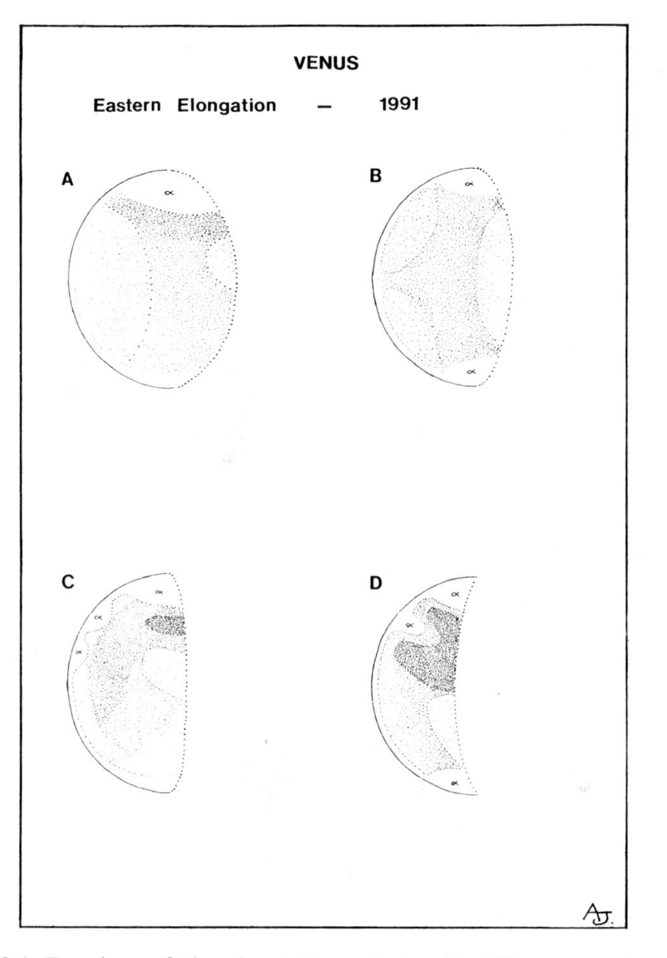

Figure 8.4. Drawings of the planet Venus during its 1991 eastern elongation, made by Andrew Johnson using his 8½-inch (210 mm) Newtonian reflector, at ×195 and using a Wratten 15 (yellow) filter. A – March 21; B – April 27; C – May 22; D – June 27.

and any other factors that effect the accuracy of your observation, such as telescope windshake, etc). If the accuracy of your drawing is at all suspect, then include a note indicating this.

Observed features

Sometimes the planet will appear to have bright *cusp-caps*, and each is often bordered by a dark collar. Figure 8.3 shows this effect well. These tend to be of somewhat variable visibility and they do not always lie exactly at the cusps. Bright patches can be visible in other positions. The whole of the limb of the planet often has a particularly brilliant appearance, though this is usually due to contrast with the darker sky-background.

Grey streaks and shadings are often faintly seen crossing the disc and these tend to be especially prominent near the terminator. Take particular care to record the general terminator shading, as well as any additional darkenings. Also watch out for terminator irregularities – sometimes the smooth curve is disturbed by one or more corrugations. In the same way the cusps of the planet may appear extended well beyond the normal north and south limbs of the disc during the crescent phase. At other times they may appear blunted short. As always, it is important to go to the telescope with an open mind. It is all too easy to deceive oneself into 'seeing' what one expects to see, merely becauses one expects to see it!

The use of coloured filters

Used with care, coloured filters can be a valuable aid to the observer. They must be of good quality and free from scratches. They are normally mounted in a small cell which screws into the eyepiece barrel, though I use 135 film canisters to mount my filters. I cut away most of the bottom of the canister and lay the filter inside the canister over the hole. Then I use a small plastic collar (cut from an insulation tape roll insert) to hold the filter in position. The 1¼-inch drawtube of my telescope is removable and I simply push the canister filter-first into the end of the drawtube opposite to the eyepiece.

I recommend that you obtain the 'Wratten' range of filters manufactured by Kodak. These have become adopted as the standard filter set of the planetary observer. Filters are obtainable from many photographic shops (though not usually unmounted) or direct from Kodak in the form of small gelatine sheets, ready for mounting. Kodak also produce a booklet (available direct from them) in which the spectral passband curves of all their filters are listed. Glass filters are also useable but tend to be expensive. When cutting gelatine filters to shape it is as well to sandwich them between two stiff cards and then cut through all three. In this way the filter will be protected from damage while cutting. As always when you introduce any potential obstruction into the light path of your telescope, do make sure that the clear aperture of the filter is sufficient to prevent vignetting.

As far as Venus is concerned, Kodak's W47 (violet) filter tends to enhance the dark shadings in the atmosphere, whereas terminator irregularities are often most clearly seen using a W25 (red) filter. A yellow (W15) or orange (W23A) filter can also be useful for observing Venus or Mercury in full daylight, as these suppress the apparent brightness of the sky. Always include details of any filters you use in your report.

Schröter's effect

The anomaly between the theoretical and actual phase values has been known for about two centuries. The observed phase is always a little less than the theoretical, so dichotomy occurs a few days early when the phase is waning and late when it is waxing. Moreover, the phase is often slightly different when observed through filters of different colour and the amounts of anomaly present do seem to differ with time. Phase estimates are particularly valuable, especially around the time of dichotomy, when the phase is most easy to estimate accurately. For instance, on 1989 November 6^{d} I observed the waning crescent and found the observed phase to be 49 per cent. The slight concavity of the terminator was very easy to see. However, theoretical dichotomy was still about 2 days away. Measurements with a micrometer are even better than estimates but, of course, require an accurate micrometer and a clock-driven telescope.

The ashen light

One of the most enigmatic of Venus's phenomena is the occasional visibility of the night-time hemisphere of the planet. Whether the effect is real or illusory was a matter of some controversy. However, an extensive study was conducted by Dr John Phillips of the Los Alamos National Laboratory and Dr Christopher Russell of the University of California. They organised a world-wide observing campaign for the 1988 elongations of the planet. Their report is published in the January 1990 issue of *Sky & Telescope* magazine, suffice to say that their overall conclusion is that the effect is real. I took part in the programme myself and had one definite sighting of the ashen light and several suspect ones.

During my positive sighting, the dim glow of the night-time hemisphere reminded me of Earthshine on the Moon, though it was less easy to see. At the time I was using three telescopes at the Royal Greenwhich Observatory: a 7-inch (178 mm) *f*/24 refractor, a 30-inch (0.76 m) coudé reflector and a 36-inch (0.91 m) Cassegrain reflector. The ashen light was visible through all three telescopes. I saw the glow as grey in colour and another observer present saw it as blue-grey. Most people who see the ashen light report it as grey but other colours are sometimes seen, particularly blue, brown and violet.

If you think you see the ashen light do be careful to rule out scattered light as the cause. Of course it is necessary to observe the planet against a dark sky and this usually means an unfavourable altitude, with its attendant poor seeing and spurious colour effects. Try blocking out the bright portion of the planet (either move the telescope so that the bright portion is beyond the edge of the field of view or use an occulting bar in the focal plane of the eyepiece). Is the effect still visible (more so, or less easily seen)?

In particular, beware of the diffraction bars which extend from the bright portion of the planet caused by the secondary mirror support vanes in compound reflectors. These bars will overlap at the same position as the dark part of the planet and may give the illusion that the dark hemisphere is glowing. If you see these bars (there will be four visible if your telescope has a four-vaned spider and six if it has a three-vaned spider), try using an off-axis stop (like that shown in Figure 8.1). Is the ashen light still visible? Try using coloured filters – is the effect enhanced or reduced?

Observing Mars

Being a *superior planet* (orbiting further out from the Sun than the Earth), Mars is very much more convenient to observe than Mercury or Venus. Instead of playing celestial hide-and-seek with the Sun, the red planet comes closest to the Earth at nearly the same time as it appears highest in a midnight sky. Successive *oppositions* of the planet occur, on average, 26 months apart. This period is subject to a large variation because of Mars's highly eccentric orbit. Its mean orbital radius is 228 million km but the perihelic distance is 208 million km and the aphelic distance 250 million km.

Oppositions of Mars that occur in late summer (such as those for the years 1971 and 1988) coincide with Mars near perihelion. The Earth–Mars distance can be as small as 57 million km and the 6 800 km globe of the planet subtends an apparent angle of nearly 25 arcseconds. At these times a magnification of about ×80 is enough to make Mars appear as large as does the Moon when the latter is seen with the naked eye. Oppositions that occur in late winter happen when Mars is close to aphelion and our distance from the planet is then 101 million km. At these times (for instance 1980 and 1995) the planet subtends less than 14 arcseconds and a power of about ×135 is needed to enlarge the disc to the same extent as before.

As with all the superior planets, oppositions occurring near the time of the summer solstice happen when the planet has a maximum southerly declination and so are least favourable for observers in the northern hemisphere. The period when the planet subtends an apparent angular diameter larger than 7 arcseconds, spans anything from 7 months for a time roughly centred on the date of an aphelic opposition, to 10 months for a perihelic one. During these periods even a moderate (6- or 8-inches aperture) telescope can show a fair amount of detail on the planet.

When the disc appears smaller than about 7 arcseconds (this is a somewhat arbitrary figure) particularly good seeing conditions are necessary to show much detail, no matter how big the telescope. At superior conjuction the planet can be over 400 million km from us and it then subtends a tiny disc of only 3.5 arcseconds. However, the planet is then effectively unobservable, being so close to the Sun's glare.

Drawing Mars

Most observing groups adopt a scale of 50 mm to the planet's full diameter and circular outlines should be pre-prepared to this size. In many ways the procedure for making drawings of Mars is the same as that for drawing Venus. Initially spend some time getting your eye used to the planet. Use a range of magnifications. Low powers should allow low contrast shadings to be more easily seen, though fine details require higher magnifications. You may find that glare swamps out the fine details, especially when using low magnifications, but do try using neutral density or coloured filters (treated in more detail shortly) before reaching for the off-axis stop.

Begin work on the drawing by outlining the positions of any sharply defined features, such as the polar caps and any gibbosity of phase (at extremes the planet can appear with a phase like the 11 or 18 day old Moon). Then shade in the overall hue of the planet, rubbing with the end of the finger to produce a smooth finish. Build the drawing by adding darker and darker layers of pencil to represent the darker features visible on the planet. Take special care with the positional accuracy of the details. The apparent centre of the disc and its limb are good reference points when positioning details. Use a clean rubber when making alterations. A pointed rubber can be used to create highlights. When you have recorded all the major features, check to see if there are any minor details that you have left out.

With practice, you should be able to make a fairly accurate sketch of the planet, taking no more than 20 minutes to complete the job. You should note the orientation of the image on your drawing (but not on the disc of the planet). If you are unsure of the orientation of the image produced by your telescope, nudge the telescope towards Polaris slightly. The image moves towards the **south**. With the telescope drive switched off the planet heads in the *preceding* direction. Mark this direction *p* on your drawing. The other direction is termed *following* and should be marked *f*. The same convention is followed when making drawings of all the other planets.

As is the case for any record of an observation, all relevant details of telescope, date, time, magnification, filters used, etc., should be included with the drawing. Though not strictly essential, it is a convenience for those who analyse your work if you record the Martian longitude of the central meridian of the planet at the time of your observation. This can be readily found from an ephemeris. A few written notes are also useful, especially where these concern details that you have had difficulty in representing accurately. For instance 'South polar cap glistening white', or 'Meridiani Sinus shown a little too far north on my drawing'. You can also comment on the colours visible.

Figure 8.5 shows a drawing of Mars I made in October 1988. Though I was busy with other observational projects, I did manage a few drawings of the red planet during its very favourable opposition that year. The telescope I used was, admittedly, rather larger than the amateur usually has at his disposal, though the seeing at the time of the observation was quite poor (ANT.IV) and all the details shown would have been visible to an observer using a much smaller telescope.

This highlights a point that is often glossed over, and yet I think is important: seeing conditions can dominate as the limiting factor for seeing planetary details, even for rather small apertures. Though some authorities state that an 8-inch (203 mm) aperture telescope is the smallest that can produce useful results when used to observe Mars, I am not so sure. Certainly not much is visible through a small telescope but the seeing conditions and the apparent size of the planet at the time of the observation also have a lot to do with it. Figure 8.6 shows what a skilful observer can record when using a telescope of moderate aperture.

One alternative to a monochrome shaded drawing is a sketch marked with dotted boundaries and intensity estimates, like that shown in Figure 8.2 for Venus. With Mars, the Association of Lunar and Planetary Observers (ALPO) and the British Astronomical Association (BAA) recommend a scale of 0 = absolutely black to 10 = brightest. Another possibility is the making of a coloured drawing, though I would suggest first making either the monochrome shaded drawing, or the intensity estimate

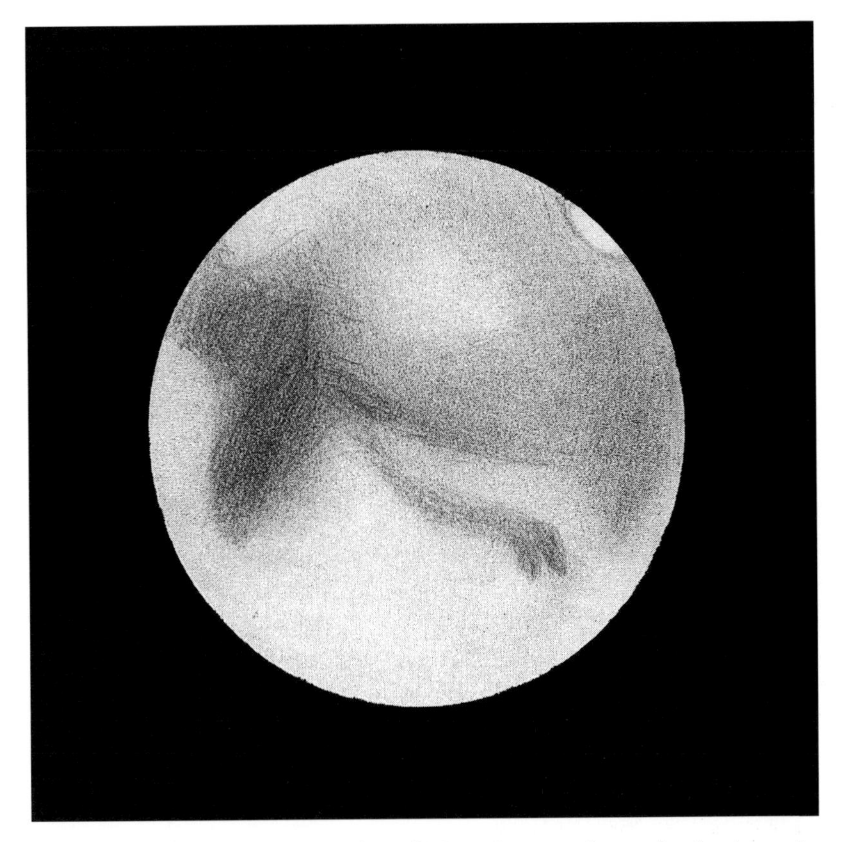

Figure 8.5. A pencil-shaded drawing of Mars, from an observation by the author on 1988 October 28^{d} 00^{h} 20^{m} UT, using the 36-inch (0.9 m) Cassegrain reflector of the Royal Greenwich Observatory, Herstmonceux. The Martian longitude on the central meridian at the time was 325°.5. The seeing was ANT. IV and so a smaller telescope would have shown the planet just as well. The smallest available magnification, ×312, was used but a lower power would have been preferred had one been available.

drawing. Then, if you desire, go ahead and attempt a sketch using coloured pencils. Do be careful that the light you draw by does not tend to distort the colours. If you draw by a red light, you may receive a shock when you view your sketch in daylight! As with a black and white sketch, the aim is accuracy. Colour perception is somewhat subjective but unless you are sure that your drawing is true to what you see do not send it in for analysis. It will do more harm than good.

The following notes detail specific features and lines of observation that are especially useful targets for the amateur.

Views of MARS Around Opposition.

Andrew Johnson

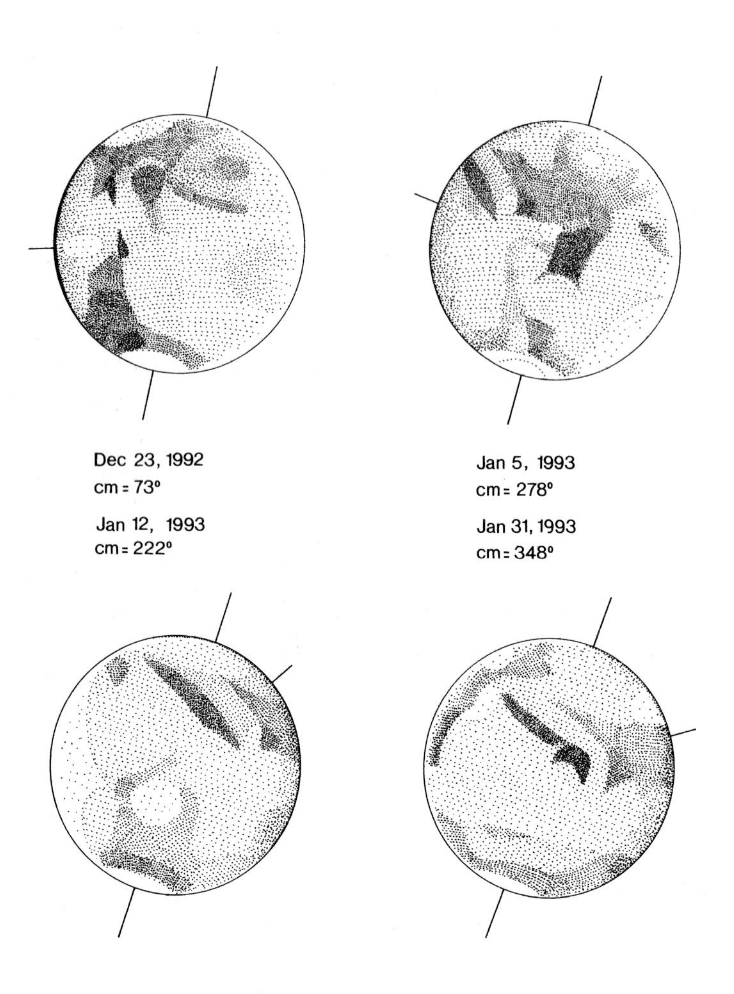

AJ.

Figure 8.6. Drawings of the planet Mars by Andrew Johnson.

The polar caps

One, or sometimes both, polar caps will be visible, depending mainly on the orientation of the globe as we see it from the Earth. The Martian south pole is presented to us at the times of favourable opposition. Both ice caps are subject to extensive changes of size and shape as the Martian seasons progress. Look out for detached portions and rifts in the caps. If you have an adequate telescope and an eyepiece micrometer you can make actual measures of the extent of the caps, though estimates based on measurements of drawings are also valuable.

Surface details

Figure 8.7 shows a computer-generated map of Mars obtained by Dr Tony Cook. He used a small CCD at the focus of an 8-inch telescope to obtain the raw images. These were later computer processed to generate the Mercator projection shown. Figure 8.8 shows an albedo map of the planet I made using Dr Cook's map as a basis and adding details obtained from my own observations and Terry Platt's CCD images. I have included the IAU approved names of the major features.

These features do undergo slight variations of intensity and shape over a period of time and so their long-term monitoring is very useful. The observer using red, orange, and yellow (W25, W23A, and W15 respectively) filters will see the dark markings more clearly, since Mars's atmosphere is more transparent to red light.

The actual colours of the dark markings are greys and browns. However, they can appear as greens, or even blues, owing to the contrast with the vivid red-ochre areas that dominate the planet. It is easy to understand how previous generations of observers were fooled into thinking that Mars was covered in vegetation. The fact that they also recorded a 'wave of darkening' and a colour change from brown to green, sweeping down from the pole, that nicely coincided with Martian spring for that hemisphere, (and the reverse happening in the Martian autumn), I think provides a lesson to all of us. Another example is provided by the saga of the Martian canals. We should consciously make every effort to avoid recording what we think we ought to see and go to the eyepiece with an open mind.

White clouds and hazes

Especially prominent in green, blue, and violet filters (W57, W38 and W47), white clouds often cloak either small portions of the planet, or vast areas.

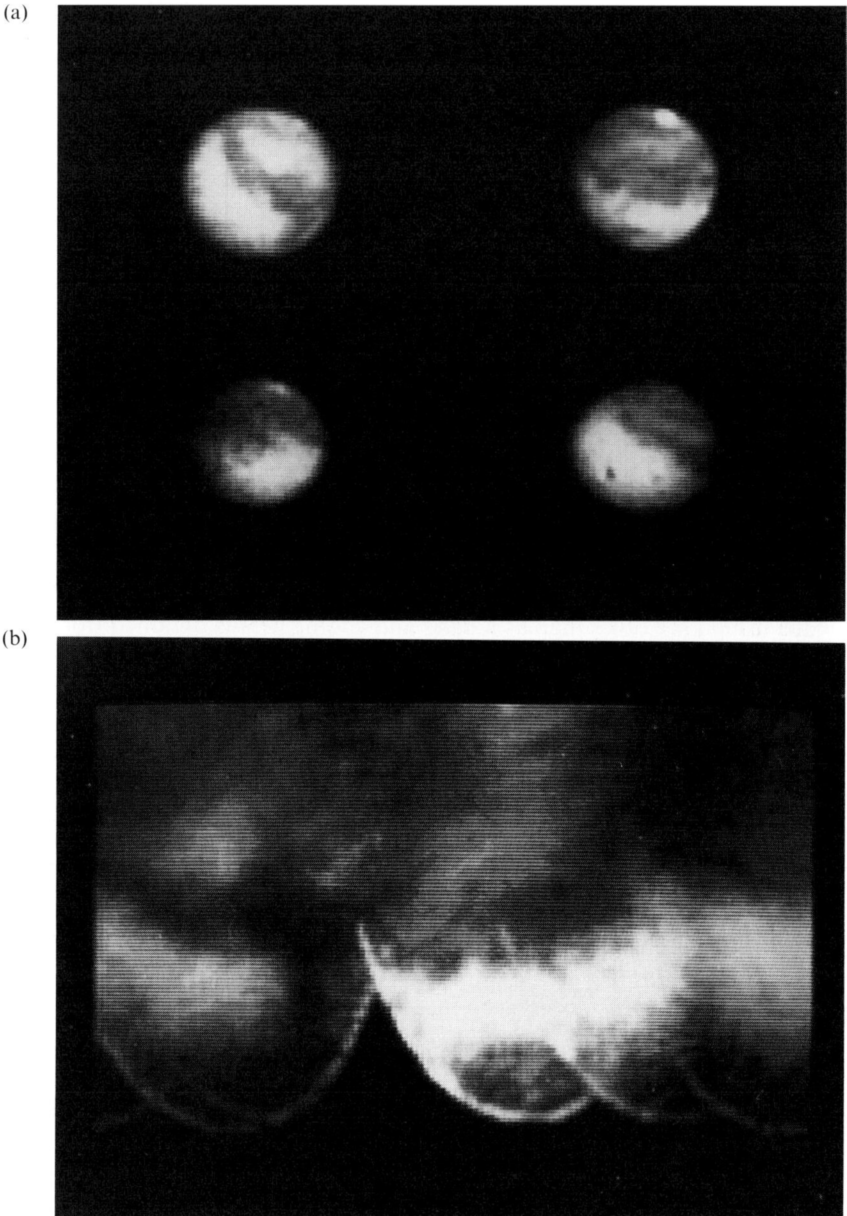

Figure 8.7. (a) Images of Mars obtained by Dr Tony Cook, using a 128 × 76 pixel MA328 CCD camera, attached to an 8-inch (203 mm) Newtonian reflector. (Upper left) 1988 September $23^{d}\ 22^{h}\ 40^{m}$ UT. Longitude of central meridian = 224°. (Upper right) 1988 September $11^{d}\ 00^{h}\ 30^{m}$ UT. Longitude of central meridian = 19°. (Lower left) 1988 September $4^{d}\ 00^{m}\ 19^{m}$ UT. Longitude of central meridian = 86°. (Lower right) 1988 October $1^{d}\ 22^{h}\ 32^{m}$ UT. Longitude of central meridian = 172°. (b) Dr Cook has combined the images to produce this Mercator projection.

As you watch the planet, it slowly rotates, the length of the Martian day being 41 minutes longer than ours. The two CCD images of Mars shown in Figure 8.9 clearly demonstrate Mars's rotation. The planet's morning terminator lies on the following side of the disc. Many features often emerge into sunlight shrouded in cloud, mist, or fog which then rapidly dissipates. The highest clouds show up best in violet light.

Blue clouds

Visible only in blue or violet light, these features are very poorly understood. Consequently their detection and study is particularly valuable.

Violet clearing

Usually Mars's atmosphere is translucent to violet light but occasional clearings occur. The surface markings then become quite apparent when viewed through a violet filter. Scientists are not sure why this happens.

Yellow clouds and dust-storms

Yellow clouds are so-called because they appear brightest when seen through a yellow filter. Unlike the white clouds, which are composed of tiny crystals of water-ice, yellow clouds are primarily composed of suspended dust particles. Occasionally, the windforce in Mars's tenuous atmosphere can whip up a dust-storm of global proportions. At these times Mars can be transformed into a virtually blank yellow disc, as seen through the telescope!

Surface frosts

Appearing as small bright patches, these can be distinguished from clouds by their appearance through coloured filters. If the patch is most clearly and sharply defined through a green filter (and still well seen in an orange filter), rather than blue, it is likely to be a surface frost (or at least a ground-hugging fog).

Mars's satellites

Mars's two natural satellites are far from easy to see. Phobos has a magnitude of $11^{m}.6$ around the time of an average opposition of Mars, Deimos's

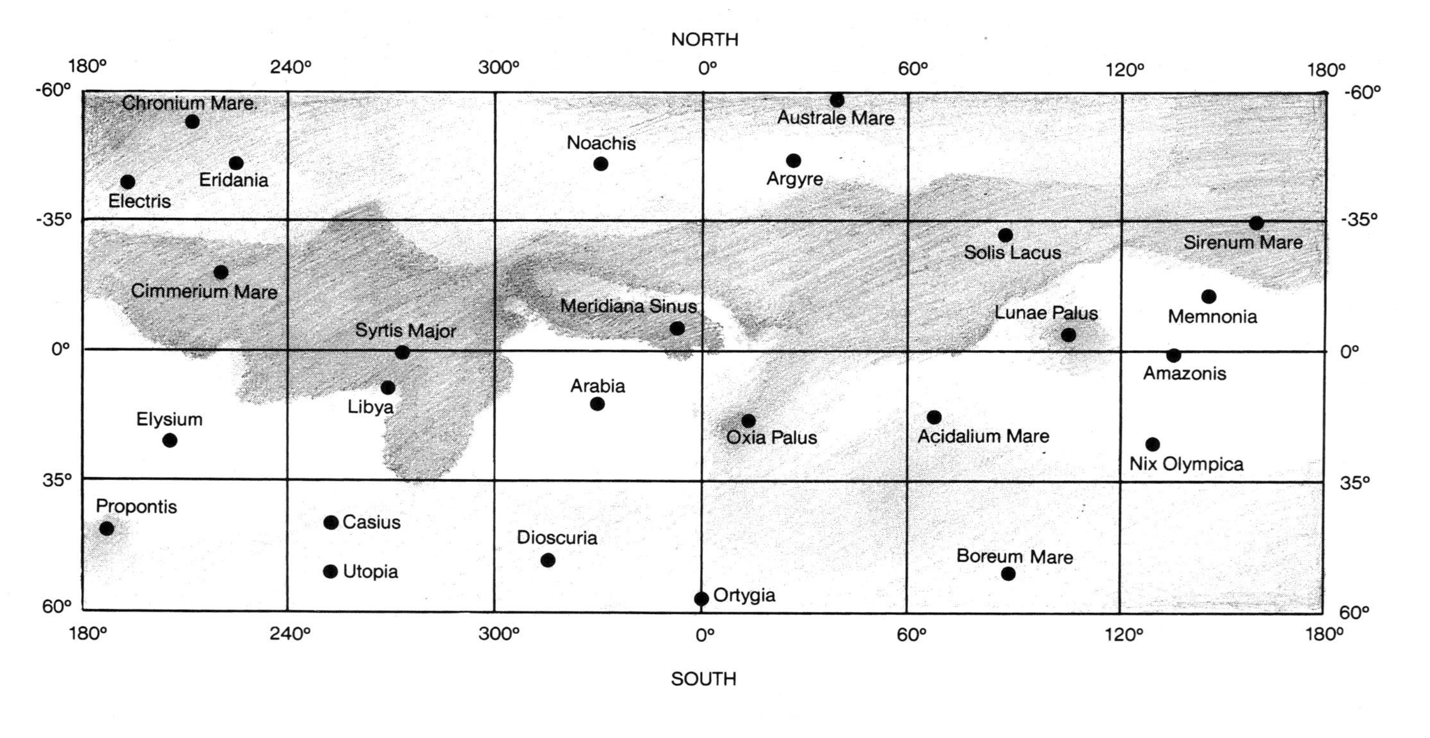

Figure 8.8. A Mercator projection map of Mars hand-drawn by the author from his own observations and the CCD images generated by Dr Tony Cook and Terry Platt, as well as Dr Cook's computer-generated Mercator map. The IAU approved names of some of the major features are included on this map.

(a)

(b)

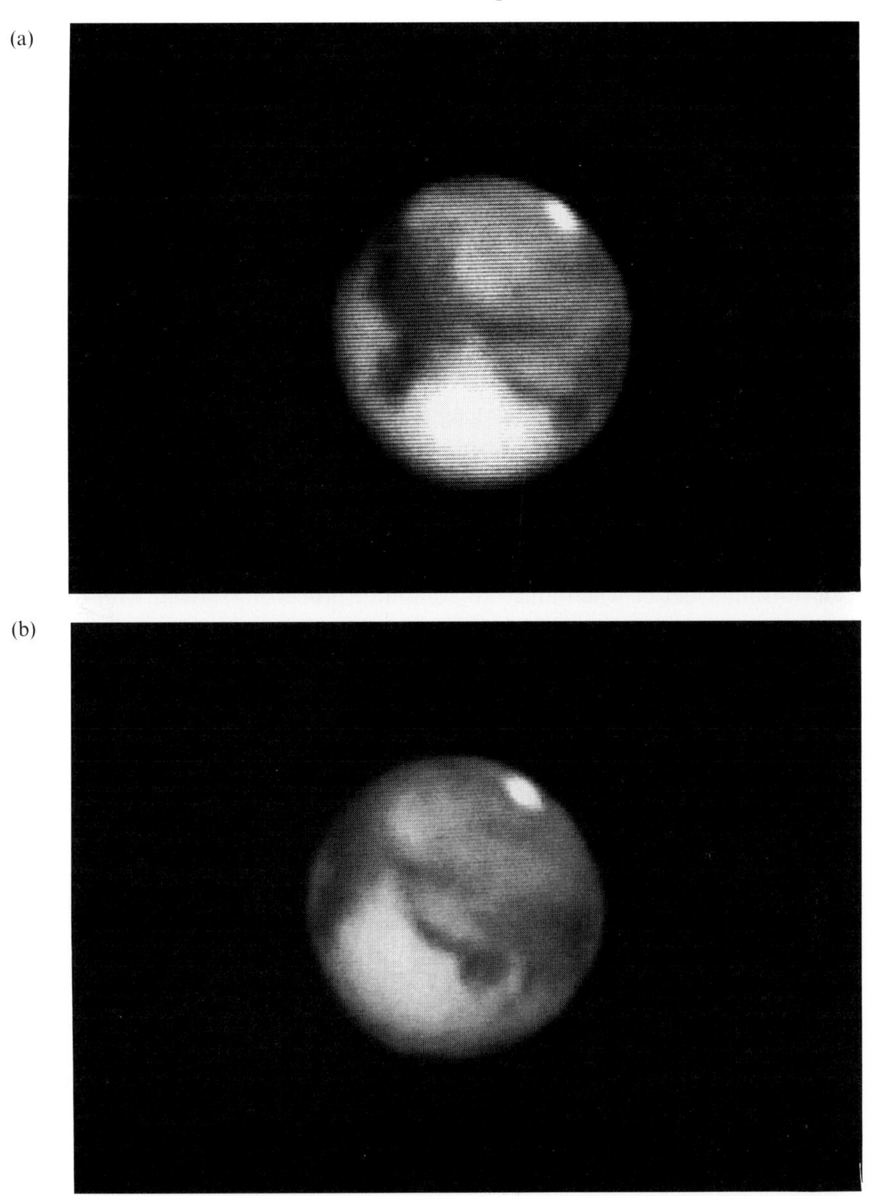

Figure 8.9. CCD images of Mars obtained by Terry Platt with his 12½-inch (318 mm) tri-schiefspiegler reflector on 1988 September 19^d (a) shows the planet at $0^h\ 05^m$ UT (longitude of central meridian = 309°.0), and (b) shows it at $1^h\ 59^m$ UT (longitude of central meridian = 336°.8). The rotation of the planet in this time interval is obvious.

magnitude then being $12^m.8$. Both satellites are always involved in the bright glare of the planet, though with care they can be detected with a 12-inch telescope under very favourable conditions. Using an occulting bar to hide the bright disc of Mars will help in detecting them. Deimos is often rather easier to see than Phobos, since it orbits further from the planet. Deimos moves at a mean distance of 23 500 km from Mars, taking $30^h\ 21^m$ to orbit it, while Phobos orbits at 9 380 km, on average, taking only $7^h\ 39^m$. There is nothing really useful that the amateur can do when observing these tiny bodies.

Photography/CCD imaging of the terrestrial planets

All the foregoing notes describing the types of visual work that the amateur can undertake apply almost equally well to photographic/CCD observation, though the average amateur's photographs are unlikely to show each planet as well as a good drawing would, made at the same time (but CCDs can do rather better when allied to computer processing). Even with the best of techniques, atmospheric turbulence causes the image to smear during the exposure. For more details on high-resolution photography read Chapter 5, and for electronic imaging also read Chapter 6. Getting a sharp image of Mercury is a challenge in itself, let alone recording any surface details.

The best photographs taken through a near-ultraviolet passband filter show some of the dark shadings on Venus. Otherwise, photographs of Venus taken with integrated light, or through coloured filters, can be useful for phase measurements (but do make sure that the film exposure time is long enough to record **all** of the bright portion of the planet – it is best to take a series of exposures of length bracketing the estimated value). There is at least one reported near-ultraviolet photograph of the ashen light effect, taken by Bernd Flach-Wilken in Germany on 1988 May 12^d. Other occasional anomalies are photographed from time to time in integrated light and through coloured filters.

Though Mars never presents a large disc, amateur photographs can show a lot of detail. Kodak's Technical Pan 2415 seems to be a particularly good film for planetary photography, but colour photographs of Mars are also valuable, as are black and white photographs taken through coloured filters. Figure 8.9 (a) and (b) shows two views of Mars obtained by Terry Platt with his 12½-inch (318 mm) tri-schiefspiegler reflector and CCD detector. These images, like those shown earlier in this book, were computer-enhanced.

9

The gas-giant planets

Out beyond the terrestrial planets and the asteriods, the gas-giant planets are the dominant bodies of the Solar System. As is the case for Venus, we can never see the surfaces of Jupiter, Saturn, Uranus and Neptune. We see just the upper layers of their atmospheres. Consequently, most of our observations of these worlds are concerned with their atmospheric phenomena.

Observing Jupiter

Jupiter is the easiest target for planetary observation. It is also one of the most fascinating. Orbiting the Sun at a distance of 778 million km, its vast globe spans 49 arcseconds at opposition. At such times a magnification of about ×40 will show the planet through the telescope as large as the full Moon looks when seen with the unaided eye. Even at conjuction the disc still spans 32 arcseconds, when a power of ×62 will enlarge it to the same extent as before. Moreover, the planet comes into opposition every 13 months and it is observable somewhere in the sky virtually all year round, apart from a month or two near the time of conjunction.

Nomenclature

Jupiter is a planet in turmoil. Even more exciting, from the point of view of the amateur astronomer, small telescopes show abundant detail. A 6-inch or 8-inch (152 mm or 203 mm) reflector is adequate for the serious study of the planet, anything larger being a bonus. As always with planetary observation, instrumental quality is to be preferred over sheer size.

Figure 9.1 shows the standard scheme for the nomenclature of Jupiter (and, incidently, for Saturn). The light regions are known as *zones*. We now know that these are the upper clouds in Jupiter's (and Saturn's) atmosphere. They are chiefly composed of ammonia crystals, hence their glistening whiteness. The belts, contrary to the visual impression they give, are just gaps between the clouds, where we can peer a little further into Jupiter's gaseous envelope. It is in these deeper levels that (depending on temperature and pressure) most of the coloured compounds reside.

Jupiter's outflowing internal heat causes the convection which generates the formation of the ammonia clouds. At the edges of the clouds, the ammonia is sinking once again into the lower, coloured, layers. The planet's rapid rotation generates the coriolis forces which wrap the clouds around Jupiter's globe, so producing the belts and zones. The shearing which occurs between the various belts and zones causes instabilities which manifest in a variety of observable phenomena. Loops, swirls, spots and various other transient deformations are always visible to the gaze of the observer.

Making whole-disc drawings of Jupiter

The general procedures are the same as those detailed for making disc drawings of the other planets. However, a circular outline cannot be used, since in reality Jupiter is markedly flattened at the poles. Jupiter's equatorial diameter is 143 800 km, while it's polar diameter is 133 500 km. Most coordinated observing groups will supply their members with blanks: the usual scales adopted are either 50 mm or 60 mm to the equatorial diameter of the planet. Photographs that show the limb regions of the planet well can also be used to make tracings of the outline.

After the usual period to allow your eye to adapt to the scene, begin your drawing by faintly sketching in the outlines of the largest features. Then use these as a basis for adding in finer details (once again, just in outline). Then note the time. Jupiter rotates rapidly on its axis (a day on Jupiter is less than 10 hours long) and so your initial sketch must be completed within a few minutes, otherwise severe distortions will occur in your representation. Then some more time can be spent filling in the shadings to their correct level of darkness and adding the finest details. Even so, the time taken to finish the drawing should not be much more than 10 or 15 minutes.

Take special care with the widths and the latitudes of the belts. These are quite difficult to show accurately. It is, I find, very easy to show the belts as too narrow and crowded too close together near the equatorial region of Jupiter's globe. Become accustomed to critically comparing the finished result with the view through the telescope. I find that the limb darkening on

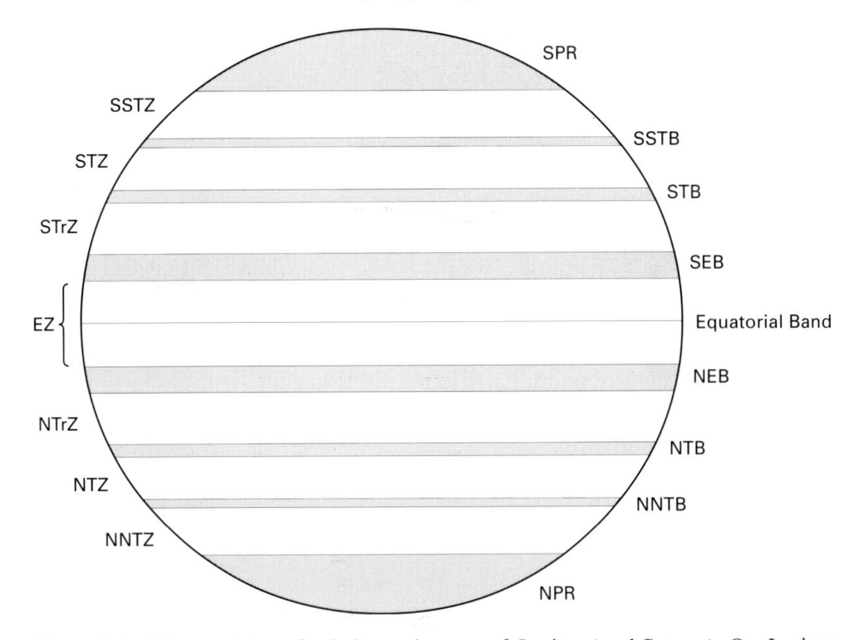

Figure 9.1. Nomenclature for belts and zones of Jupiter (and Saturn). On Jupiter a north north north temperate belt and a south south south temperate belt are sometimes seen, though they are usually continuous with the polar regions. These are never seen on Saturn. Even the south south temperate and north north temperate belts are not usually seen. The whole of the region between Saturn's STB and the SPR is known as the south temperate zone. Similarly, for the region between the NTB and the NPR (the north temperate zone).

Key:

SPR = south polar region
SSTB = south south temperate belt
STB = south temperate belt
SEB = south equatorial belt
NEB = north equatorial belt
NTB = north temperate belt
NNTB = north north temperate belt
NPR = north polar region
SSTZ = south south temperate zone
STZ = south temperate zone
STrZ = south tropical zone
EZ = equatorial zone
NTrZ = north tropical zone
NTZ = north temperate zone
NNTZ = north north temperate zone

Figure 9.2. Drawing of Jupiter by the author, using his 18¼-inch (464 mm) Newtonian reflector, magnification ×288, on 1977 December 30^{d} 23^{h} 25^{m} UT. At the time of the observation the transparency was poor, due to haze, but the seeing was a fairly steady ANT. III.

the globe is really quite difficult to appreciate when seen visually at low magnifications. It becomes more apparent at higher powers. Yet it is extremely obvious in photographs. This is probably due to the contrast of the bright limb of the planet against the dark sky and is a sharp reminder of just how subjective vision really is. I always try to represent the amount of limb darkening I see with the magnification mainly used for making the drawing.

As usual, remember to include **all** the relevant details with your observation (I do not possess the audacity to list them yet again!). Figure 9.2 shows a drawing of Jupiter I made using my 18¼-inch (464 mm) Newtonian reflector under hazy, though fairly steady, seeing conditions.

Central meridian transit timings and strip-sketches

A particularly useful project is the determination of the longitudes of the various ephemeral features on Jupiter's visible disc. This is done by timing the transits of the features across the north–south line, known as the *central meridian*, of the planet. One complication is that the apparent rotation rate of Jupiter varies with latitude. In order to define Jovian longitudes, two coordinate systems, *System I* and *System II*, are used. System I operates only in the equatorial zone of the planet and is based on an average rotation period of 9 hours 50 minutes 30 seconds. System II is taken to operate for the rest of the planet and is based on an average rotation period of 9 hours 55 minutes 41 seconds.

The transit time of a given feature is converted to Jovian longitude using the tables in an ephemeris. With care the transits can be timed to an accuracy of about one minute, even without the use of a crosswire to bisect the planet along the central meridian. The longitudes will then be accurate to within about 0°.6. The information thus obtained allows a detailed quantitative study of Jupiter's atmospheric phenomena.

In the same vein *strip-sketches* are particularly useful. Instead of making a full disc drawing, the observer draws just that north–south strip of the planet which is crossing the central meridian. However, the observer continues drawing as more and more of the planet crosses the meridian. The new details are added from left to right on the drawing, the position corresponding to the time that part of the drawing is made. The result is a band of Jupiter drawn free from the longitude distortions caused by the rotundity of the planet. The whole range of latitude from pole to pole could be shown, or maybe just a small range centred on some feature of interest. Recording the times on the strip-sketch allows the System I and System II longitudes to be subsequently calculated (see Figure 9.3).

The Great Red Spot

The most famous of Jupiter's visible features has to be the Great Red Spot (GRS). It resides in the planet's south tropical zone. The first person to definitely record the GRS was the astronomer Cassini in 1665, though an observation by Robert Hooke made in the previous year might refer to it. We now know that the GRS is a *soliton* (a solitary wave). It is a vast anticyclonic whirlwind dragging up the coloured materials low in Jupiter's atmosphere to well above the level of the ammonia clouds. It (and the smaller spots like it) are formed as a result of the counterflowing atmospheric jet streams and currents. Some authorities think that its colour is due to free red phosphorus.

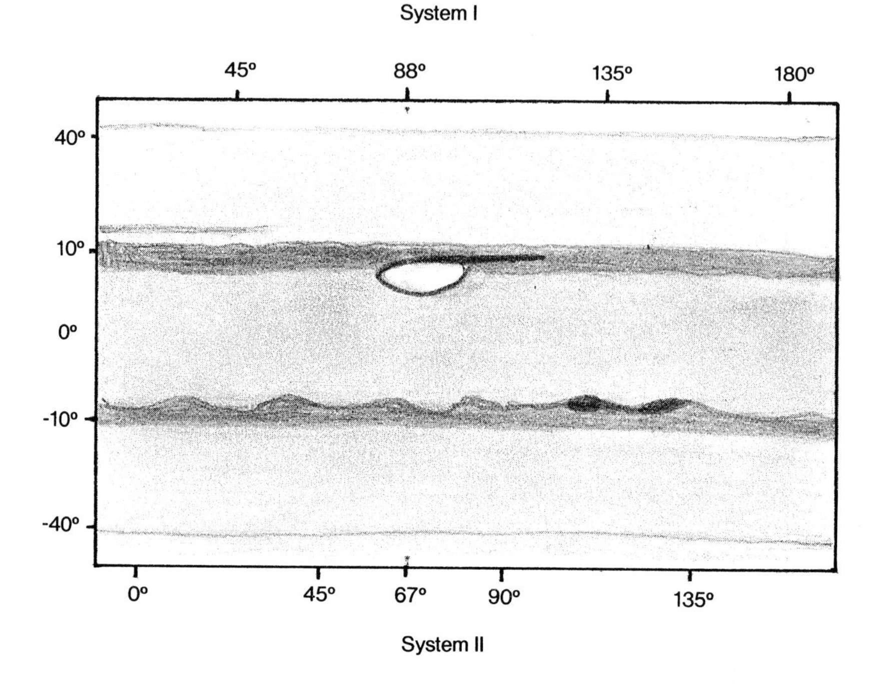

Figure 9.3. Strip-sketch of Jupiter, made from an observation by the author on 1977 November 28^d, using the 8-inch (203 mm) refractor of the Jeremiah Horrocks Observatory in Lancashire. At 22^h 11^m UT the System I longitude of the central meridian (ω_1) was 88°. At the same instant the System II longitude of the central meridian (ω_2) was 67°. The latitudes shown are merely estimates – no actual measures were made.

The Great Red spot undergoes extensive changes in longitude and slight changes in latitude. It has effectively wandered round the planet several times (with respect to System II) since its discovery! It also undergoes changes in appearance. In the 1880s the spot was at its largest, being about 50 000 km in length. Now it is about half that size. It's width has stayed fairly constant at about 14 000 km and so it is now much less elliptical than it used to be. It also varies its colour. At times it can be the darkest feature on the planet, then being a vivid red. I started observing right at the beginning of the 1970s and I well remember its salmon-pink hue at that time. In the 1980s it faded almost to invisibility, leaving only a white 'hollow' in its place (obviously less of the coloured material was being dragged up from the lower atmospheric layers). At other times the GRS can be seen partly separated from this 'hollow'. Recently, the Great Red Spot has shown signs of recovering its previous colour and intensity.

Terry Platt's CCD image (Figure 6.9(b)) and Martin Mobberley's photographs (Figure 9.4 (a) and (b)) show views of the Great Red Spot in the late 1980s.

Observations of the small-scale atmospheric features as they interact with the GRS are particularly valuable. For instance, what happens when a small white spot encounters the GRS? Does it get taken rapidly round the GRS – if so in which direction? Does it remain attached to the GRS, continuing to go round and round, or does it accelerate over the perimeter of it and then break free? Does the small spot survive, or is it destroyed by its interaction? To successfully observe such small-scale details one needs both a large aperture telescope and very good seeing.

Other features

Even the major belts and zones are subject to apparent changes of colour, intensity and width. One never knows quite what to expect when going to the eyepiece. Sometimes major disturbances are present. At other times the planet can seem relatively quiescent. A major disturbance erupted in 1989, with all the southern hemisphere of the planet apparently invisible under a white pall. As Figure 9.4(b) shows, the planet then looked very strange. Various loops and swirls disturbing the belts and zones are commonplace, as are small spots, both light and dark in hue. They may be much more transient than the Great Red Spot but they are, nonetheless, just as interesting.

A truly awesome event occurred in July 1994: the impact of the fragments of the disintegrated comet Shoemaker–Levy 9 with Jupiter. Planetary specialists are still learning from the data gathered by the world's major observatories (and the Hubble Space Telescope) of this remarkable happening. Amateur astronomers also played their part. Unfortunately conditions were not good in Britain because Jupiter was only seen rather low in the sky. The problem was exacerbated by the planet being not far from conjunction with the Sun.

I could not see the planet from my largest telescope and, instead, used my 6¼-inch (158 mm) Newtonian reflector. Figure 9.5 shows one of the sequence of observations I made of this once in a millenium event.

Colours on Jupiter

The observing of colours through the telescope is a somewhat controversial subject. Some observers maintain that only shades of grey are visible and they cite the low levels of illumination as the reason. Others equally

(a)

(b)

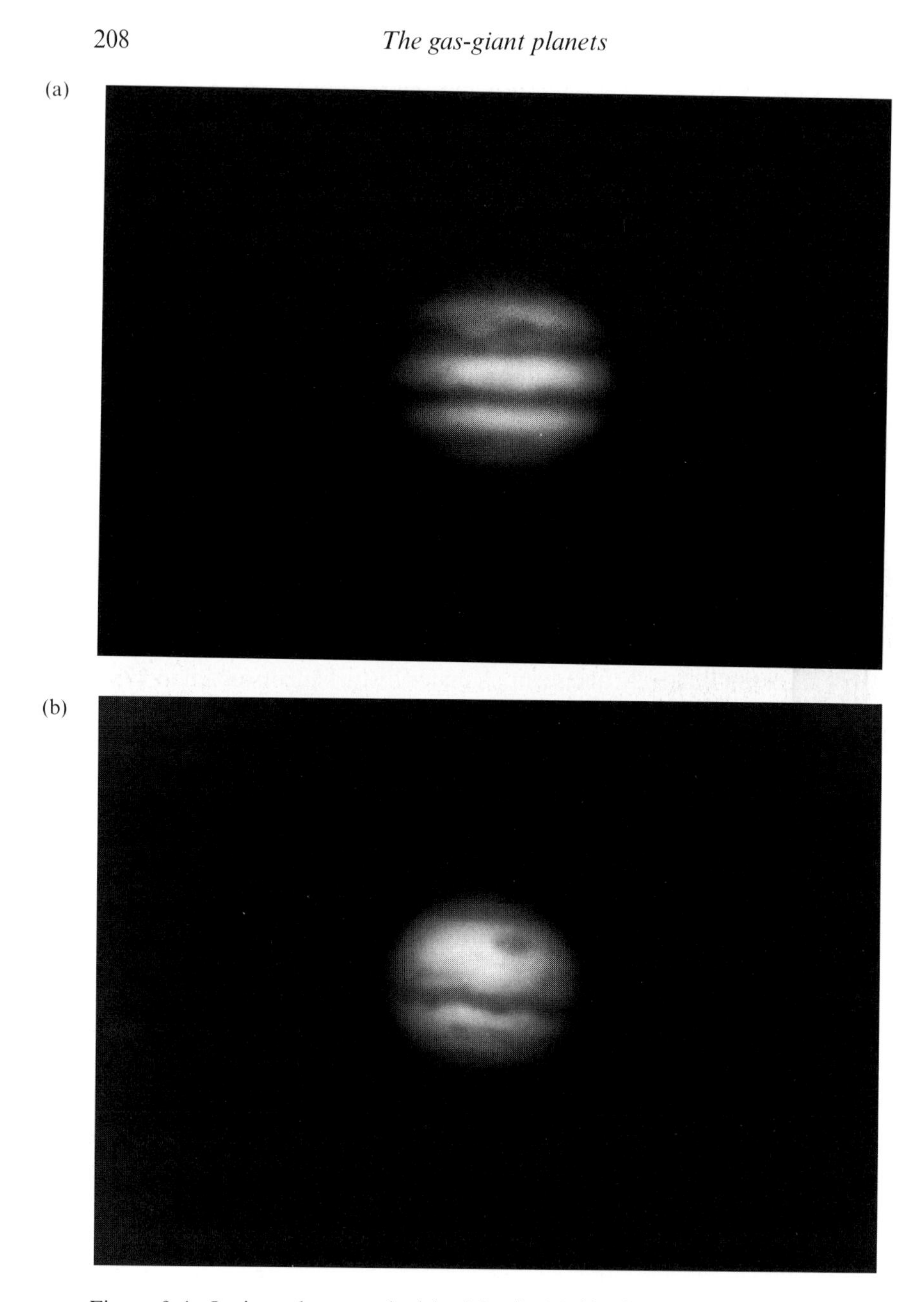

Figure 9.4. Jupiter photographed by Martin Mobberley. (a) was taken on 1987 August 20^{d} 02^{h} 43^{m} UT. ($\omega_1 = 139°.8$ and $\omega_2 = 24°.0$). Martin used eyepiece projection at the Cassegrain focus of his 14-inch (356 mm) reflector, the effective focal ratio being *f*/80. The 1 second exposure was made on Kodak's TP2415 film. (b) was taken on 1990 January 13^{d} 23^{h} 17^{m} UT. $\omega_1 = 347°.6$ and $\omega_2 = 13°.5$. This time he projected the image to *f*/55.

Figure 9.5. The most dramatic event to occur on Jupiter for many centuries! The author made this observation of Jupiter with his 6¼-inch (158 mm) Newtonian reflector, at ×203, on 1994 July 20^{d} 20^{h} 21^{m} UT. Unfortunately the seeing was a bad ANT.V, since the planet was very low over the horizon at the time. Also the sky was still brightly twilit and the transparency was rather poor, due to low-level haze, but it was worth persevering to observe the unique spectacle of the impact scars of the doomed comet Shoemaker–Levy 9 that had explosively ploughed into the giant planet's atmosphere.

firmly say that colours can be seen. I find that I can see colours in many celestial bodies. In the case of Jupiter I find the colours particularly apparent. However, I must also say that my experience has indicated that my eyes may be more colour sensitive than most people's. I find that the belts are generally brown in colour, with dashes of other colours – even blues, evident. I usually see the polar regions as yellowish, beige, or even slightly greenish at times.

Note that I am not asserting that the colours I see are very accurate. Colour perception is notoriously subjective. For instance, the perceived colour of a feature depends heavily on the colours and intensities of its surrounds. As with the observation of Mars, there is nothing wrong with preparing coloured drawings of Jupiter as an addition to the main work of monochrome drawings, strip-sketches and transit timings. Even if one does not go to the trouble of making coloured drawings, one can still include notes on any colours seen in the observation report.

Perhaps better (certainly of more scientific value) than coloured drawings are monochrome renditions of the appearance of the planet as seen through coloured filters. A yellow filter (Wratten 15) will selectively darken any blue features present, while a blue filter (say, a W38) darkens orange, brown (which is really just dark orange) and red features.

Latitude measures

Very few amateurs make latitude measures of planetary discs. In years past this was largely because a well-mounted and accurately driven telescope was as necessary as a good filar micrometer. Nowadays, amateurs can obtain good high resolution photographs, and especially CCD images, and make the measures from these, instead. Whether using an eyepiece micrometer or a photograph, the basic method for making the measure is the same. Here I will give the procedure to be used with the eyepiece micrometer, since it can easily be adapted to making measures on a hard-copy image, while the reverse is not so obvious.

The highest possible magnification should be used that still shows the feature to be measured distinctly. The use of the filar micrometer is more fully described in Chapter 14. Figure 9.6 shows the sequence of steps necessary to measure the latitude of one of Jupiter's belts. First align the normally horizontal fixed reference wire to planet's central meridian. This can be precisely ascertained by bringing up one of the movable wires (which are perpendicular to the fixed reference wire) to a cloud belt and checking that the wire runs exactly along it, as illustrated in Figure 9.6(a).

Next set the movable wires (obviously, moving the telescope in declination if your micrometer only has one movable wire) onto the extreme north and south limbs of the planet (9.6(b)), so measuring the planet's polar diameter. By moving the wires, measure the distance from the south limb to the belt (9.6(c)), and repeat for the distance of the north limb to the belt (9.6(d)). Let us call these measured distances X and Y, respectively.

In making the calculation, we first check that the distances X and Y add up to the total polar diameter of the planet, which we can denote D. If not, then one, two, or all three of the measures are in error. Assuming that our feature lies in the southern hemisphere, we use the distance X in the following equation (if it lies in the northern hemisphere substitute Y):

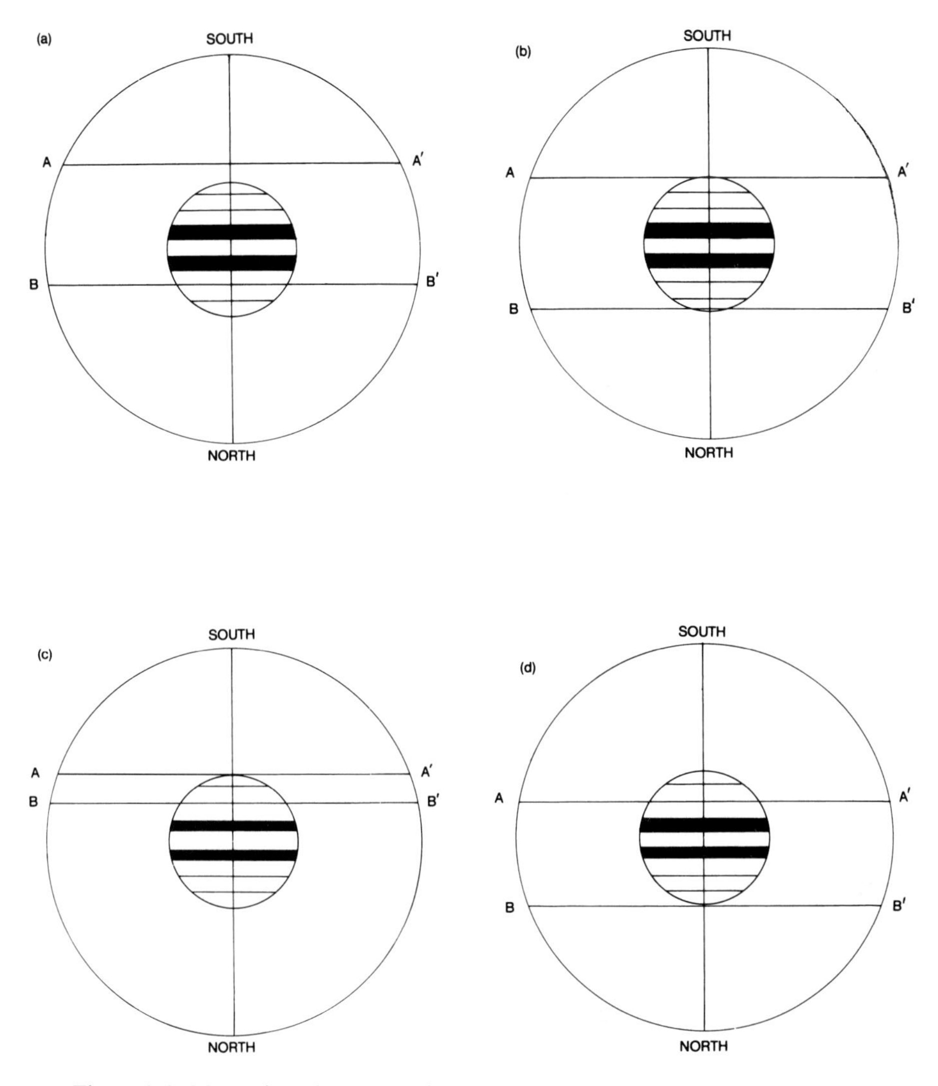

Figure 9.6. Measuring planetary latitudes. In this example the feature whose latitude is to be measured lies in the planet's southern hemisphere. See text for full details. (a) the position angle of the micrometer is adjusted until the crosswire BB′ runs parallel to one of the planet's belts (for Jupiter, or Saturn). (b) Crosswire AA′ is set on the extreme southern limb of the planet and BB′ is set on the extreme northern limb. The distance between the crosswires is the polar diameter, D. (c) The crosswire AA′ is set on the extreme southern limb of the planet and crosswire BB′ is set on the feature whose latitude is to be determined. The distance between the crosswires corresponds to the distance X. (d) Crosswire AA′ is now set on the feature whose latitude is to be determined and crosswire BB′ is set on the extreme northern limb. The distance between the crosswires now corresponds to the distance Y.

$$sin\theta = \frac{2\left(\frac{D}{2} - X\right)}{D + 0.07D\cos\theta_{(est)}},$$

where θ is the required latitude. $\theta_{(est)}$ is your initial estimate of the latitude of the feature. Unless your initial estimate is wildly in error, the derived latitude should, after one final correction, be accurate to better than 1° (if your micrometrical measures are this good).

The final correction is to take into account the tilt of Jupiter's axis. The latitude of the sub-Earth point is found from an ephemeris. If this latitude is 0°, then your calculated value is correct as it stands. If this latitude is positive, then the south pole of the planet is tilted away from us by the given amount. Your value of θ should be decreased by this amount if the feature you are measuring lies in the southern hemisphere and increased by this amount if it lies in the northern hemisphere. The reverse is the case if the latitude of the sub-Earth point is negative.

If you are troubled that the value of $\theta_{(est)}$ you put in might be too inaccurate (it is a minor correction which makes very little difference to the final result), try re-doing the calculation but this time inserting your freshly calculated value of θ. Your new answer should differ very little from that of your first calculation (any difference is likely to be very much less than the instrumental errors in making the measurements at the telescope.

Observing Jupiter's satellites

The four major satellites of Jupiter are visible through the smallest telescope, or a good pair of binoculars. They are numbered in order of increasing orbital radius from Jupiter : (I) Io, (II) Europa, (III) Ganymede and (IV) Callisto. Of these four, Io is undoubtably the most interesting. The Voyager 1 space-probe revealed this little world (diameter 3600 km) to have a sulphur covering and active volcanoes. From Earth its 1.1 arcsecond diameter disc appears distinctly yellow. Some observers suspect that it is a little variable in brightness and colour. Obviously long-term photometric measures (in B and V) would be a particularly valuable line of study for amateurs with the necessary equipment (see Chapter 13). Io normally appears of visual magnitude $5^{m}.4$.

Do note that different authorities quote different magnitudes for Jupiter's satellites. I have 'averaged' the figures from a number of sources. Perhaps a definitive study of the brightnesses of Jupiter's satellites would make a worthy project for the suitably equipped amateur?

The other three major satellites are much more quiescent bodies, mainly composed of ices with a 'dirty' covering. Europa appears of magnitude $5^m.6$ and its 3000 km globe subtends 1 arcsecond from the Earth. Ganymede, diameter 5270 km, subtends 1.7 arcseconds and appears of magnitude $5^m.0$. Callisto's 5000 km globe spans 1.6 arcseconds, as we see it. This satellite appears to us rather fainter than the rest, at $6^m.1$, because of its lower albedo.

The only other of Jupiter's retinue to be visible through amateur-sized telescopes (and then only with difficulty) is thirteenth magnitude Amalthea. The transits and occultations of the various satellites with Jupiter (and very occasionally with each other) are interesting to watch but it is probably fair to say that no scientifically useful work can be performed.

Observing Saturn

The outermost of the bright planets known to the ancients, Saturn takes 29½ years to complete one orbit at 1400 million km from the Sun. Since Saturn moves rather slowly round the Sun, oppositions occur only about a fortnight later each year. That for 1995 occurred on 14 September, and for the fifteen years following the planet will be positioned north of the Celestial Equator and so well placed for observers stationed in the northern hemisphere of the Earth.

At opposition the planet reaches a visual brightness of $-0^m.4$. Its 121 000 km diameter globe then spans about 20 arcseconds and a power of about $\times 100$ is necessary to enlarge the disc to the same apparent size as the full Moon, when the latter is seen with the naked eye.

Like Jupiter, Saturn is observable somewhere in the sky for most of the year. Its variations in apparent size are even smaller than Jupiter's. Even when close to conjuction, Saturn's disc still subtends 16 arcseconds. Of course, the real spectacle of the planet is provided by its ring system. The rings that are visible through a telescope span 270 000 km, an apparent angular extent of 45 arcseconds when the planet is at opposition.

Observing Saturn's globe

Saturn is very much more quiescent in appearance than Jupiter. In part this is due to a haze layer lying well above the planet's upper clouds. This haze is responsible for the muted contrasts and colours of the planet. The nomenclature of the visible features follows the same scheme as for Jupiter (Figure 9.1), though I have only ever seen the equatorial belts, the temperate belts and polar regions with any clarity. Other observers have shown additional belts in their drawings. As far as the zones are concerned, the whole of the region south of the SEB and north of the SPR is known as the South Tropical Zone (STZ). Similarly for the region north of the NEB and south of the NPR, (the NTZ), irrespective of whether additional belts are visible in these regions.

Occasional irregularities are seen in Saturn's belts and zones. Jupiter-type spots are only very rarely observed. Timing the transits of any prominent features is an especially valuable exercise, as this allows the various circulating currents to be studied. The procedure is exactly the same as for Jupiter. However, the central meridian longitudes are not generally published, unlike the System I and System II values for Jupiter. The accepted figure for Saturn's rotation period is 10 hours 40 minutes, though this figure varies with latitude. Nonetheless, the 'raw' timings can be submitted to the coordinator of an observing group for further analysis.

Intensity estimates are valuable and can be carried out with a smaller telescope than the minimum of about 10-inches (254 mm) aperture required for detecting spots and irregularities. The standard 0 = brightest to 10 = black scale is adopted, with fractions used where necessary.

Observing Saturn's rings

Figure 9.7 shows the nomenclature of the main rings. I have seen the crêpe ring clearly defined against the black sky with my 6¼-inch reflector. On the other hand, it can be difficult even in my 18¼-inch. Much depends upon the aspect of the rings and the freedom from haze. However, I can easily see the darkening on the globe caused by the crêpe ring crossing in front of it, with a 60 mm refractor.

Despite occasional reports to the contrary, rings other than A, B and C are not visible at visual wavelengths, even with fairly large telescopes. The same cannot be said for the ring divisions. Cassini's division is easily seen, even with a small telescope, when the rings are wide open. Encke's division, near the outer edge of ring A, has been reported from time to time (though I have never seen it myself), though prior to the encounter of the spaceprobe Voyager 1 most observers placed it running through the middle of ring A!

Once in a while observers reported many other ring divisions present and we now know that the rings are, indeed, heavily grooved. However, I am not sure that the fine divisions present really could have been seen from Earth (the Earth's atmosphere imposing a resolution limit). I am not questioning

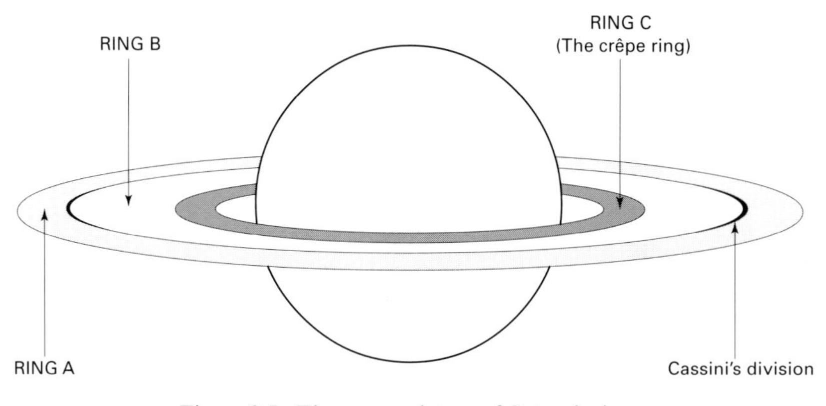

Figure 9.7. The nomenclature of Saturn's rings.

the honesty of the observers concerned but it is all too easy to be deceived when one's eyes are straining to see features that may be just out of reach. Perhaps these observers were really recording the boundaries of sections of the rings of slightly differing brightness? One can usefully keep a watch on the intensities of the various rings (using the same standard number scale as for intensity estimates of globe details) and look out for any divisions or irregularities that may be visible.

The yaw and tilt of Saturn's rings changes constantly over a 29½ year cycle. The times when the rings are edgewise are particularly interesting as they are then reduced to a fine line of light when seen with a large telescope. Through small apertures Saturn may then appear ringless!

Making drawings of Saturn

A pre-prepared outline is essential. The globe of the planet is very flattened at the poles and the rings are especially difficult to draw. One solution is to get pre-prepared blanks from your coordinating group or society. Another is to make your own blanks. Here a computer with graphics facilities and a printer is a considerable aid. If D is the equatorial diameter of Saturn, the other quantities follow:

Polar diameter of globe $= D - 0.12D\cos B$
Outer edge of ring A (major axis) $= 2.26D$
Outer edge of ring A (minor axis) $= 2.26D\sin B$
Inner edge of ring A (major axis) $= 2.01D$
Inner edge of ring A (minor axis) $= 2.01D\sin B$
Outer edge of ring B (major axis) $= 1.95D$

Outer edge of ring B (minor axis) = $1.95 D \sin B$
Inner edge of ring B (major axis) = $1.53D$
Inner edge of ring B (minor axis) = $1.53 D \sin B$
Inner edge of ring C (major axis) = $1.21D$
Inner edge of ring C (minor axis) = $1.21 D \sin B$

Note that the gap between the outer part of ring B and the inner part of ring A is Cassini's division. Also, the inner edge of ring B is the outer edge of ring C. In all cases the quantity B is the apparent tilt of Saturn's polar axis (numerically the same as the latitude of the sub-Earth point, though ignore any negative sign) and is obtainable from an ephemeris. The ellipses representing the rings and the planet can be generated (either using the computer or plotting out by hand) using the following equation:

$$y = b\sqrt{1 - \frac{x^2}{a^2}},$$

where a is half the major axis and b is half the minor axis, respectively. I recommend using 40 mm as the value of D.

If you are plotting out the ellipses by hand, I would suggest just doing this once on graph paper. After drawing your freehand curves through all the points of each ellipse you can then ink over them to make the shapes show through for tracing. Apart from near the dates of edgewise ring presentations, the tilt of the rings does not vary very much over a period of just a few months. The one 'master' can then be used to generate many blanks. However, the coordinators of observing groups may well supply you with the necessary blanks.

When making your sketch, begin by adding in any shadows of the globe on the rings or visa versa. The procedure then continues in the same way as for drawing Jupiter. Remember to include all the relevant data in your report. Figure 9.8 shows a drawing I made of Saturn back in 1975, when the south face of the rings were well presented to us.

Observing colours on Saturn

My comments here follow on the same lines as those given for Jupiter. Saturn's NEB and SEB usually look brownish-pink to me, the rest of the globe being various shades of beige. Ring B usually looks a creamy-white colour and I find that ring A looks grey to bluish-grey. Monochrome drawings made using colour filters are of more scientific value than coloured drawings, though the latter are far from worthless.

Figure 9.8. Saturn, a drawing made by the author, using his 18¼-inch (464 mm) Newtonian reflector, ×260, on 1975 March 7^{d} 22^{h} 00^{m} UT.

Measuring latitudes on Saturn

The general procedure is exactly the same as for Jupiter, though one difficulty is that Saturn can be tilted towards or away from us by as much as 27°. Hence aligning the fixed crosswire to the central meridian is a little more difficult than is the case for Jupiter, since the belts can appear very curved. Aligning one of the movable wires so that it runs from one extreme of the planet's ring system to the other is the best procedure. The measurements are carried out in the same way as outlined for Jupiter, the corresponding reduction formula being:

$$sin\theta = \frac{2\left(\frac{D}{2} - X\right)}{D + (0.12D\sin\theta_{(est)}(\cos\theta))}$$

where B is the Saturnian latitude of the sub-Earth point, or axial tilt, (given in an ephemeris) and all the other quantities are as previously defined for the measurement of Jovian latitudes.

Observing Saturn's satellites

Of Saturn's extensive retinue of satellites, nine are within the visual range of amateur telescopes. They are, in increasing radius of orbit from Saturn:

Mimas (I, $12^m.1$), Enceladus (II, $11^m.9$), Tethys (III, $10^m.4$), Dione (IV, $10^m.4$), Rhea (V, $9^m.8$), Titan (VI, $8^m.3$), Hyperion (VII, $14^m.2$), Iapetus (VIII, variable brightness – 10th magnitude to 12th magnitude), and Phoebe (IX, 16^m approx.). In each case, the first number in parenthesis is the official number of the satellite. The second is its visual magnitude.

Don't be too surprised if you see slightly differing magnitude values assigned to the various satellites in other books. Authorities differ. I have tried to give an 'average' of the quoted figures. There is surely an opportunity for a suitably equipped observer to carry out extensive photoelectric observations. Particularly so, in the case of the peculiar Iapetus, which has half of its icy surface laid bare but the other half covered in a dark brownish material. The brownish surface leads as the satellite orbits Saturn (it has a captured rotation) and this causes its wide variations in apparent magnitude. All of Saturn's satellites are fairly diminutive objects, except Titan which is over 5100 km across. The second largest is Rhea, with a diameter of 1530 km. Even Titan subtends an apparent diameter of less than an arcsecond from the Earth.

Observing Uranus, Neptune, and their satellites

Uranus was the first major planet to be discovered after the advent of the telescope, though it can be just discernible to the naked eye on a night of excellent transparency. It moves at a mean distance of 2870 million km from the Sun, taking 84 years to complete one orbit. This planet was at opposition on 20 July in 1995 and subsequent oppositions occur roughly 4 days later each year. As with all the planets, summer oppositions are very unfavourable for mid-northern hemisphere observers, since the planet always has a very low altitude. Uranus sports a small (3.9 arcsecond diameter) cyan coloured disc. A power of about $\times 500$ is needed to enlarge this disc to the same apparent size as the full Moon seen with the naked eye.

I have never been able to see any features on the disc of Uranus, though others report seeing bands and spots variously positioned. The Voyager 2 space-probe images showed that the orb was bland and only after computer enhancement could definite details be seen. Even so, it is surely worth keeping a watch on this planet as large changes might occur. Uranus normally has a magnitude $+5^m.6$ at opposition. There have been reports of changes of brightness and so it is worth making magnitude estimates from time to time. In order to do this use a very low magnification so that the planet appears 'starlike'. Its brightness can then be assessed by comparison with nearby stars of known brightness (see Chapter 12). Better still, the observer equipped with a photoelectric photometer can get very accurate brightness measures (see Chapter 13). Uranus's five major moonlets are all of the fourteenth magnitude, or fainter, and the amateur can do little to further our knowledge of them.

Neptune appears even smaller than Uranus. It's 2.5 arcsecond blue disc came into opposition on 17 July in 1995. This planet orbits at 4497 million km from the Sun and takes 165 years to journey once around the Solar System. Consequently it comes into opposition very roughly two days later each year. We shall have to wait even longer for oppositions of Neptune to occur with the planet at a reasonable altitude from mid-northern latitudes, than we will for Uranus.

Voyager 2 showed us that Neptune is a far more showy spectacle than Uranus. However, with a power of about $\times 770$ needed to show the disc of the planet the same size as the full Moon seen with the naked eye, it is hardly surprising that no reliable observations of albedo features have ever been obtained from the Earth. Apart, that is, from some near infrared images which revealed a patchiness due to clouds in Neptune's atmosphere. As with Uranus, the amateur can usefully keep a watch on the overall brightness of the planet. Neptune is usually of magnitude $7^{m}.9$. Of the planet's two major moons, only Triton, of the thirteenth magnitude, is bright enough to be visible in an amateur's telescope.

Planetary occultations

Watching a planet, or a planetary satellite, passing in front of a star can be an exciting affair. It can also provide information of real scientific value. Accurate timings and descriptions of the change in appearance of the star can be made, perhaps using a tape or cassette recorder to leave the observer free to watch the event uninterrupted. Those with CCD or photometric equipment can make an extremely valuable contribution. A recent example was the 1989 July 3^{d} occultation of 28 Sagittarii by Titan and Saturn's rings. Hundreds of observers all over the world contributed the results of their observations. Cloud layers in Titan's atmosphere caused the star to flash and fade a number of times during immersion and egress, which took about a minute from start to finish in each case. UK observers saw the star hidden from view for about 4 minutes.

Unfortunately, it was still daylight in Britain and Europe when the transit of 28 Sagittarii behind Saturn's rings took place. By all accounts the event was truly spectacular. The star winked and flashed as it traversed through the multitude of grooves and divisions. The data obtained will take years to analyse – just another example of cutting edge science that the amateur can perform!

Accounts of the 28 Sagittarii event can be found in the October issues of the *Journal of the British Astronomical Association* and *Sky & Telescope* magazine. The unexpected can often happen. Remember, it was because of the light variations of the star SAO 158687 in March 1977, before and after occultation by Uranus, that the rings around this planet were first discovered.

Photography and CCD imaging of the gas-giant planets

Chapters 5 and 6 cover the techniques of high-resolution photography and CCD imaging; suffice it to say here that recording the image through coloured filters is particularly useful. Jupiter is the easiest of the gas-giant worlds to photograph as it is relatively bright and subtends a large apparent diameter. At the other extreme, it is difficult to get anything approaching a sharp disc on photographs of Uranus and Neptune. So much depends on the steadiness of the Earth's atmosphere.

Figure 9.4(a) and (b) shows two fairly typical photographs of Jupiter obtained by an experienced astrophotographer. Note how apparent the limb darkening is. Limb darkening can be suppressed during printing by a 'dodging' technique. A small central hole is made in a large card and this card is held in the enlarger beam a little above the photographic paper. Of course, the size of the hole and its height above the paper must depend upon the size of the image of the planet. Angling the card slightly will take into account the ellipticity of Jupiter's outline. In this way the illumination to the limb regions of the planet is severely restricted (and the illumination of the sky-background part of the image is almost totally cut off). After part of the exposure time the enlarger is switched off and the card is removed. The enlarger is once more switched on to complete the exposure. Slightly gyrating the card when it is in position above the photographic paper prevents any artificial-looking boundaries from being generated in the final print.

CCD imaging and computer enhancement have already been referred to. It is most definitely in this direction that the very best possible results lie.

10

Asteroids, comets, meteors, and aurorae

For many people the Bohemian asteroids, ghostly comets, flaring meteors and ethereal aurorae are what astronomy is all about. These objects and phenomena rarely fail to strike up feelings of wonder in astronomers and lay persons alike. These are also areas where amateur endeavours are pre-eminent. The last time an amateur discovered a major planet was way back in 1781. I am probably not sticking my neck out too far in stating that the chances of a modern-day amateur doing the same are nil. However, there is **every** chance that **you** might discover a new comet or asteroid, or observe a particularly spectacular fireball-meteor, or a vivid aurora. Even if you do not make the actual discovery, you can still provide observations of real scientific value.

Observing asteroids

Known asteroids can be located from their ephemerides. Thousands have been discovered so far and many more remain yet to be found. They can be hunted down visually, though there is little to distinguish their appearances from stars, apart from their relative motions. Undoubtably the best way to discover new asteroids is to use wide-field photography. A Schmidt camera is the ideal tool to use but, failing that, use any camera capable of giving a fairly wide field of view (several degrees) and with the largest possible aperture. Figure 10.1 shows a photograph of the asteroid 5 Astraea as it passed through the Beehive star cluster. Long exposures tracked at the stellar rate might reveal asteroids by their trails. Alternatively, negatives obtained on different nights can be compared, to reveal any interlopers.

The easiest way of carrying out this comparison is by holding the two negatives together, so that corresponding images of the stars are very slightly separated. One could hold the negatives up to the light and view

Figure 10.1. The asteroid 5 Astraea passing through the Beehive star cluster. This 10 minute exposure on Tri-X film was taken by Martin Mobberley, using his 14-inch (356 mm) reflector, at its *f*/5 Newtonian focus, on 1987 March 29^{d} 20^{h} 31^{m} UT. The asteroid is identified by the arrow.

with a magnifying glass but projecting though both with a photographic enlarger is much easier. The projected image will show all the stars paired except for any asteroids which will show single images and so will be immediately obvious.

Another way of making the comparison is to set up two projectors (they must have lenses of identical focal length) side by side. Each projector produces an image of one of the negatives onto a screen. The images are made to coincide as nearly as possible. Alternately covering one of the lenses with a mask allows the negatives to be 'blinked'. Any object not in the same position on both negatives will appear to hop back and forth (or appear and disappear if it is only shown on one negative) as each projector lens is alternately covered. The other star images will remain virtually stationary. This device has been given the acronym *problicom*, standing for *pro*jection *bl*ink *com*parator.

Some enthusiasts go to the trouble of building a *blink microscope comparator*, or a *blink stereoscope comparator*. In the first instrument an optical arrangement sends the light from each of the two negatives alternately into the same single eyepiece. In the second, simpler, type of instrument the two negatives are viewed through two lenses, one for each eye.

Switching on and off the illumination behind each negative produces the blinking effect.

Clyde Tombaugh used a blink microscope comparitor to examine several hundred photographic plates, in the course of which he discovered the ninth planet, Pluto. This was the main reason for his search, though he also discovered a globular star cluster, several clusters of galaxies, several star clusters, a comet – and nearly eight hundred asteroids!

In many ways Pluto can be considered more like an asteroid than a major planet. This is certainly true from the point of view of the observer. Pluto takes 248 years to move once round its highly eccentric and highly inclined orbit, the mean Pluto–Sun distance being 5900 million km. It appears like a star of the fourteenth magnitude and a telescope of at least 8-inches (203 mm) aperture is needed to show it visually. Much of what is said in this section about observing asteroids is also applicable to observing Pluto.

If you find a new asteroid (and you **must** be sure that it is a new one) then you can have your discovery registered by sending a telegram to:

IAU Central Bureau for Astronomical Telegrams,
Smithsonian Astrophysical Observatory,
60, Garden Street,
Cambridge
Massachusetts 02138,
U.S.A.,

giving all the details (especially the positions at given times). There is a preferred (and rather involved) format for setting out the data in your telegram. I would urge interested readers to write to the Central Bureau asking for guidance. They will send you examples of how to report the discovery of an asteroid, a comet, a nova, a supernova, or a variable star.

Alternatively, you could communicate with the Bureau via the Internet. The address is:

http://cfa-www.harvard.edu/cfa/ps/cbat.html.

Select the section titled 'How Do I Report a Discovery' and fill in the questionnaire according to the instructions. However, do be warned that most apparent discoveries turn out to be completely bogus and instigating a false alarm will only serve to annoy the already overworked clearing-house staff and will, furthermore, destroy your credibility as a competent observer. If you think that you have a discovery to report, stop and consider. Are you **really** sure?

In any case, the asteroid will not be considered to be new and distinct from one of the thousands of others until it is recovered and its position is measured on at least two nights. You might not be able to measure its position accurately but someone, somewhere, will have to.

Asteroid designations

Firstly, asteroids are also known as *minor planets*. The first asteroid, Ceres, was discovered in 1801 and for the next half century they were referred to by name. Later numbers were added, indicating their order of discovery. Various other schemes have been used and are still valid. However, the latest scheme is IAU approved and should now be considered the standard. In this the asteroid is designated by a four part number which refers to its date of discovery.

The first part is simply the year of discovery: 1995, etc. Following the year is a space. After that goes a letter to indicate the half-month of discovery, according to the following:

A	January 1–15	B	January 16–31
C	February 1–15	D	February 16–29
E	March 1–15	F	March 16–31
G	April 1–15	H	April 16–30
J	May 1–15	K	May 16–31
L	June 1–15	M	June 16–30
N	July 1–15	O	July 16–31
P	August 1–15	Q	August 16–31
R	September 1–15	S	September 16–30
T	October 1–15	U	October 16–31
V	November 1–15	W	November 16–30
X	December 1–15	Y	December 16–31

Note that the letter I is omitted and, of course, Z is not needed.

After that comes another letter which denotes the order of discovery within the given half-month. It starts off with A for the first discovery and runs, in alphabetical order, through to Z for the 25th (once again I is not used, so H denotes the 8th discovery, J denotes the 9th and K denotes the 10th).

In the event of more than twenty five discoveries in a given half-month a number '1' is added to the end of the designation and the sequence of 'order of discovery' letters is re-run. For more than fifty discoveries the final letter is re-run once more and a final number '2' added; and so on.

For example: if an asteroid discovered on 1997 May 9^{d} is the tenth to be discovered in that half-month, its designation would be: 1997 JK.

As another example: if a particular asteroid discovered on 1998 December 22^d is the 53rd to be discovered in the half-month beginning 16 December, then its designation would be: 1998 YC2.

Preparing a finder-chart

Whether it be for tracking down predicted asteroids or comets, or locating faint nebulae and galaxies, finder-charts are a very powerful aid to the astronomer. The first step is to select a good star atlas. It should show the faintest possible stars and have the largest possible scale. Tirion's *Sky Atlas 2000.0* is an excellent atlas for stars down to eighth magnitude. 43 000 stars are plotted and so an eyepiece giving a 1° real field of view should show at least one plotted star in view, no matter at which part of the sky the telescope is pointed. The scale of the 26 charts are about 3° per inch (1°.2 per centimetre). Even better is another Atlas by Wil Tirion (with Barry Rappaport and George Lovi), *Uranometria 2000.0*. Nearly a third of a million stars are plotted (down to the tenth magnitude) on nearly 500 charts in this two-volume work. The scale is about 1°.4 per inch (0°.6 per centimetre) and there is a very convenient grid superimposed.

You should locate the predicted position of the asteroid (or other object) on the appropriate chart and determine the scale of the atlas at that position (use the grid reference marks on the Celestial Equator if this is not known). Then either photocopy that section of the chart, or obtain a small overlay sheet (acetate plastic will do, or very transparent tracing paper). Your next step is to scribe a circle of radius equivalent to the real field of your eyepiece either on the overlay, or on the photocopy. The circle should be centred on the predicted position of the object.

Figure 10.2(a) shows a circle of diameter 1°.5 centred on an R.A. of 5^h $29^m.8$ and a declination of $-1°.0$. Divide up the circle into a grid of squares as shown in the diagram. The number of squares is not important but do try to make them all equal in size. On a separate sheet of paper draw a large circle and divide it up into the same number of squares as the one on the photocopy, or overlay. The final step is to use the grids to reproduce the pattern of stars contained in the circle on the photocopy or overlay, onto your large circle. There is no need to make precise measures. Eye estimates are quite good enough. The purpose of the grid is to help you do this. Figure 10.2(b) shows the result. The whole process should only take you a few minutes and will save you a great deal of time at the telescope.

If your circle has been drawn to the correct scale (the procedure for

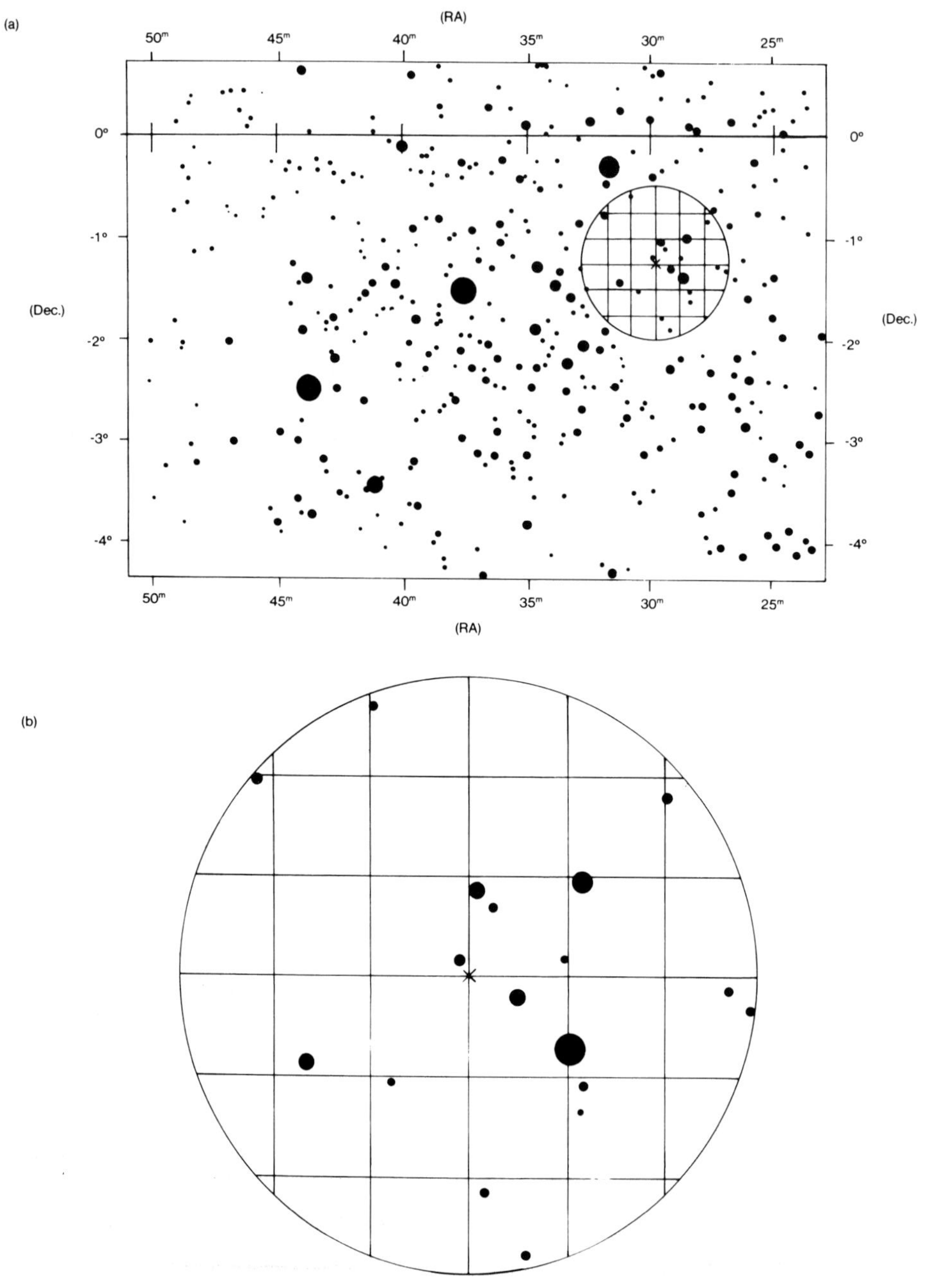

Figure 10.2. The preparation of a finder-chart. See text for details. In this case an eyepiece real field of 1°.5 is represented, centred on a position corresponding to $\alpha = 5^h\ 29^m.8$ and $\delta = -1°$.

measuring the real field of an eyepiece is explained in Chapter 2) the view through the eyepiece will coincide with that represented on your finder-chart. The patterns of stars should be easy to recognise. In difficult cases you can prepare two finder-charts. One is for the view through the finder-scope. The other is for the view through the main telescope.

Allowing for precession

If the predicted coordinates of the asteroid, or other object, are expressed in an epoch more than a few years different from that of the atlas (for instance, comet positions are often expressed in 1950 coordinates, while your atlas may be of 2000 epoch), then the effects of precession will have to be allowed for in calculating the correct position for the centre of your 'eyepiece field' circle. Use the following equations:

$$\Delta\,\mathrm{RA} = 3^{s}.073 + 1^{s}.336\ \sin\alpha\ \tan\delta$$

$$\Delta\,\mathrm{Dec} = 20''.04\ \cos\alpha$$

In each case α is the right ascension of the object and δ is its declination. Δ RA and Δ Dec are the annual amounts of precession in right ascension and declination, respectively.

As an example, suppose that the predicted position of an asteroid, in epoch 1950.0 coordinates, is given as:

$$\alpha = 10^{h}\ 00^{m}\ 00^{s}$$

$$\delta = +5^{\circ}\ 00'\ 00''$$

and you want to know where to put the centre of the circle on your epoch 2000.0 chart.

The right ascension value has to be converted into an angle. Since 24^{h} of RA is equivalent to 360°, the conversion is simple:

$$10^{h} \equiv 360^{\circ} \times 10/24 = 150^{\circ}.$$

Most electronic calculators will accept angles larger than 90° and display the correct sines and cosines. If not, then $\sin 150^{\circ} = \sin 30^{\circ}$ and $\cos 150^{\circ} = -\cos 30^{\circ}$.

Using the equations, Δ RA $= 3^{s}.13$ per year, and Δ Dec $= -17''.36$ per year. In precessing the 1950.0 coordinates forward 50 years, the values of annual precession are multiplied by 50. These amounts of precession are then added (taking account of the sign of the correction) to the 1950.0 coordinate values.

Hence, the epoch 2000.0 coordinates of the asteroid become:

$$\alpha = 10^h\ 02^m\ 37^s$$

$$\delta = +4°\ 45'\ 32''.$$

If you have to precess backwards (say from epoch 2000.0 to 1950.0) remember to reverse the signs of the corrections.

Photometric and colormetric studies

Precision photometric measures of asteroids can reveal subtle light curves which betray rotation. If observed for long enough the precise rotation period of the asteroid can be determined. Filter photometry allows colour-index determinations to be made. The procedure is the same as applied to stars and is outlined in Chapter 13. Such colour measurements provide information on the composition of the asteroid's surface. However, it must be said that only two asteroids (Vesta and Ceres) ever get brighter than the seventh magnitude and not many more will be within the range of amateur photoelectric equipment.

Astrometry

Rough positions can be measured on any photograph (or projected negative) simply using a ruler. Knowing the image-scale and orientation, measuring from a star of known coordinates allows the approximate RA and declination of the object (be it an asteroid, comet, or other object) to be determined.

However, achieving sub-arcsecond accuracy in your positions is very much more difficult. It requires that the photographs be obtained with a telescope of at least 1 metre focal length. It also requires a precision instrument for measuring the negative. This *measuring engine* should be able to reliably record the positions on the negative to an accuracy of about a micron. Even then, a rigorous mathematical technique has to be applied to the measurements in order to get reliably accurate measures.

Very few amateurs undertake such positional work, though the results would be extremely valuable; for instance in determining comet and asteroid orbits. So, rather than spend a great many pages outlining the procedures and equipment requirements, I would point any reader who might be interested to two articles in the September 1982 issue of *Sky & Telescope* magazine: 'Constructing a measuring engine' and 'How to reduce plate measurements'. An article in the February 1985 issue of the *Journal of the British Astronomical Association* 'Quick-look photo-astrometry with a linear micrometer' might also be of interest.

Potentially the most accurate way for the amateur to perform astrometry is by utilising the appropriate software on CCD images. I commend those interested to read a paper by Nick James in the August 1994 edition of the *Journal of the British Astronomical Association*, entitled 'Some applications for amateur CCD cameras'. At the time of writing more and more software is becoming commercially available and by the time this edition is published I would hope that the reader will be spoiled for choice.

Observing comets

Hunting down previously undiscovered comets, recovering comets on their predicted returns, studying their physical nature and changes, measuring their positions accurately for orbit determination – these are all areas open to the amateur.

Comet hunting

It is just within the bounds of possibility that you might come across a comet when looking at some other celestial body. However, if you set your mind to go comet hunting then it is as well to maximise your chances of success. The first thing to do is provide yourself with suitable equipment. Comets are usually faint things (even if not, someone else is likely to discover it before it becomes bright). So, the largest possible aperture is needed. On the other hand, the largest possible field of view is also highly desirable and this demands a short focal length. Together these requirements lead to a moderate aperture telescope of low focal ratio. In practice a telescope of aperture 4, 5, or 6 inches (102 mm, 127 mm, or 152 mm) and of focal ratio lower than *f*/6 is the ideal for comet hunting.

I ought to deal briefly with one common misconception (most disgracefully, I have even seen this perpetuated in adverts for commercial telescopes). Low focal ratio telescopes do **not** produce brighter images when used visually, than those of larger *f*/number. The two main factors responsible for the perceived brightness of an extended object are the aperture of the telescope and the magnification used. The other factors involved are the transmission/reflection efficiency of the optics, the absence of scattered light, and the size of any obstruction in the telescope's light path.

A power of ×30 on a 6-inch *f*/4 telescope will produce an apparent surface brightness in the image of any extended body identical to that produced by a power of ×30 on a 6-inch *f*/10 telescope. Where the low focal ratio scores is in the possibility of using an eyepiece with a larger field of view.

The image scale at the focus of the 6-inch *f*/4 telescope is 339 arcseconds mm^{-1}. At the focus of the 6-inch *f*/10 telescope this is 136 arcseconds mm^{-1}. If we suppose that both telescopes are fitted with eyepiece focusing mounts which can take 1¼-inch (31.7 mm) standard barrel eyepieces, then we can pick a figure of 28 mm for the maximum diameter of the field-stop aperture. (It obviously has to be less than the diameter of the eyepiece barrel). For the *f*/4 telescope this diameter at the focal plane corresponds to 339×28 arcseconds, or 2°.6. This is the quantity known as the real field of the eyepiece (see Chapter 2). For the *f*/10 telescope this size of field-stop produces a real field of 1°.1.

Taking our calculations forward a step (and rounding figures where appropriate), to achieve a magnification of ×30 and a field of view of 2°.6 on the *f*/4 telescope we would plug in an eyepiece of focal length 20 mm and 78° apparent field. One of the complex wide-field eyepieces would be a suitable choice. A 50 mm focal length eyepiece, will produce the same magnification on the *f*/10 telescope but the maximum value of apparent field could only be 33°. Even a Ramsden eyepiece would be sufficient for this.

To summarise, for successful comet hunting one desires a telescope of the largest possible light grasp but also the largest field of view. Something in the range of 4 to 6 inches aperture and a focal ratio of less than *f*/6 is the ideal. Even so, you could still achieve success with a telescope outside these limits, or even with just a pair of binoculars. The magnification should be as low as possible, without wasting any of the light gathered by the objective/primary mirror – ideally ×4 or ×5 per inch (×1.6 or ×2.0 per centimetre) of aperture. However, do not use a magnification much lower than ×16, or ×20, even for very small apertures.

Of those using purely visual techniques, the most successful comet hunter this century is William Bradfield. He continues to observe and make new discoveries today. He found his thirteenth comet (this was the one that put his name in the record books) in August 1987, after 17 years of searching. Eleven of these thirteen comets were found with his 6-inch short focus refractor. Though he did find one using a 10-inch (254 mm) reflector and one with a pair of 7×35 binoculars!

A short-focus Newtonian reflector is almost as good as a short-focus refractor and will cost a lot less. A 'comet-catcher' telescope need not have an equatorial mount (indeed, sweeping may be easiest with an altazimuth) but it should be sturdy. Most of the discoveries are made when the comets are at an elongation of between 40° and 80° from the Sun, either to the east, or to the west of it.

For evening hunting, it is best to begin by sweeping the telescope in azimuth along the western horizon just as the sky is getting dark. The sweep will probably cover about 90° of azimuth centred on the setting point of the Sun. Then raise the telescope in altitude by about half a field diameter (allowing for the fact that the stars are setting) and sweep back in the opposite direction. Maintain this sweeping. You should gradually beat the diurnal setting of the stars and be working with the telescope pointing at an ever increasing altitude. However, don't sweep so fast that you miss out on seeing any faint objects. After a couple of hours, or so, you will be sweeping a strip of sky about 90° away from the Sun. Of course, you could continue longer, though with the knowledge that most of the reasonably bright comets are found when they are somewhere near 60° from the Sun.

Not many people get up early enough to hunt for comets in the morning sky. If you do you will have few competitors! Search in the east starting at an altitude of about 50° and sweep east–west as before, gradually getting lower to meet the morning twilight.

If you see a fuzzy blob do first check that it is not a deep-sky object. Obviously a good star map is essential. I recommend '*Sky Atlas 2000.0*' by Wil Tirion. Also beware of ghost images caused by reflections in the optics. As you move the telescope does the image really stay with the stars? Try changing the eyepiece to really make sure. Try using higher powers to see if the object can be resolved into a faint grouping of stars. Estimate the brightness of the object (how this is done is covered later in this chapter).

Make a careful note of the position of the object with respect to the star background (best of all, make a sketch). If there is time enough before the object sets below the western horizon, or is engulfed by the morning twilight, you might be able to detect some motion. If so, draw another sketch. Your sketches can be used to determine the approximate position(s) of the object by comparison with a star map. Even better if you have a telescope (not necessarily the one you use for comet sweeping) with accurate setting circles. You can set on the comet and simply read off the circles. Better still is a telescope with a computer-aided (CAT) readout and best of all is securing a CCD image or photograph of the suspect new arrival.

If you are really sure that you have found a comet (it may not be a new one – just a predicted return of a period comet) then you can lay claim to it by sending a telegram, or E-mailing, the IAU Central Bureau for Astronomical Telegrams, as described previously in connection with asteroids, giving all the details (especially the positions at given times).

If your claim of discovering a comet is confirmed and you were the first to see it, then that comet will bear your name. If others independently observed it at about the same time, then the comet will bear their names as well as yours. Previously, a comet was also given a designation signifying the year of its discovery and the order in which that particular comet was discovered in that year. For instance, William Bradfield's thirteenth comet discovery was designated 1987s, because it was the nineteenth comet to be discovered in 1987.

When the orbit of a comet has been firmly established it is given yet another designation, comprising the year of discovery and the order of perihelion passage that year. For instance, the famous short-period Comet Encke is known as 1786 1 because it was the first of the comets discovered that year to pass perihelion. The prefix P was used in front of a comet's name to denote those that are *periodic* (have known periods and can be predicted). Hence P/Encke, P/Halley, and P/Brorsen-Metcalf.

You might be wondering at my use of the past-tense. The reason is that in January 1995 the preferred designation scheme had changed again, becoming uniform with that for asteroids. See the earlier section on asteroid designations for guidance. The only difference is that a number rather than a letter indicates the order of discovery in each half-month. There is one addition for comets: a letter C or a letter P, either being followed by a slash, precedes the designation. P/ denotes a comet of period less than 200 years and C/ is for all others.

At the time of writing a particularly impressive comet is gracing our skies. This is Comet Hyakutake C/1996 B2; the designation meaning that it was the second comet (2) discovered in the second half of the month (B) of 1996.

All but the most recent books, scientific papers and other publications will, of course, use the older scheme. You will find examples of both the old and the new schemes of comet nomenclature in this chapter.

Recovering a periodic comet

Many organisations (such as the B.A.A. and the I.A.U.) issue predictions for the returns of periodic comets, listing the expected RA, declination, and magnitude on given dates. Certain magazines (such as *Sky & Telescope*) also provide this information. Recovering a comet as early as possible before perihelion allows for a full and detailed study of its physical development and changes as it nears the inner regions of the Solar System.

If your telescope has accurate setting circles (or is computer-controlled) then you can set on the expected position. You will probably find that the comet is not there but methodically searching the patch of sky out to a degree or so of the predicted position should be enough to find it. Your measurement of the actual position of the comet (covered later in this chapter) is extremely useful and should be sent to your comet observations coordinator. If your telescope has no accurate setting circles then you will have to make a finder-chart to help you track it down. The making of finder-charts is covered earlier in this chapter. Suffice it to say here that the predicted positions are plotted on the chart and area of the sky is located by matching the charted and actual star patterns as seen through the eyepiece.

Photography/CCD imaging of comets

Reference should first be made to Chapters 4 and 5, where the basic principles of astrophotgraphy are outlined, and then to Chapter 6 where the specialism of electronic imaging is covered. Comet photography is usually carried out at the telescope's principle focus; that is, without any amplification of the image at the focal plane. A comet can appear as a fuzzy smudge just a few arcseconds across. At the other extreme, rare ones can stretch tens of degrees across the sky. A telescope is most appropriate for the former type, but a camera fitted with a standard lens may be best for the latter. In either case low focal ratios are to be preferred.

The general procedure follows that for photographing any faint and extended body, such as a nebula or a galaxy. However, there is one important difference: the comet moves against the star background during the exposure. A photograph carefully guided on the stars will usually show the comet smeared out unless it is sufficiently bright to allow for a very short exposure.

One remedy is to guide on the nucleus of the comet. However, that is not always easy as the nucleus may not be sufficiently bright and well-defined. If your telescope is computer-automated, then the computer can be programmed to track the comet (the rate of change in RA and declination can be worked out from the daily change in the predicted position of the comet).

Alternatively, your camera could be mounted on its own drive unit at the focus of your telescope. Figure 10.3(a) shows such a unit designed and built

(a)

(b)

Figure 10.3. (a) Ron Arbour's automatic comet-tracker camera unit. Note the stepper motor, which drives the camera on the cross-slide so that the moving image of the comet remains stationary on the film. (b) Ron Arbour's automatic comet-tracker camera unit in position at the Newtonian focus of his 16-inch (406 mm) reflector.

Figure 10.4. Halley's Comet, 1985 December 7^d 20^h 22^m UT, photographed by Martin Mobberley, using his 14-inch (356 mm) reflector at its *f*/5 Newtonian focus. The 18 minute exposure was made on 3M Colourslide 1000 film, developed as a colour negative. North is uppermost in this photograph.

by Ron Arbour. A computer-controlled stepper motor drives the camera in a transverse direction across the focal plane, the position angle being determined and set beforehand (once again, from the predictions). The camera is made to move at a rate which keeps the image of the comet stationary on the film. Meanwhile the telescope is guided on the stars in the ordinary way. Figure 10.3(b) shows the unit mounted at the focus of Ron's 16-inch (406 mm) Newtonian reflector.

If you have a filar micrometer which you can plug into the guide-scope then you could change the setting every minute, and then alter the pointing of the telescope (using the slow motions) to re-set the guide star on the crosswires. The amount you alter the micrometer setting should obviously correspond to the amount of relative motion of the comet in the interval. The position angle of the micrometer is set to that of the direction of relative motion of the comet. All in all, this is a very tricky procedure, especially as the adjustments have to be carried out in the dark!

Figures 10.4, 10.5 and 10.6 show the sort of photographs that can be obtained using a moderate-sized telescope. Telephoto or Schmidt camera photographs (especially the latter) can be used to hunt for comets; the

Figure 10.5. Comet Bradfield (1987s) surpassed Halley's Comet in spectacle for UK observers. Martin Mobberley took this photograph on 1987 November 14^{d} 18^{h} 03^{m} UT. The 17 minute exposure was made on Tri-X film at the *f*/5 Newtonian focus of his 14-inch (356 mm) reflector. Note the use of the old designation scheme, also used in Figures 10.6 and 10.7.

(a)

Figure 10.6. Three photographs of Comet P/Brorsen-Metcalf (1989o). All were taken by Martin Mobberley on T-Max 400 film, at the *f*/5 Newtonian focus of his 14-inch (356 mm) reflector. (a) [above] 1989 August 13^{d} 02^{h} 08^{m} UT. 13 minute exposure. (b) [right, above] 1989 August 28^{d} 03^{h} 02^{m} UT. 12 minute exposure. (c) [right, below] 1989 September 3^{d} 03^{h} 14^{m} UT. 9 minutes exposure. In all of these photographs north is approximately uppermost.

(b)
(c)

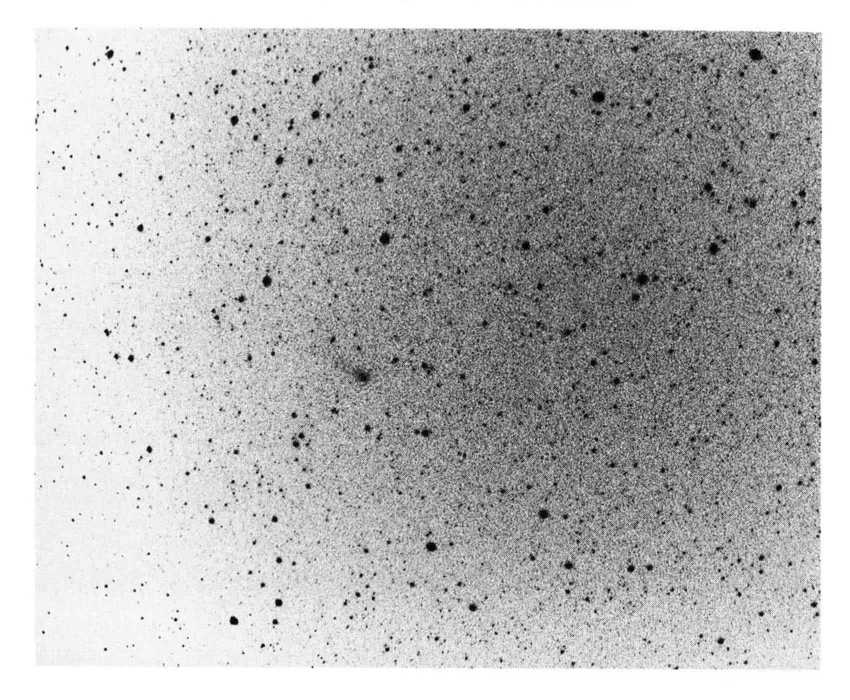

Figure 10.7. Faint extended objects are much easier to see on a negative, than on a positive print. The very small fuzzy object shown is Comet Bradfield (1987s). The photograph was taken by the author on 3M Colourslide 1000 film, using an ordinary 135 mm telephoto lens. The 10 minute exposure was made on 1988 January 14^{d} 18^{h} 45^{m} UT, when the comet was well on the wane. The image-scale produced by the telephoto lens permitted the camera to be tracked at the stellar rate, as can be seen by the round star images.

procedure being to cover as much of the pre-dawn or post-sunset sky as possible with overlapping photographs. A fast film and a low focal ratio enables one to use exposures of just a few minutes. Negative (black stars on a white background) images are best for detecting faint and fuzzy objects (see Figure 10.7). The negatives are not printed but are carefully examined under the enlarger, or through a magnifying lens.

The possibilities for an amateur astronomer who is equipped to carry out electronic imaging, and subsequent computer processing are virtually boundless. Precision astrometry and brightness measurements, intensity profiles and the extraction of delicate structural details (see Figure 10.8), quantitative chemical analysis from image brightness measures through specific waveband filters; the list goes on. The power of computer processing can even be used to do away with the need for special guiding arrangements, as shown earlier in this book, in Figure 6.13.

Figure 10.8. Comet P/1995 S1 (De Vico) imaged by Nick James on 1995 September 29^{d} 04^{h} 33^{m} UT, using a 'Starlight Xpress' SX camera at the Newtonian focus of his 12-inch (305 mm) *f*/5.25 reflector. After the six minute integration, Nick used his own software to radially filter the image and so preferentially enhance details in the tail. Note the new style designation.

Observing the physical details of comets

Figure 10.9 illustrates the structure and nomenclature of a hypothetical comet. Of course, this is very much an idealised representation. Any given comet may possess only some of these features. Also the shapes, sizes and brightnesses of the various parts of comets differ enormously.

When out in the depths of space a comet consists just of a small solid object, the *nucleus*. However, as the comet wings its way towards perihelion, the Sun's heat vaporises materials from its surface. It is these vaporising materials that produce the visible spectacle of the comet. The nucleus is at most a few kilometres across and appears point-like even at high powers. We think that it is a very loose conglomeration of ices of various chemical compounds. Every once in a while a comet is seen whose nucleus may suddenly break into two or more pieces. This happened to one of the most spectacular comets in recent years, Comet West of 1976. Such was also the case for Comet Shoemaker–Levy 9, of Jupiter impact fame.

Through the telescope *jets* of matter can sometimes be seen spiralling from the nucleus before they are swept back into the tail. When a bright comet is close to the Sun it can display a bright *central condensation*,

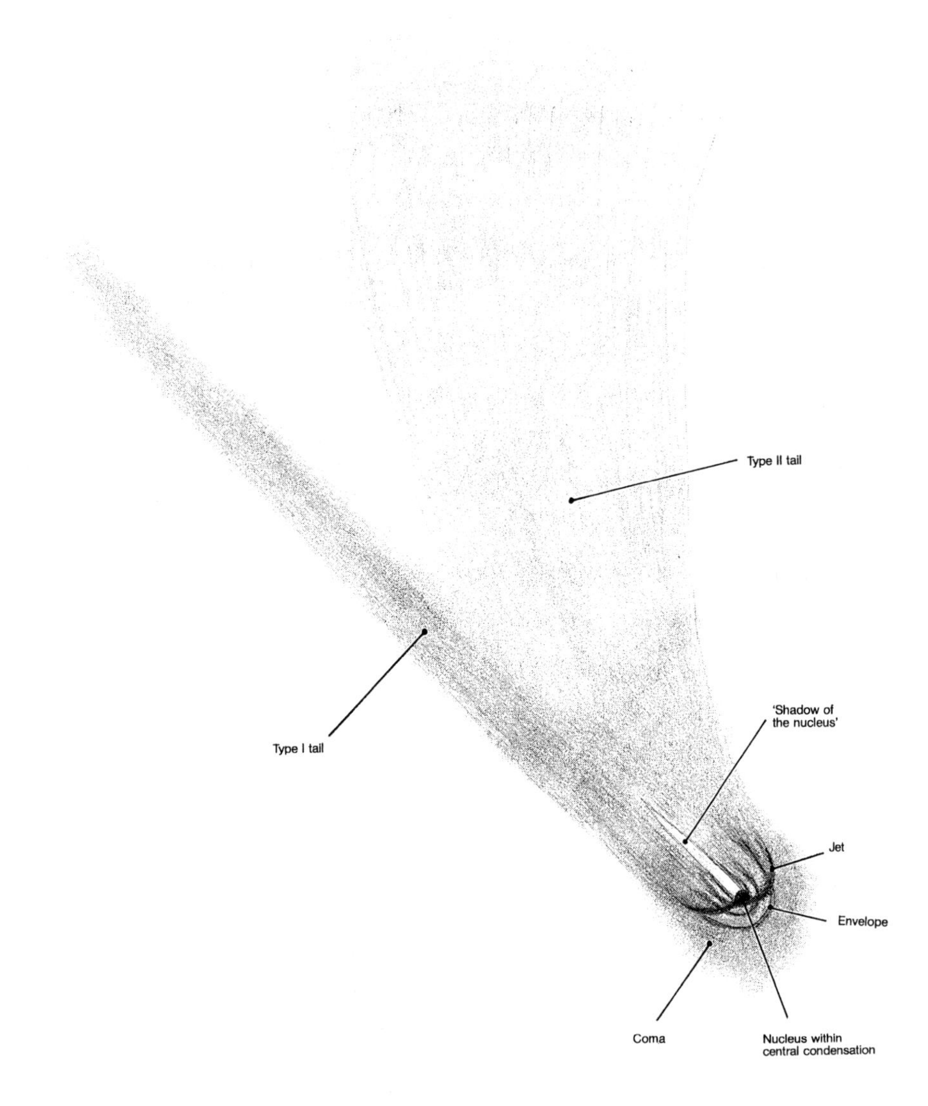

Figure 10.9. The nomenclature of the visible parts of a hypothetical comet are shown in this idealised representation.

centred on the nucleus. This can appear as a bright disc several arcseconds across and represents a region of fairly dense matter issuing from the heated nucleus. Sometimes distinct shells, or *envelopes* of matter can be seen surrounding the nucleus. They can even be seen to puff out of the nucleus and expand away from it over timescales of several hours.

Sometimes a dark *'shadow of the nucleus'* effect is seen trailing from the nucleus and into the tail of the comet. The region at the head of the comet is known as its *coma*. The actual size of the coma can be collosal – over one million kilometres across in some instances. This region is effectively the comet's 'atmosphere', containing mainly neutral atoms and molecules, such as: H, OH, NH, NH_2, HCN, CH_3, CN, CS, S, O, CO, C, C_2, C_3 and traces of metals, including sodium and iron.

Small, faint, comets tend to have approximately spherical comas and may not develop much of a tail. The larger and brighter examples tend to have ellipsoidal comas and may develop extensive tails. *Type I tails* are made of ionised gas stripped back from the coma by the solar wind. Spectra reveal such ionised molecules and radicals as: CO^+, OH^+, CH^+, CN^+, and many others. These species are chemically very reactive and can only survive uncombined due to the extremely tenuous nature of the tail (far more so than the best vacuum we can produce on Earth). They are usually bluish in colour, caused by the ionised gases absorbing short-wave solar radiation and re-radiating it at visible wavelengths. Type I tails are usually fairly straight, pointing almost exactly in the anti-solar direction. However, bright knots are sometimes seen to form and then thread along the tail, being caught up in the complex magnetic fields frozen into the solar wind.

Type II tails are composed of colloidal-sized 'dust' particles. They appear yellowish in colour because they shine by reflected sunlight. I should make it clear, though, that the colours in comets only show up visually in the very brightest examples. The tints show up best in colour photographs. Owing to the comet's transverse motion when close to perihelion, type II tails are curved. Some comets sport very highly curved fan-shaped tails. These are classed as *Type III*, though in essence they have the same nature as type II tails. A few comets show *anti-tails*, short sunward pointing spikes extending from the head of the comet. This is an illusion caused by us seeing part of the highly curved dust tail projected a little forward of our line of sight to the head of the comet.

Nowadays it is almost solely down to amateur observers to record the physical changes that take place during a comet's apparition. The best way of doing this is by photography/CCD imaging. Various exposures can be given to record parts of the comet of differing brightness. Drawings are also useful. They can be made with a black pencil on white paper (a negative representation), or as white on black. As always the most important factor is accuracy. A telescope, binoculars or the naked eye can each play a complementary part, depending on the size and brightness of the comet.

Monitoring a comet's brightness is another valuable exercise. Binoculars will prove best for this, unless the comet is very faint. The procedure is to defocus until nearby stars are expanded into similar sized discs as the comet (which will also appear blurred, and a little expanded). Ideally, the field of view will contain at least one star disc which is just a little brighter than the comet and one which is a little fainter. We will discount the very remote possibility of getting a nearby star disc of matching brightness. The two comparison discs are then used to estimate the brightness of the comet.

As a simple example, let us suppose that the out-of-focus comet disc appears one third of the way between the brightnesses of two nearby out-of-focus star discs, and is nearer the brighter one. If the brightness of the fainter star is $7^m.8$ and the other is $7^m.2$, then the magnitude of the comet is:

$$(7^m.8 - 7^m.2)1/3 + 7^m.2 = 7^m.4$$

It is important to include the details of the equipment used when reporting brightness estimates. Magnitude estimates made with larger apertures generally come out a little lower than those made with smaller apertures.

Estimates of the coma's size can be made, given the value of the real field of view of the eyepiece. Eye estimates will do but a better method is to time the transit of the coma into or out of the field of view (or across a cross-wire) with the telescope drive switched off. The angular diameter of the coma, in arcseconds, is then given by:

$$\text{Diameter} = 15t \cos\delta,$$

where δ is the declination of the comet (in degrees, ignoring the negative sign of southerly declinations) and t is the transit time in seconds. The length and position angle of the tail can also be read off, using an eyepiece micrometer for small comets. Position angles are dealt with in Chapter 14, as is the use of the micrometer. For larger comets, noting the positions of the head and the end of the tail on a star map will allow the apparent length and position angle to be scaled off.

Observing meteors

Shoals of particles, called *meteoroids*, orbit the Sun. When the Earth passes through one of these shoals, its atmosphere sweeps many of them up. These particles then become *meteors*, as they hurtle to a fiery destruction. The visible effect comes not from the particle itself but from the trail of ionised air that the rapidly vaporising particle creates. The vast majority of these projectiles from space are, fortunately for us, little bigger than grains of sand. Meteors that briefly shine out their demise as brightly as a zero magnitude star are not uncommon, though they are still far brighter than the average. Such a meteor might well have a mass of about a gram, enter the atmosphere at about 50 km/s and vaporise while still at something like 100 km above the ground.

Meteors that glow brighter than -4^{m} (brighter than the planet Venus) are rather uncommon, and are termed *fireballs*. Some fireballs explode. Those that do are termed *bolides*. Fireballs and the brighter meteors tend to leave visible tracks, or *trains*, which can persist for several seconds (very rarely, even several minutes!). Any solid object which reaches the ground is termed a *meteorite*.

Perspective causes the meteors in a given shower to appear to originate from a small patch of sky, called the *radiant*. Meteor showers are each named according to the constellation within which the radiant of the shower lies. For instance, all the paths of the meteors in the prominent shower which occurs in mid-August each year appear (when tracked backwards) to intersect at a point in the sky close to the star ϵ Persei. Hence this particular shower is known as the *Perseid* meteor shower. Owing to the rotation of the Earth, meteors are more common, and are faster moving, when seen after midnight. This is because a given location is on the Earth's trailing face before midnight and meteoroids have to 'catch us up', so to speak. After midnight, the location is on the Earth's leading face and we meet them 'head on'.

Showers vary considerably in character. Some, like the Perseids, produce nearly identical displays each year. Others, like the *Leonids* (mid-November), produce just a trickle of meteors most years but every once in a while put on a real celestial fireworks display. Some showers, like the *Taurids* of late October through to late November, produce slow moving meteors. Others, such as the late April *Lyrids*, produce meteors which rapidly dash across the heavens. There are many showers, some rich but others consisting of just a few meteors per hour at best. Full details can be found in the various ephemerides, such as the annual *Handbook of the British Astronomical Association*. Some meteors will not seem to belong to any particular shower. These are termed *sporadic meteors*.

Visually recording meteors

The only optical equipment needed is the unaided eye. The procedure is to recline on a sun-bed or deck-chair and simply watch the sky. It is only sensible to make oneself as comfortable as possible. Warm clothing and a woollen hat are a must for cold winter evenings. Perhaps even a thick blanket or a sleeping bag is best. Getting too cold is dangerous. After passing through the uncomfortable stage, hypothermia tends to creep up unnoticed. The end result could be tragic – **please do be careful**.

You should make a note of each meteor you see. Using a tape recorder will avoid you having to take your eyes away from the skies. However, do warn your neighbours. If they see you lying on a sun-bed in the garden on a cold and frosty night, apparently chatting merrily to some unseen entity, you might just be faced with having to explain your actions to visiting officials!

For each meteor you see record the time (UT is always preferred), type of meteor (whether it is a shower meteor – and if so which shower, or a sporadic – determined by noting its direction), the visible magnitude at its brightest (comparing with the background stars to aid you in making an estimate to the nearest half magnitude if possible), the altitude (or range of altitude) of the meteor, and any other pertinent details (visible description where appropriate). Record any of these quantities as a questionmark if you are not sure, rather than giving erroneous information. If the shower is so rich that you cannot record details of the individual meteors, then stick to recording the total number of shower and sporadic meteors seen within specified times (usually 15 or 30 minutes).

As soon as is convenient, this information should be tabulated for sending to one of the meteor watching groups. The other information to be included in your report is the date and the exact time (to the nearest minute) when you started watching, which shouldn't be until after a period of at least 10 minutes in order to become dark-adapted, and the start and stop times of any breaks you took. The exact time of ending the observing session should also be recorded. Another important detail is the limiting magnitude of the sky, which can be ascertained by noting the faintest stars which can just be seen. There is another method of finding the limiting magnitude. This is by counting the numbers of stars visible in certain triangular portions of the sky, defined for meteor observers for this very purpose. If your group uses this system, the coordinator will supply you with all the information you need.

Some observing groups will supply their contributors with *gnomonic-projection* star maps. This type of map projection has all great circles, such

Figure 10.10. A bright Perseid meteor captured by Nick James with a fixed camera fitted with a 50 mm lens. The film was Tri-X processed in D19 developer. Several small explosions are evident in this photograph, before this meteor reached magnitude -6^{m} in its terminal burst. Several of the bright stars of Cassiopeia are also visible. Their trails appear broken because of occasional interuptions by passing clouds during the 16 minute exposure, taken on 1980 August $12^{d}\ 02^{h}\ 50^{m}$ UT.

as the Celestial Equator and the lines of right ascension, as straight lines. The observer can then plot the paths of meteors on these maps, noting the beginning and end of each meteor trail.

Photographing meteors

Setting up a camera on a tripod is good enough for meteor photography. The camera should be fitted with a standard lens, set fully open, and a 'B-setting' exposure given. If the film is very 'fast' it will record fairly faint meteors but the maximum time of exposure can only be a few minutes, otherwise the film will fog due to sky-glow. A slow film can be exposed for longer but will fail to record fainter meteors. Probably a film of speed about 400 ISO is best and the exposure limited to about 15–20 minutes, if the lens is open to about *f*/2.8. Remember always to record the exact time of any meteor that occurs during the exposure if you want your photograph to be really useful. Figure 10.10 shows a bright meteor Nick James captured using an ordinary camera.

If money is no object, one can have several cameras pointing at different parts of the sky, so covering a much wider field of view. Otherwise, set your camera pointing directly at the zenith or inclined a little towards the shower's radiant. A rotating shutter mounted in front of the camera lens will allow the angular speed of the meteor to be determined by measuring the trail lengths between the breaks. The rotating shutter is very simple to make. It consists of a disc with sections cut out, mounted on a central spindle and driven by an electric motor. If the motor is synchronous, the number of times the camera lens is uncovered per second can be easily determined. If not then some other means will have to be found. An infrared emitter and detector set up either side of the disc and driving suitable pulse-counting electronics will do.

If another observer, several tens of kilometres away, records the same meteor photographically, then its precise tragectory through the Earth's atmosphere can be determined. The method is to use triangulation, based on measurements of the apparent positions of the meteor against the star background. If one of you is using a rotating shutter then its actual speed can also be determined.

Finding the ZHR from your own observation

The predicted *zenith hourly rate* (ZHR) of the shower is the number of meteors that would be observed by a person whose eyes are fully dark-adapted on a night when stars of magnitude $6^{m}.5$ are visible, with the radiant at the zenith. It would be rare indeed to have all these conditions met.

If the actual altitude of the radiant is A, then the predicted rate must be multiplied by $\sin(A+6)$, A being measured in degrees. The resulting figure x should be inserted in: $\text{ZHR} = x^{(6.5-M)}$, where M is the actual limiting magnitude of the sky. If you wish to correct your own observed rate of shower meteors to the zenith hourly rate, divide your figure by the two factors previously given.

Meteor spectrography is covered in Chapter 15 and the radio observation of meteors is dealt with in Chapter 16.

Observing aurorae

The night of 1989 March 13^{d} began ordinarily enough. I was busy at the Royal Greenwich Observatory, using the 30-inch coudé reflector there for lunar research. At 19^{h} 19^{m} UT I left the dome and immediately noticed a peculiar glow in the northern sky: which looked like faint beams from a row of searchlights extended upwards, fanning out from the horizon. I had

Figure 10.11. The brilliant aurora of 1989 March 13^{d}, photographed by Peter Strugnell. His camera was loaded with Kodacolor Gold 100 film and was fitted with a 35 mm wide-angle lens, set at *f*/2.4. The exposure was 30 seconds. This display was particularly bright, though for most auroral photography a faster film is needed.

observed the Sun that day and had seen an enormous and complex sunspot group. It was obvious what was happening.

I ran to alert the only other observer there, Peter Strugnell, who was using the satellite-tracking camera in an adjacent dome. We climbed out onto the roof of the adjoining buildings. We watched the streamers glow and then die away. I did not have my camera handy. Even if I had, I had no colour film. Peter did. He set up and photographed as we watched an amazing auroral display. Brilliant patches of cyan light growing into vast curtain-like streamers extending over the zenith and becoming a rich vibrant red. Later vivid arches of light, reds and greens dominating, with tints of yellows and orange.

At one point the whole sky became a vivid blood red colour. Later a white patch developed at the zenith and then broke into moving ripples, reminiscent of waves lapping at the sea shore. About the same time darts of light, like tickertape streamers, criss-crossed in the southern part of the sky. During that evening Peter and I had several visitors and the telephone lines to the observatory were virtually jammed. We had witnessed a once in a lifetime auroral display. Yet it is rare to see the aurora at all from a latitude of 51°N!

Figure 10.11 shows one of the photographs Peter Strugnell took that night. All he had in the camera was 100 ISO colour negative film. We decided that 30 second exposures would be required, with the 35 mm wide-angle lens set wide open at *f*/2.4 and this estimate proved to be about right. Ideally a very fast (400 ISO or more) colour film would have been preferred, with exposures of ten seconds or less, so that the aurora's structure was not smeared too much by its movements and changes.

Monitoring changes in the Earth's magnetic field during an aurora is interesting. It may also give an early warning of an event that is about to happen. A simple device to do this, the *jam-jar magnetometer*, is described in the October 1989 issue of *Sky & Telescope*. A more sophisticated device is the *flux-gate magnetometer*, described in the February 1984 issue of the *Journal of the British Astronomical Association*. Visual observing consists of a description of the appearance of the display at given times. Virtually the only clearing house for observations with professional links is the auroral observing section of the *British Astronomical Association*. I would strongly advise any reader interested in auroral observing to contact this group.

11
The Sun

It is true that the Sun is monitored daily by several solar observatories, using specialised equipment largely beyond the resources of the amateur. However, there are significant gaps in the professionals' monitoring programmes; especially so in the current climate of budget cuts. It could just be that you might watch the unfolding of an eruption on the Sun's surface at a time when no professional telescope is trained on our daytime star!

Methods of viewing the solar image

Sorry, I must say it: **On no account look through any ordinary telescope, binoculars, camera viewfinder, or any other optical equipment not specially designed for the purpose, which is pointed at the Sun. Also, do not be tempted to use one of the dark filters which screw into the eyepiece barrel – they may well be supplied by the manufacturer of the telescope but they are NOT safe**. I mean no disrespect to you, the reader, for throwing such an elementary warning in your direction. It is a sad fact that several people each year damage their eyesight by taking risks with the Sun. Why? I don't know. There are plenty of published warnings, when even common sense should be sufficient – and yet it still happens. If my warning rescues just one person from the temptation of having a peek, then I am sure you will agree that it is well worth including.

The classic safe way of observing the Sun is by projecting it's image onto a screen (Figure 11.1). A shield can keep the direct light off the screen. Better still, the screen can be enclosed in a box which has a large hole in its side to allow viewing of the image. If your telescope is of large aperture, stop it down to 6-inches (152 mm) at most. Otherwise the instrument will collect so much heat radiation that the optics could be damaged. Provided

(a)

Figure 11.1. (a) [above] and (b) [right]. Projecting the Sun. The telescope shown is the author's 6¼-inch (158 mm) Newtonian reflector.

no additional 'star diagonal' is used, the orientation of the projected image is with north uppermost and west (the direction of drift of the image if the telescope is undriven) to the left.

I must admit that I used an orthoscopic eyepiece when I used to project the Sun's image with my 6¼-inch Newtonian reflector because it delivered sharper and better colour corrected images than my Ramsden or Huygenian eyepieces. However, on the 'don't do as I do, do as I say' principle, I would caution against using complex eyepieces. The concentrated heat might well damage the optical cement between the lens elements (and differential expansion is more likely to crack one of the lenses). Also, the simpler eyepieces are much cheaper to replace should damage occur. Stopping the telescope down so that it has an effective focal ratio of greater than *f*/10 will allow the simpler eyepieces to deliver crisp projected images. In any case I would recommend using an eyepiece of focal length at least 18 mm to do the projecting in order that the concentrated heat at the focal plane is as far from the lens elements as possible.

One filter which is safe, **if properly made and fitted**, is a full-aperture 'Mylar' filter. It fits onto the front end of the telescope tube (which has to be light-tight. You cannot use this method with open-tubed telescopes). These filters are commercially manufactured by telescope suppliers (see Chapter 17). You can make your own but do make sure that you use the correct material. It should be 12 μm thick polyester, metallised with aluminium on both sides. Don't attempt to use anything else, even if it looks about the same. You can't be sure what the transmission characteristics are for the

(a)

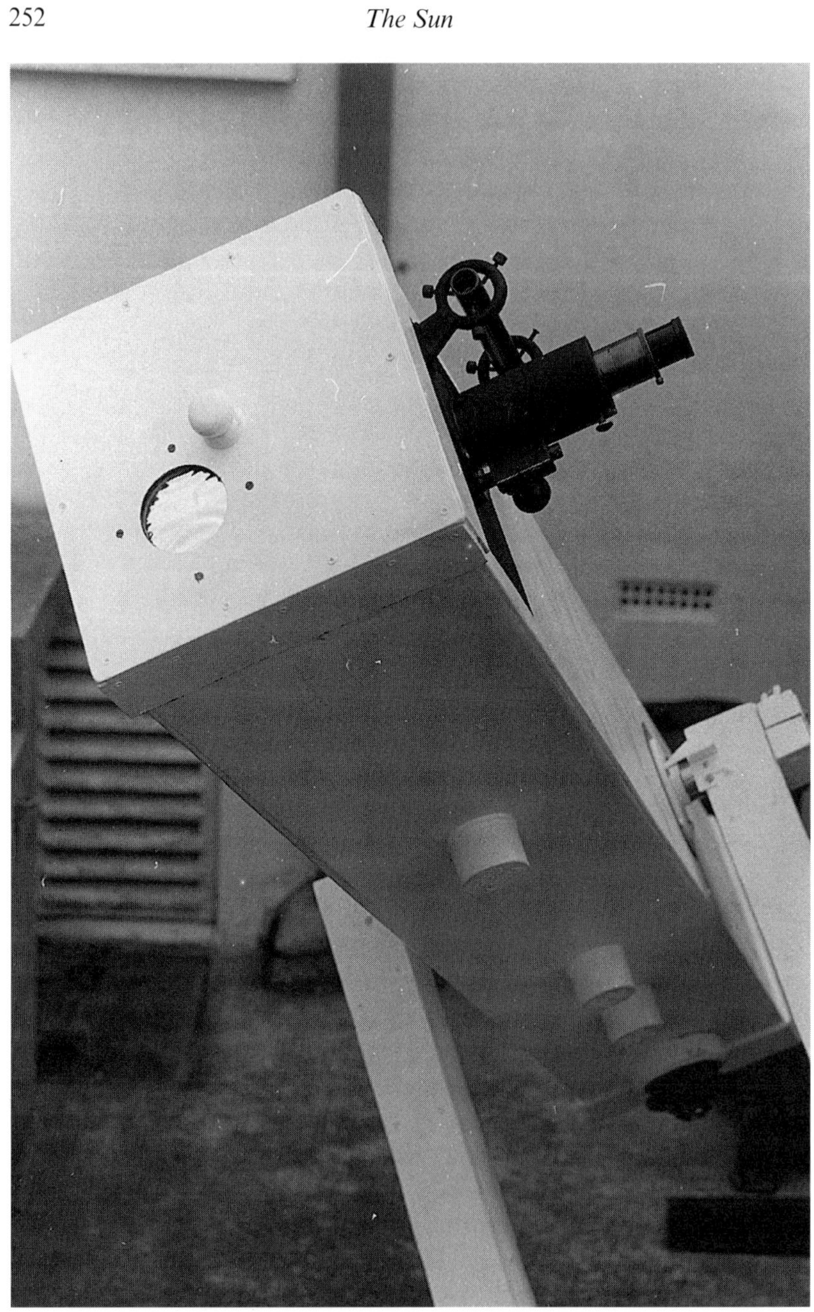

Figure 11.2. (a) [above] A Mylar filter is mounted in the lid that closes the tube of the author's 8½-inch (216 mm) Newtonian reflector for solar observation. A 62 mm off-axis aperture was considered adequate, given the daytime seeing conditions usually experienced. (b) [right] Showing how the filter is securely mounted inside the lid.

(b)

infrared and ultraviolet wavelengths. 12 μm 'Mylar' is safe. You can purchase this material in rolls.

To mount a filter cut a ring of cardboard to the correct size of the cell for fitting to the end of the telescope tube. Cover one side of this ring in adhesive (or double-sided adhesive tape). Then unroll some of the Mylar onto a clean table surface. Lightly drop the ring, adhesive side down, onto the Mylar. When the adhesive has set, the Mylar can be trimmed to the edge of the ring. Inspect the filter to make sure that there are no pinholes or other imperfections in it. The filter should be slightly slack and wavy in its cell. If all is O.K., then the ring is set into the mounting cell.

Do make sure that it is properly secured. Don't risk it falling off! Unless you can be absolutely sure that you have mounted the filter securely, don't use this method. Stick to projecting the Sun's image, instead. I now do my own solar observing using a Mylar filter attached to one of my telescopes (see Figure 11.2).

Mylar transmits more blue light than red, so the Sun's image looks bluish when seen through a Mylar filter. If this is considered undesirable, a yellow filter can be added at the eyepiece. Actually, an additional filter will be found helpful when using magnifications of less than about ×8 per inch (×3.2 per cm) of aperture, in order to reduce the image brightness to a comfortable level. The effects of sky-haze are also reduced, and contrast consequently improved, using a yellow, orange, or red filter.

A device known as a *Herschel wedge* was, at one time, popular with solar observers. It is basically a star diagonal with an unsilvered (and wedge shaped – to avoid problems with reflections from it's rear face) mirror. The back of the mirror is left open to let most of the heat and light pass straight through. Even so, about 4 per cent of the radiation is still sent to the eyepiece and a very dark filter (incorporating a heat-rejection filter) is still required. I regard Herchel wedges as safe only when used on small telescopes (say, less than 4-inch aperture) and many observers have had accidents with them because of the heat streaming out of the back of the device. They also mirror-reverse and change the orientation of the image. These devices are now little used – probably a good thing!

A number of amateurs have contructed a more specialised type of solar telescope by using a full aperture metallised glass filter to double as a Newtonian secondary mirror (see Figure 11.3). The filter is coated on both sides to reduce the incident radiation to less than 0.2 per cent of its intensity. There are several advantages to this design of telescope. One is that should any disaster befall the secondary mirror/filter the light going to the eyepiece is quite likely to be cut off straight away. Another is that the light path is free from the secondary obstructions inherent in most other types of telescope. So this telescope is capable of providing maximum contrast images. Also most of the Sun's heat radiation is kept outside the telescope and the optical system is totally enclosed. Both these factors tend to restrict the tube currents which can be a problem in 'open' systems, such as the Newtonian reflector used for projection.

The only real drawback of this type of telescope is its cost. The diagonal glass plate has to be made of 'lens quality' optical glass and both its surfaces have to be figured to be optically flat and parallel to one another. A much cheaper telescope could be made by leaving the primary mirror of a standard Newtonian reflector uncoated (in which condition it will reflect 4 per cent of the incident light). The secondary mirror should also be left uncoated. Little of the Sun's light and heat should then reach the eyepiece, though a heat-rejection filter, and possibly a dark filter, may still be needed.

Although I must give a mention to *coelostats*, where a driven plane mirror directs the sunlight into a fixed and (usually) horizontal telescope, very few amateurs use these for making white light observations of the Sun.

Photographing the Sun

The basics of astrophotography are covered in Chapters 4 and 5 in this book. I recommend reading those chapters first. Add chapter 6 to your reading list if you wish to do electronic imaging. The main thing to be careful of is the Sun's intense radiant heat. Telescopes and cameras can be easily damaged. Projecting the solar image onto a screen is the safest

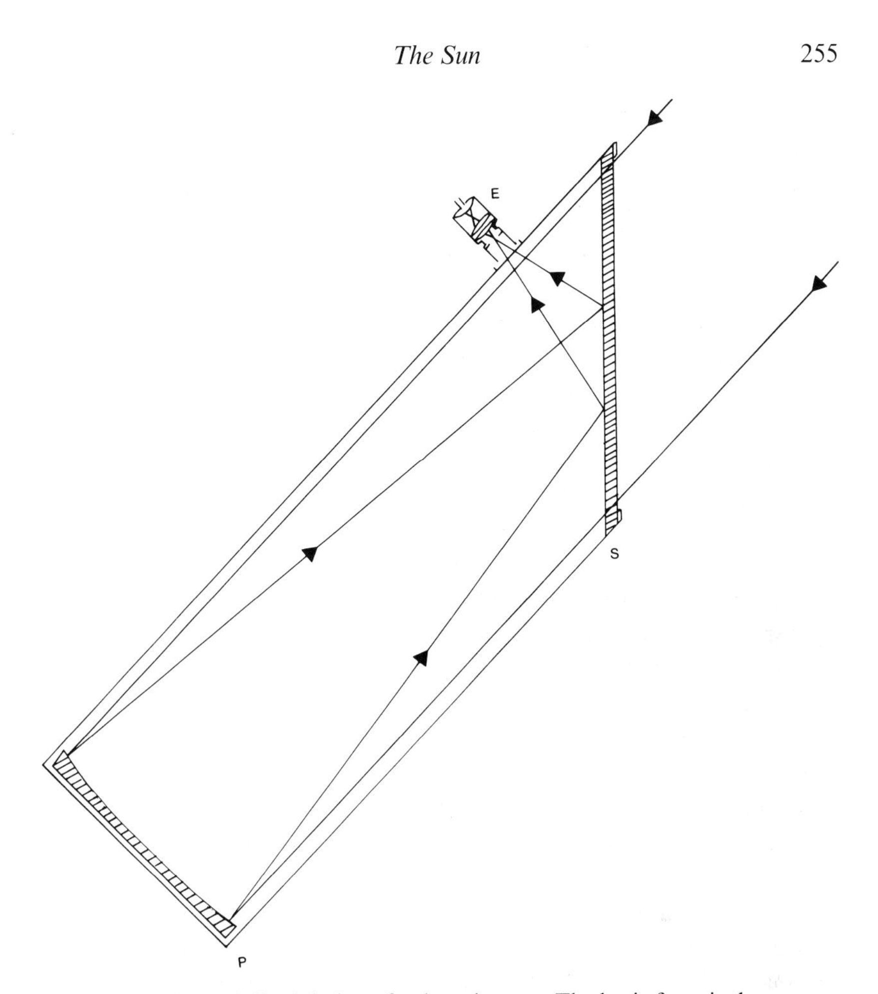

Figure 11.3. A specialised design of solar telescope. The basic form is the same as that of the Newtonian reflector. The primary mirror, P, is aluminised in the usual way but a full-aperture plane parallel glass plate is placed at the end of the tube and at 45° to the incoming light. Both sides of the glass plate are aluminised to admit the minimum of light and heat into the telescope.The underside also functions as the Newtonian secondary mirror.

method of visual observation. It is also the safest way of photographing the Sun. The lengths of exposures given are a little less than that required for ordinary photography of the sunlit scene. Hence a drive, while convenient, is certainly not necessary.

Figure 11.4 shows an excellent example obtained by Martin Mobberley. He used a 5-inch (127 mm) *f*/14 refractor (the guiding telescope on his 14-inch reflector) to project the image onto a white card. He then photographed the card using a 50 mm 'macro' lens. The high contrast of the

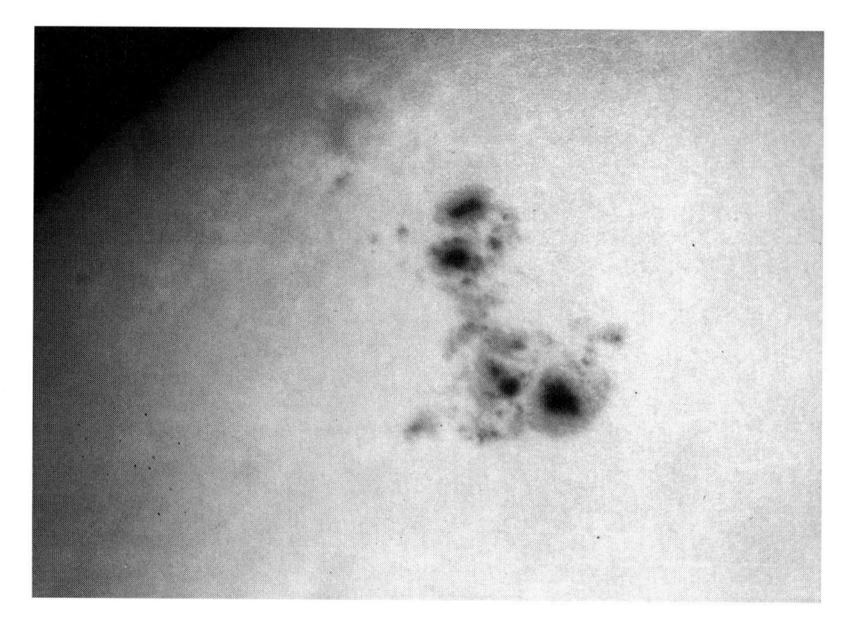

Figure 11.4. Giant sunspot group, photographed on 1989 June 11^{d} 13^{h} 30^{m} UT by Martin Mobberley. He used a 5-inch (127 mm) *f*/14 refractor to project the solar image onto a card and photographed the card using a 50 mm 'macro' lens. 1/125 second exposure on Kodak Technical Pan 2415 film. As is usual with solar projection (the telescope being in the Earth's northern hemisphere, and no eyepiece diagonal used) north is uppermost and west (the direction of the diurnal drift) is to the left.

TP2415 film he used has allowed Martin to produce a print which shows the 'sky' (actually the projection screen) background as virtually black. Obviously prints should be made on the highest contrast paper. Details on the photographed image then stand out with maximum contrast.

I would recommend using an eyepiece of long focal length to do the projection for a reason additional to the one given earlier. The required size of the solar image (15 cm to 30 cm) is then achieved with the projection screen at a fairly large distance from the eyepiece. This allows one to position a camera close to the eyepiece and so photograph the Sun 'straight on' without the distortion involved in an angled shot.

On the subject of distortion, I would not recommend trying to restore the image to circularity by tilting the projection screen. Doing so would cause the image to be thrown out of focus everywhere except for one particular strip across the image. It is better to do this while printing. If the enlarger lens is stopped down the easel can be tilted enough to make a slightly oval view of the Sun circular while still maintaining sharp focus across the image. Before inserting the photographic paper, use a plain sheet of paper with a compass-drawn circle of the required size to set the tilt. When correct, the Sun's image will exactly fill the circle.

Of course, taking the photographs direct through the telescope avoids this type of distortion. Once again, I would urge against using a dark filter mounted close to the focal plane. A Mylar filter mounted at the end of the telescope tube is certainly the best of the cheap solutions (coated optical end windows may be better if accurately made, though they are much more expensive). The procedure is the same as for photographing the Moon and planets, except that the highest enlargement factors are seldom used. Contrast may be improved by using a yellow, orange, or red filter, as for visual observation.

If desired, the whole of the solar disc can be fitted onto a frame of 135 format (35 mm) film provided the effective focal length is no more than 2.5 m. Probably the best film to use is Kodak's TP2415. Figure 11.5 shows a photograph taken by Nick James using this method. The image is remarkably detailed considering that the 50 mm refractor he used was effectively stopped down to 35 mm!

Whether for photography, or for visual observation, conditions are usually best in the morning. At most sites, the atmosphere becomes increasingly turbulent as the day progresses.

'White light' features and nomenclature

The term 'white light' features denotes those observable in integrated light. Monochromatic observations are covered later in this chapter.

Even when the Sun's surface is devoid of transient features, the *limb darkening* is always visible. Its cause lies in the fact that the outermost tenuous layers of the Sun are partially transparent. When we look directly at the centre of the Sun's visible disc we can see down to a level where the temperature is about 5800 K. However, when we look at a region close to the limb our line-of-sight ploughs into the Sun's outer layers at an angle and we would require a much longer path length in order to be able to see down as deep as before. The result is that we now can only see down to a level where the temperature isn't so high. The radiation from this level is redder and less intense: hence the limb darkening, which Figure 11.4 shows especially well.

The Sun's surface is seething with convective motion. When viewed with sub-arcsecond resolution the solar surface is seen to be covered in *granules*. These are the tops of rising columns of gas. The dark gaps in between are the regions where the gas is falling once again into the deeper layers.

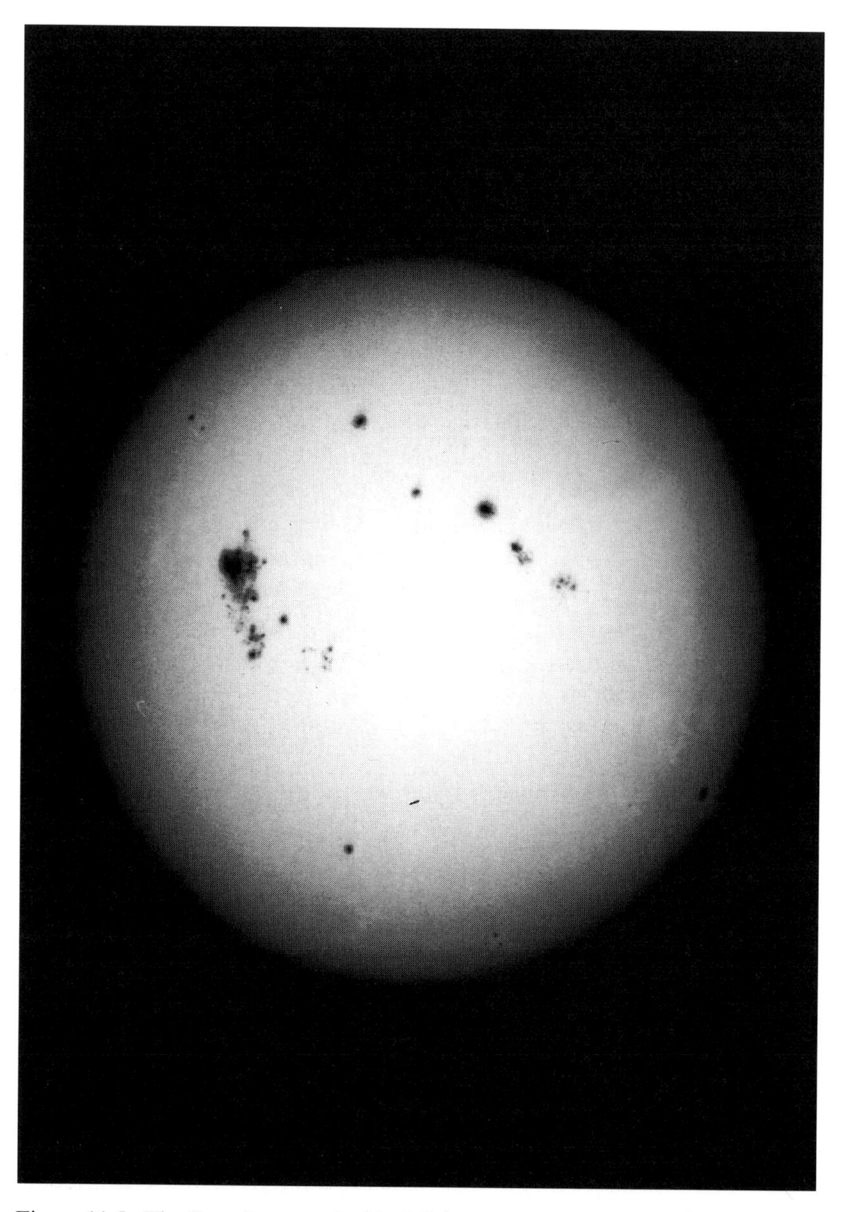

Figure 11.5. The Sun photographed by Nick James on 1989 June 16^{d} 06^{h} 34^{m} UT. The telescope used was a 50 mm refractor, stopped down to 35 mm, with a full-aperture Mylar filter. The image was projected to *f*/70, using a 25 mm eyepiece. The 1/500 second exposure was made on Kodak's Technical Pan 2415 film, processed for 3 minutes in D19 developer at 20°C. The large spot group visible produced a number of flares. South is approximately uppermost in this telescopically normal view.

The rest of the observable features on the Sun result from interactions of its powerful and complex magnetic field with the hot plasma. The most obvious are the *sunspots*, regions where the Sun's magnetic field loops out of the surface. By a coupling mechanism that is still imprecisely understood, energy is conducted away from this region to leave a patch of the *photosphere* (the Sun's visible surface) at a lower temperature than average. Sunspots usually consist of a dark region, the *umbra*, where the temperature is about 4000 K, surrounded by a lighter *penumbra* where the temperature is around 5500 K. Sunspots can range from tiny *pores* to vast and complex *sunspot groups*. This range is well shown in Figure 11.5.

White cloud-like features, known as *faculae*, are dotted over the surface of the Sun. They are regions of higher temperature than the average for the photosphere. Faculae tend to cluster around sunspots and sunspot groups and are often associated with *plages*, clouds of hot material higher up in the chromosphere (the region of the Sun's atmosphere overlaying the photosphere). At low resolutions faculae and plages look similar, though at high resolutions plages look smoother. Both faculae and plages appear to be caused by local intensifications of the Sun's magnetic field.

The equatorial regions of the Sun take about 25 days to complete one rotation but this period increases with latitude, being about 27 days at $+30°$ and $-30°$. It increases further towards the poles. The general development of features on the photosphere and the effects on them of differential rotation make the monitoring of the Sun's surface an especially fascinating study.

Recording solar disc details

Obviously, the method which produces the greatest positional accuracy is photography. However, photography may not always be the best way of recording the **finest** detail. At any given observing site atmospheric turbulence is usually more severe in the daytime than at night. Taking multiple exposures, and then selecting just the best for printing, will improve your chances of recording fine details.

Drawing directly on the projected image is also a good way of achieving high positional accuracy. However, this method does require that the telescope and projection board are sufficiently rigid. It is no good if every touch of the pencil shakes the image. I have used this technique myself with the 8-inch (203 mm) refractor of the Jeremiah Horrocks Observatory in Lancashire, England. However, that particular telescope is a mid-nineteenth century refractor of typically massive construction for the period.

Few 'off-the-peg' instruments of today are that rigid. Figure 11.6 shows one of my observations with that telescope recorded on a standard BAA solar section report form (or, observation *blank*).

In order to set the east–west orientation of the blank, pencil lines are drawn to join the marked cardinal points (as shown in Figure 11.6). With the blank mounted on the projection board, the telescope is moved so that the extreme southern or northern limb of the Sun is just in contact with the east–west pencil line (alternatively, a sunspot will do just as well). With the telescope drive switched off the limb of the Sun should just remain in contact with the east–west line as the image tracks across the blank. Adjust the orientation of the projection board until it does so.

When the orientation is correct the telescope is then moved to centre the Sun in the outline of the blank. Of course, the Sun's image should exactly fill the outline. If it does not, then adjust the eyepiece–projection board distance as necessary. The standard size adopted for full-disc drawings is 6-inches (152 mm) to the Sun's diameter. When all is correct you can draw in the details. Draw sunspots as they appear. Faculae and plages can be represented by pencil outlines, as shown in Figure 11.6.

Other details can be added after the observation. Referring again to Figure 11.6, The value of P is the apparent inclination of the polar axis. P is **positive** if the Sun's true north pole appears inclined to the **east**. B_0 is the *heliographic* (measured with reference to the Sun's surface) *latitude* of the centre of the image (following the usual convention for positive and negative). L_0 is the *heliographic longitude* of the centre of the disc. P, B_0 and L_o all vary and their values at the time of the observation can be found in an ephemeris. Using the value of P, the true polar axis of the Sun can be measured off with a protractor and marked on the drawing. It is indicated by the letter P on Figure 11.6.

If the telescope is not sufficiently rigid to enable one to draw directly onto the blank on the projection screen, then another method has to be used. The image is projected, not directly onto the blank, but onto a 6-inch circle covered in faint grid lines. The squares of the grid are identified by numbers and letters drawn along the bottom and up the side of the circle, respectively. For instance, a square near the middle of the grid might be M13, and so on.

An identical circle with heavily marked grid lines is placed underneath the report form and the blank (report form) circle is aligned with the heavily marked circle and grid, the grid then showing through. After the same orientation procedure as before, the projected markings can be copied onto the blank using the numbered and lettered grid squares. For instance, a

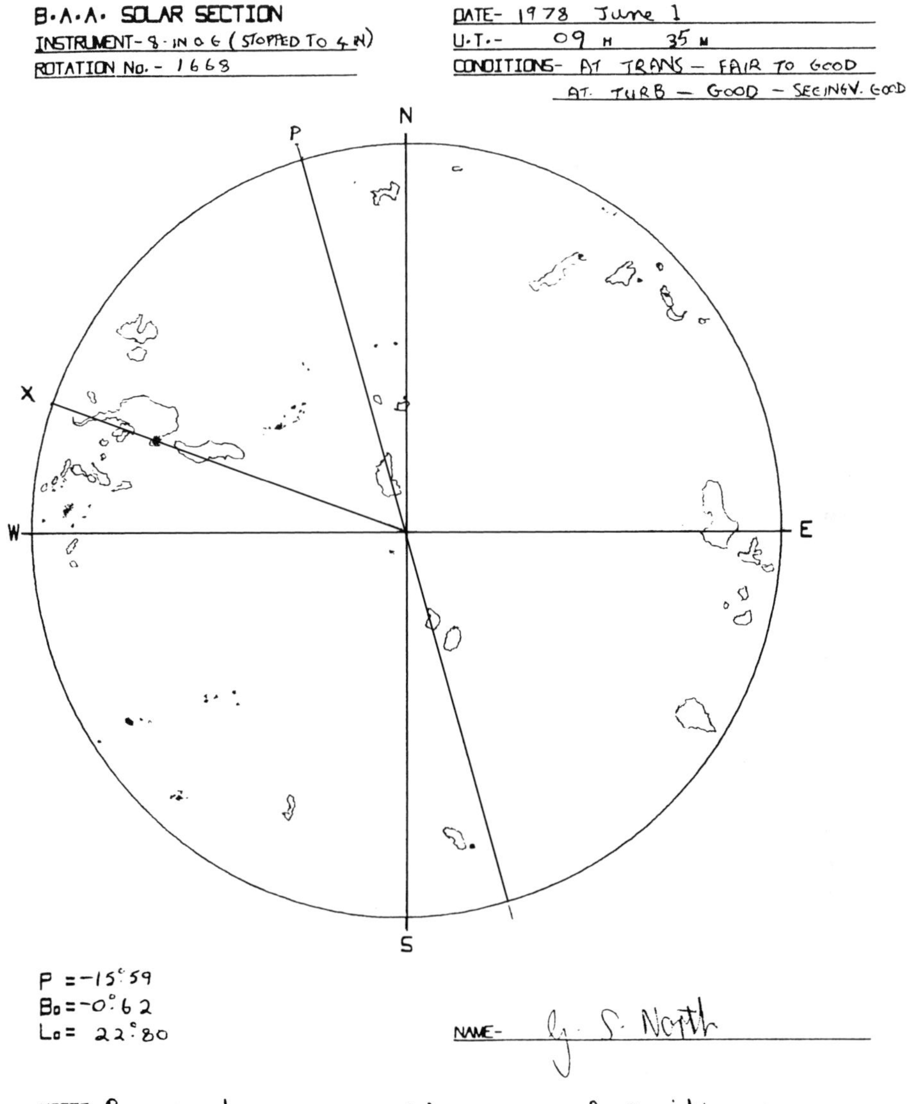

Figure 11.6. Solar observation made by the author, 1978 June 1[d] 09[h] 35[m] UT. The author used the 8-inch (203 mm) refractor of the Jeremiah Horrocks Observatory, stopped to 4-inch aperture, to project the solar image directly onto the blank report form. The details were traced over the solar image. Note the large number of faculae present, as well as several groups of small sunspots and isolated pores. The line marked X was drawn in order to calculate the heliographic latitude and longitude of the large sunspot that lies on it, as described in the text.

sunspot that appears in the middle of square D5 on the projection board is drawn in the same relative position, in the middle of square D5, on the blank.

Those able to observe directly through the telescope (using a full-aperture filter) can usefully make detailed, high-resolution drawings showing the development of sunspots, sunspot groups etc. Obviously some artistic ability is necessary for this. As always, the main requirement is accuracy.

Measuring the latitudes and longitudes of features on the Sun's surface

The easiest way of finding a feature's heliographic longitude and latitude is by using a specially prepared grid on a transparent sheet to overlay your drawing (or your photograph of the full disc). There are a number of variants. *Porter's disc* is a single grid which is used in conjunction with an auxiliary table. The other commonly used graticules: the *Thompson/Ramaut/ Ball discs*, the *Stonyhurst discs* and the *Ravenstone discs* each come in a pack of eight to cover the values of B_0 from 0° to 7° (and −7°). The observer selects the correct grid for the drawing corresponding to the value of B_0 at the time of the observation. Your national astronomical society solar observing section coordinator ought to be able to supply you with that particular observing group's preferred graticules, or at least tell you where to get them.

Solar graticules are certainly quick and easy to use but in my experience the highest accuracy is achieved by direct measurement and geometric calculation. As an example, let us find the heliographic coordinates of the largest sunspot visible on the Sun's disc in Figure 11.6.

The first step is to draw a line from the centre of the disc, through the sunspot, and to the edge. On Figure 11.6 this line is marked X. By measuring along this line the angle, R, between the sunspot and the apparent centre of the Sun, as subtended from the Earth, is found from:

$$R = Sr/r_o,$$

where S is the apparent angular radius of the Sun, as found from an ephemeris for the day of the observation, r_o is the measured radius of the Sun in millimetres (usually 76 mm if the Sun's image was projected to the standard size), and r is the measured distance of the sunspot from the apparent centre of the disc, in millimetres. A protractor is used to measure the position angle, θ, of the sunspot. θ is measured from the north point of the disc, through the east. Figure 11.7 should make all this clear. Our chosen sunspot has a measured position angle of 295°.5.

The value of S on 1987 June 1^d was 15′ 48″ (= 0°.2633) and from my original drawing $r = 53.5$ mm and $r_o = 76$ mm. This gives a value of R of 0°.1853.

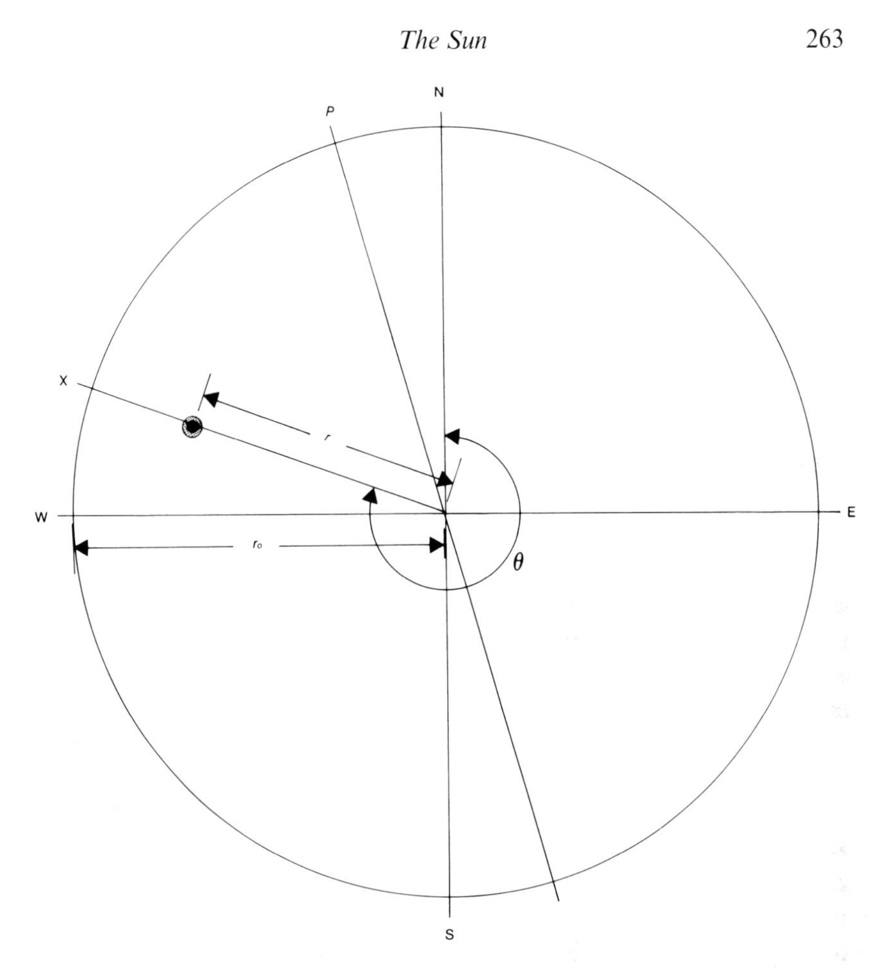

Figure 11.7. The construction necessary to calculate the heliographic latitude and longitude of the sunspot on the line X. The procedure is described in the text.

We must now calculate the angle, ρ, between the direction of the sunspot and the direction of the Earth, as seen from the centre of the Sun. We use the relation:

$$\sin\,(R+\rho)=r/r_o.$$

Putting in the figures given previously, this becomes:

$$\sin\,(0^\circ.1853+\rho)=0.7039.$$

Using the arcsine function on the calculator (or tables):

$$0^\circ.1853+\rho=44^\circ.74,$$

from which $\rho=44^\circ.56$.

At last we are now ready to calculate the heliographic latitude and longitude of our chosen sunspot.

The heliographic latitude, B, is given by:

$$\sin B = \sin B_0 \cos \rho + \cos B_0 \sin \rho \cos (P - \theta).$$

For our sunspot the value of B comes out to be 23°.3.

The heliographic longitude, L, is given by:

$$\sin (L - L_0) = \sin \rho \sin (P - \theta) / \cos B$$

If you put in the figures you will find that the heliographic longitude of our chosen sunspot is 61°.5.

The nineteenth century astronomer Richard Carrington invented a method for measuring heliographic coordinates that does not necessarily involve projecting or photographing the solar image. In this method the observer uses a crosswire eyepiece to directly view the safely filtered (see earlier) image of the Sun through an undriven telescope and time the passages of the Sun's limb and the feature(s) across the diagonally inclined crosswires. The method is just as suitable for use with an altazimuth telescope as for an equatorially mounted one and apart from the crosswire eyepiece, the only other piece of equipment needed is a split-time stopwatch.

I must straight away declare that I had never heard of Carrington's method until I read a paper by E. T. H Teague in the April 1996 issue of *The Journal of the British Astronomical Association*. In it Mr Teague describes the slightly involved reduction procedure and he points out the advantages to the modern observer who prefers to observe the Sun directly, rather than projecting its image. Unfortunately lack of space does not allow me to include the details here but I commend those interested to read Mr Teague's excellent paper.

Mr Teague also points out that with the use of a programmable calculator or a personal computer the modern amateur can execute all the reductions in a matter of moments. Of course, this is also the case for the other method and, in fact, is also true for any of the mathematical reduction procedures given in this book.

Measures of solar activity

As a long-term study, the latitudes of individual sunspots can be plotted against time (in years) on a graph to produce the characteristic *butterfly diagram* that will be familiar to readers. A typical solar-cycle may last around 11 years but there can be large variations.

Another measure of solar activity is the MDF, or *Mean Daily Frequency*. This is calculated by counting the number of *active areas* on the solar disc. The total number for any month is divided by the number of days during that month for which observations were made. Hence, a value of the MDF can be calculated for each month. Plotting the MDF against time (in years) will reveal how the activity level changes throughout the solar cycle. However, there is one snag. What counts as an active area? Every sunspot and pore, even the smallest, counts as an active area. However, that spot must be at least 10 degrees of latitude or longitude from its nearest neighbour in order to be counted as separate.

For those of you who might be interested in more detailed studies of sunspot behaviour, the December 1980 issue of the *Journal of the British Astronomical Association* carries a paper on the 'Measurement of areas on the solar disc'. Also, the November 1988 issue of *Sky & Telescope* magazine has an article 'Watching the premier star' in which one of the authors, Patrick McIntosh, describes his sunspot group classification scheme. McIntosh's scheme, introduced in 1966, has become an accepted standard in solar research.

Observing the Sun at selected wavelengths

Though most amateur astronomers are put off by the large cost involved, a few do invest in the equipment necessary to view the solar disc at certain specific wavelengths, especially the Hα line at a wavelength of 656.3 nm (6563Å, or 6.563×10^{-7} m) in the red part of the spectrum.

Solar granulation, plages and faculae are much more evident when seen in Hα light, as are *solar flares*, huge releases of energy (see Figure 11.8). The largest solar flares can radiate about as much as a thousandth of the Sun's total output. The major events last only 20 minutes or so but during this time vast quantities of electrified particles are sprayed into space. Some may give rise to auroral storms here on Earth. Occasionally solar flares are visible in integrated ('white') light as small bright patches on the photosphere.

In addition to limiting the Sun's light to Hα, if an occulting disk is used to block off the overpowering light from the Sun's photosphere, then a thin pink band can be seen showing round the edge of the occulting disc. This is the *chromosphere*, the 'lower atmosphere' of the Sun, overlaying the photosphere. Also *prominences* can usually be seen as huge, flame-like, protrusions of gas arching above the chromosphere and rooted in it.

Prominences can be of two forms. *Quiescent prominences* can survive

(a)

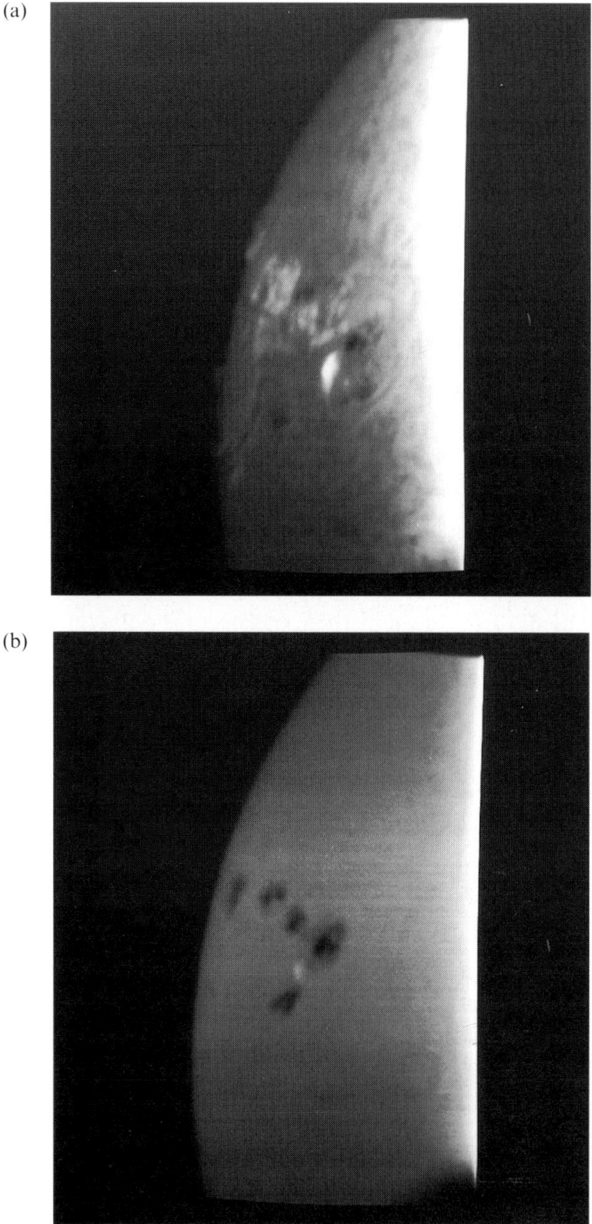

(b)

Figure 11.8. Commander (retd.) Henry Hatfield took this sequence of photographs showing the development of a solar flare on 1989 August 31^{d}. He used his spectrohelioscope tuned to Hα for all the photographs except (b) for which the instrument was set a little off Hα. All the photographs were 4 second exposures taken on Kodak SO115 film. North is uppermost and west to the right in all of these views. (a) [above, top] 16^{h} 08^{m} UT. (b) [above, bottom] 16^{h} 10^{m} UT. (c) [opposite, top] 16^{h} 12^{m} UT. (d) [opposite, bottom] 16^{h} 17^{m} UT. (e) [overleaf] 16^{h} 19^{m} UT. Note that a new flare can be seen starting to the north-east on the last image.

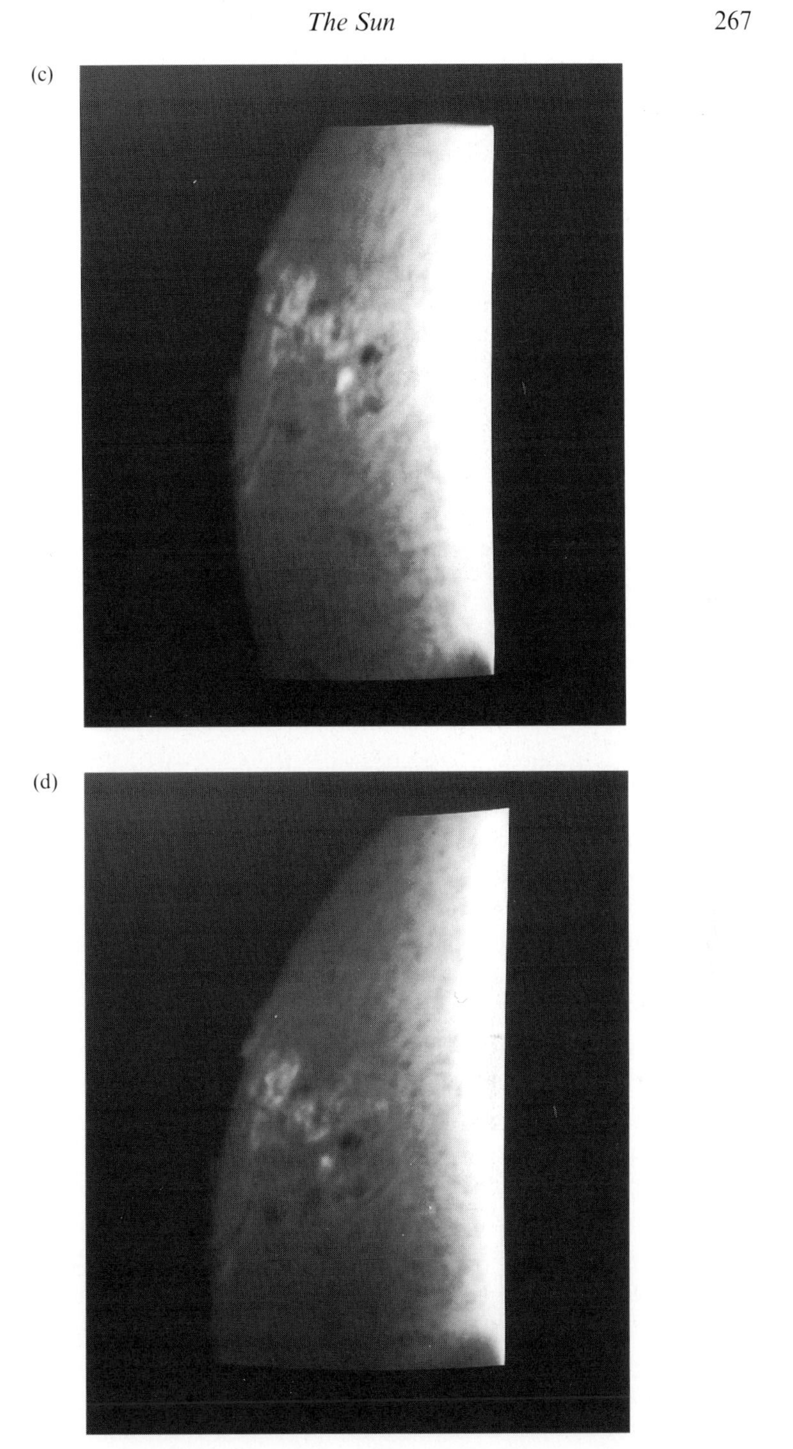
(c)
(d)

(e)

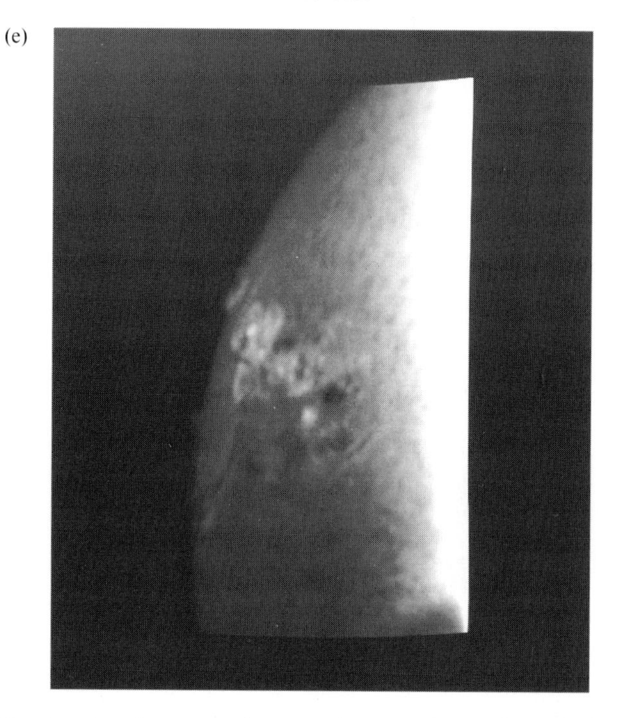

with little structural change for months at a time. Though they give the impression of upward eruptions of gas surging from the chromosphere, they are really gas condensing from the Sun's tenuous 'outer atmosphere', or *corona*. *Active prominences* are much shorter-lived and they are usually smaller than the quiescent variety. They can change in timescales of minutes. Some consist of condensing material falling back to the Sun but others are true upward surges of thousands of millions of tonnes of chromospheric gas. Without the occulting disc, dark *filaments* can be seen projected against the photosphere. These are just prominences seen as dark against the brighter background (see Figure 11.9).

One way of observing prominences is to use a *solar spectroscope*. Spectroscopes are covered in Chapter 15. Suffice it to say here that a spectroscope used for this type of work should have a dispersion high enough to easily split the twin sodium D lines. The spectrograph is mounted on the telescope with its slit coincident with the focal plane. Probably the best type of telescope for this work is a refractor of about 4-inches (102 mm) aperture and at least 60-inches (1.5 m) focal length. Larger apertures will concentrate too much heat on the slit jaws (the outer face of which should be shiny to reflect as much of the heat away as possible) and smaller apertures won't collect enough light to produce a detailed image. Shorter focal lengths will suffer from a small image scale and will be tricky to align precisely on the limb of the Sun.

The procedure is first to set the telescope with the solar image central over the slit. The adjustable slit jaws should be very nearly closed. If in doubt, close them completely and then **carefully** turn the adjusting screw until the spectrum becomes visible. Using the **spectroscope eyepiece** focuser, adjust the focus until the dark Fraunhofer lines are sharp. Adjust the spectroscope until the Hα line is brought into the centre of the field of view.

Next move the telescope until the limb of the Sun is over just half of the slit. In other words, sunlight will be pouring through half of the slit but not through the other half. You will then see half a spectrum though the spectroscope eyepiece. Then adjust the **telescope** focuser until the boundary of the half spectrum (corresponding to the edge of the solar image) is sharp.

The telescope is then moved (and the spectrograph rotated about the optical axis of the telescope as necessary) until the slit is tangential to the edge of the solar disc. This is a delicate and critical operation. I would regard an equatorially mounted and driven telescope, equipped with smooth slow motion controls, as a virtual necessity for this. At the correct position most of the spectrum goes dark and the Hα line stands out as bright. The spectroscope slit is carefully opened. The bright portion of the Sun must, obviously, be kept hidden by one of the slit jaws. If any prominences are visible you will see then as red tongues of flame. Using the same procedure you can search round the rest of the Sun's limb for prominences.

Spectrohelioscopes allow full-disc images of the Sun to be obtained. The general principle of their operation is illustrated in Figure 11.10. A moving slit samples a strip of the solar image, admitting light to a *diffraction grating* (dealt with in Chapter 15). Another slit, moving in step with the first, selects just the light reflected off the grating at an angle which corresponds to the wavelength to be studied (say, Hα). In this way a complete view of the Sun's disc is built up during one traverse of the slit. If the slits are made to oscillate rapidly (say, twenty times every second) then a fairly steady (without too much flicker) image of the Sun can be produced in any given wavelength.

In practice, spectrohelioscopes work best using long focal length lenses to collect and image the light. They are usually set up as fixed horizontal arrangements, fed with sunlight via a coelostat. They are also mechanically complicated and the moving slits have to be precisely made and operated.

(a)

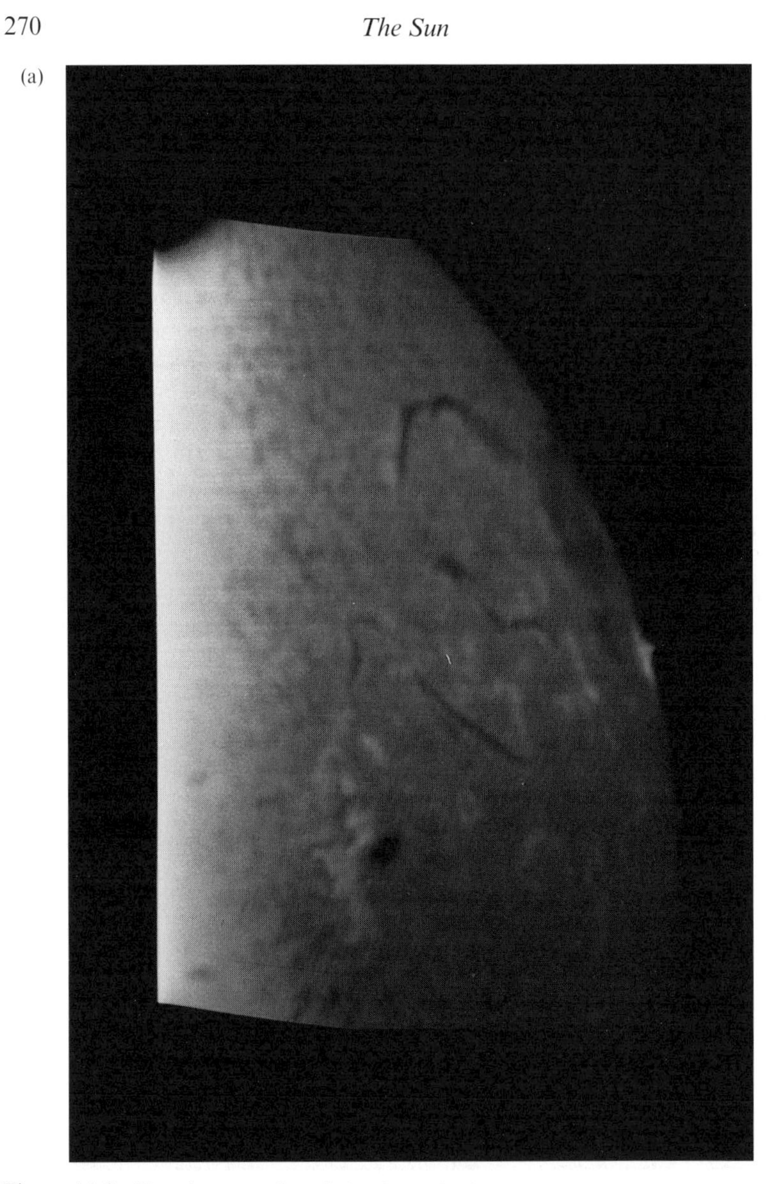

Figure 11.9. Hα photographs of the Sun obtained by Henry Hatfield using his spectrohelioscope, tuned to Hα. The 4 second exposures were taken on Kodak SO115 film on 1989 August 20^{d}. (a) [above] was taken at 14^{h} 07^{m} UT. and (b) [opposite] at 14^{h} 08^{m} UT. North is uppermost and west to the right in both these views.

(b)

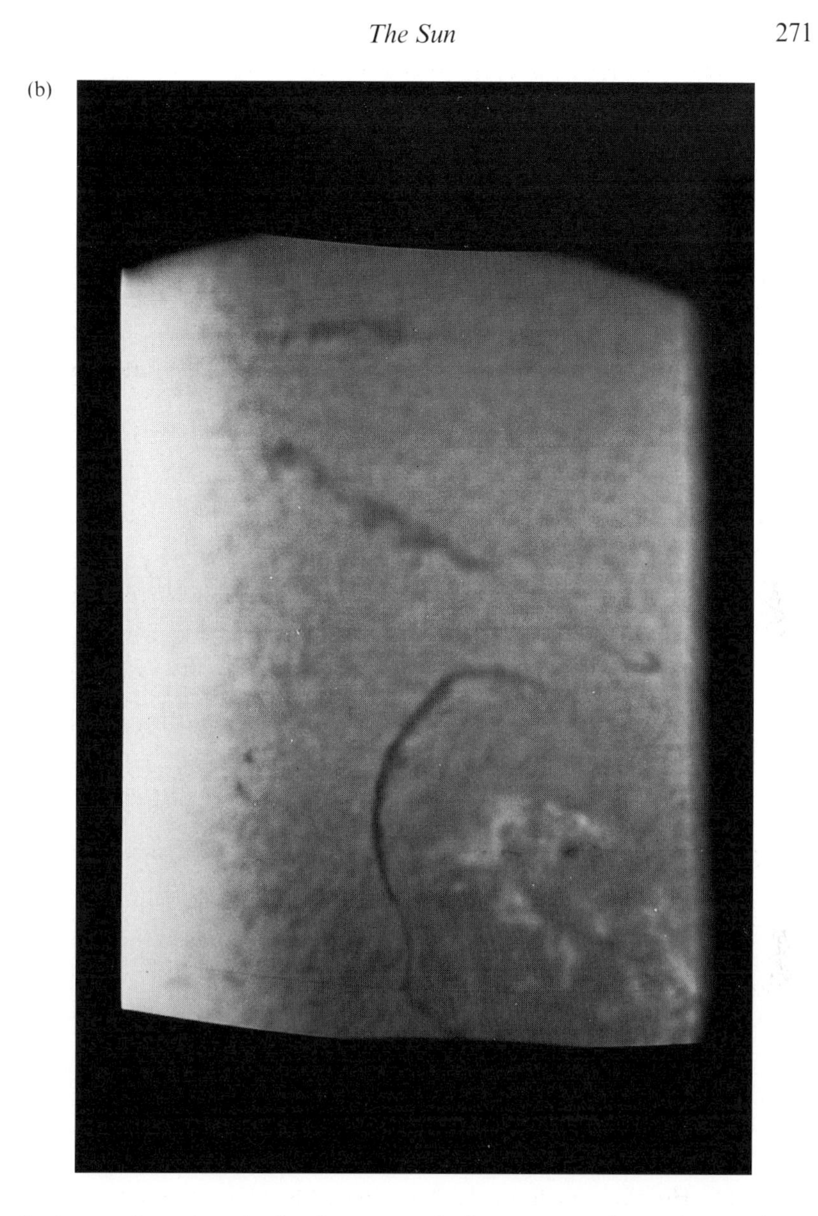

Owing to these drawbacks, few spectrohelioscopes are in use by amateurs. One notable exception is the instrument which Commander Henry Hatfield has built into his house in Kent, England. The photographs in Figures 11.8 and 11.9 were taken using that equipment.

The construction and use of Commander Hatfield's equipment is described in a long and detailed paper 'The Sevenoaks spectrohelioscope' in the December 1988 issue of the *Journal of the British Astronomical*

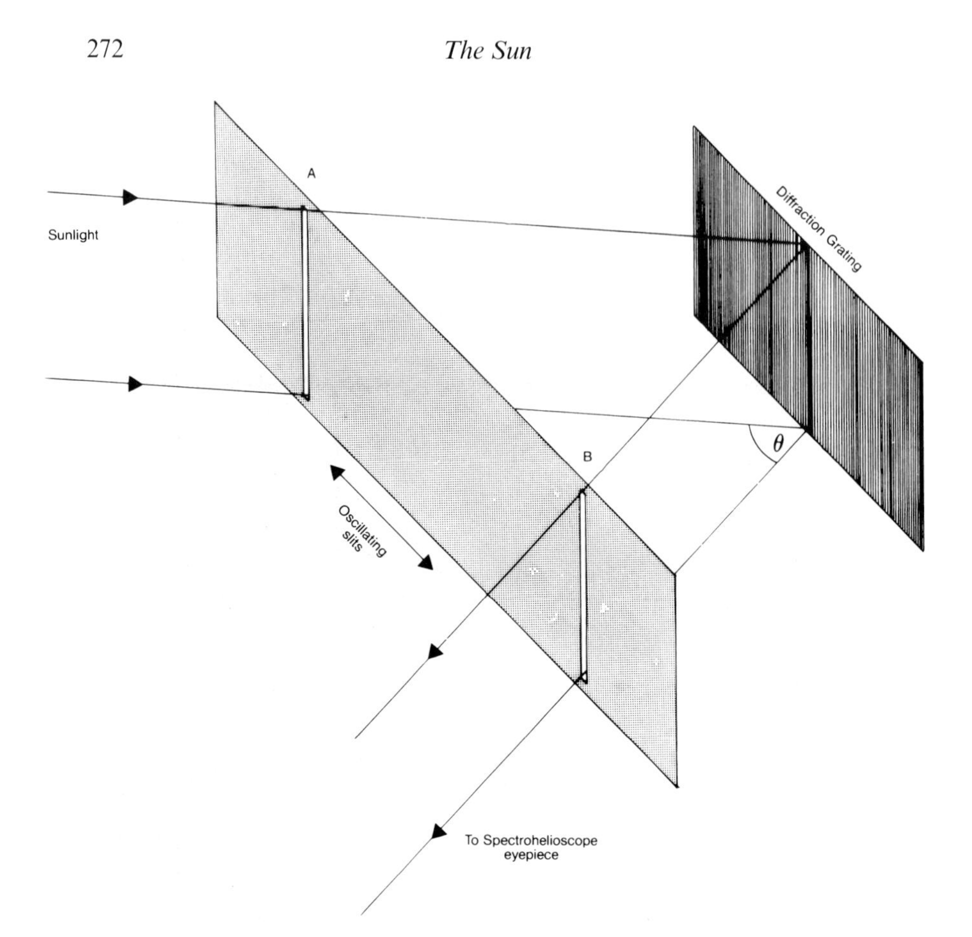

Figure 11.10. The principle of the spectrohelioscope. The Sun's image is focused over slit A. That strip of the Sun's image that passes through the slit is diffracted at the grating and the image of the slit at the required wavelength is selected by slit B. As slit A moves across the solar image, so slit B also moves in order to preserve the angle θ. In that way one sideways movement of the slit A scans the entire solar surface and the angle of diffraction, θ, is preserved. The movement of slit B then creates an entire image of the Sun at just the required wavelength. In this diagram all the auxiliary optics have been left out. In practice the light path in the instrument is many metres long.

Association. His instrument is also easily useable as a spectroscope. I can recommend no better a starting point for those interested than to read Commander Hatfield's paper.

In recent years narrow passband filters have become available. Though they are invariably expensive, they do allow high quality views of the Sun in Hα, or any of the other available wavelengths. The filter should have a bandwidth of less than 0.2 nm (2Å) if reasonable views are to be obtained.

(a)

(b)

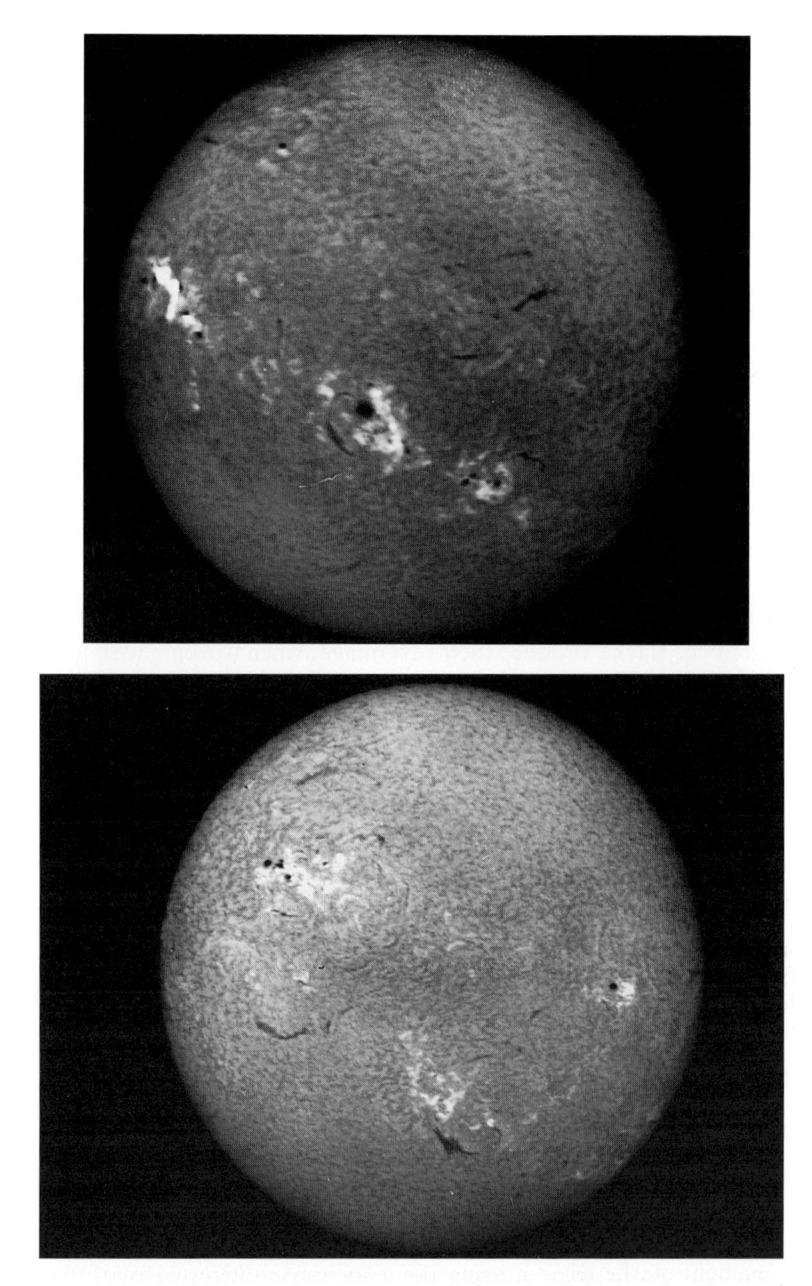

Figure 11.11. Hα filtergrams of the Sun taken by Eric Strach. He used a 0.6 Angstrom (0.06nm) passband filter centred on Hα in conjunction with his Celestron 8-inch (203 mm) Schmidt–Cassegrain telescope. Of course, the telescope was fitted with an off-axis 'energy rejection' filter of approximately 6cm clear aperture (the rest of the aperture being masked). In each case 1/30 second exposures were given on TP2415 film. (a) was taken on 1981 October 18^{d} 11^{h} 27^{m} UT and shows the end of a flare (classification 2B) amongst a group of spots on the west (left) side. Other spots with bright plages, the dark filaments and the chromospheric network (of granulation) are also apparent. (b) was taken on 1981 September 13^{d} 08^{h} 20^{m} UT and shows the granulation particularly well.

Better still less than 0.1 nm. The 'Daystar Filter Corporation', among other companies, produces a range of these filters.

However, do not run away with the idea that you can just plug one of these into your telescope and observe. For one thing, if the aperture of your telescope exceeds about 5 cm the concentrated solar radiation will quickly overheat the filter. Ideally the objective lens of your solar telescope (which can be single-element, since chromatic aberration is not a problem when viewing in monochromatic light) ought to be made from a heat-rejecting glass, such as Schott OG-590, or RG-630 (which also transmits best at Hα). Even so, an additional over-the-objective, heat-rejection filter (made from, say, Schott KG-2 glass) is highly desirable for apertures greater than 5 cm.

Ordinary telescopes can be used with these filters provided they are stopped down and a heat-rejection filter is used, preferably over the clear aperture rather than anywhere near the focal plane. Wherever you get your filter from, **do make sure that you have full instructions for its use and you are well aware of its safe limits of operation**. These filters, which are multi-layer interference stacks, produce their best images when working with incident beams of effective focal ratio greater than *f*/30. Figure 11.11 shows two Hα filtergrams obtained by Eric Strach.

Though narrow passband filters generally produce better quality images than spectrohelioscopes they are less convenient in that they cannot be set to deliver images over a wide range of wavelengths. It is true that they are often supplied in mounts which are fitted with heaters in order to control the wavelength at which they pass radiation (this is temperature dependent). Control units are usually provided which can allow one to tune the filter (by varying its temperature) to a little either side of its normal passband wavelength. It is easier to image a *Doppler-shifted* (the wavelength is altered by its motion) prominence with a spectroheliograph, than with a very narrow passband filter.

A telescope, involving a narrow passband filter and a central occulting stop, which is especially suitable for viewing solar prominences, is described in the February 1979 issue of the *Journal of the British Astronomical Association*. This paper is entitled 'Filter type solar prominence telescope for amateurs'

Many photographic emulsions are virtually useless for recording the Sun in the light of Hα, due to their poor sensitivity at this wavelength. Hence you should look at a film's spectral response curve before selecting it for this type of work. However, I have no hesitation in recommending Kodak's Technical Pan 2415. It was originally developed for exactly this type of photography and its other characteristics make it the outstanding choice, superceding another Kodak emulsion, Technical Pan SO115.

12

Variable stars and novae

Despite the seething turmoil of the Sun's surface, our daytime star's output of energy is very steady. This is very fortunate for us but many of the stars visible to the naked eye, and thousands within the range of binoculars or a moderate telescope, markedly change their brightnesses. Some change erratically. Some are very predictable. Others may be fairly predictable but can sometimes spring surprises on the observer. Noticeable changes can take years to occur or, in some cases, just minutes.

In many ways these *variable stars* are just as individual as people, each with its own characteristics. However, astronomers have classified these stars into a number of types, the behaviour of the members of each type being broadly similar. Adopting a number of variables to observe and periodically estimating their brightnesses is a very positive way of communing with our Galaxy at large. It is also a very scientifically valuable exercise. Most of the panels of astronomers that allocate observing time on professional telescopes, allow individuals or groups a few nights each to follow up some specific project. Here the amateur, with his/her telescope on hand whenever the sky is clear, scores heavily over the professional.

Though some professional astronomers are doing observational work in this field, an amateur observing group's records, for instance those of the BAA Variable Star Section or the American Association of Variable Star Observers (AAVSO) are often called upon by researchers. In the cases of many stars, amateurs are the only people watching and recording their brightness variations.

Every so often a *nova* bursts forth onto the stellar scene, or even a vastly more powerful *supernova*. Once again the amateur observer has the advantage over the professional and it is not surprising that many of these impressive outbursts have the names of amateurs associated with them, as their discoverers.

Supernovae can, and do, occur in our own Galaxy. However, most examples are found in external systems and so the subject of supernova hunting is dealt with in Chapter 14, together with observing galaxies. Also, amateurs have long been making eye-estimates of stellar brightnesses and this work still provides useful results today. However, a number of amateurs have begun to emulate their professional counterparts and use photometric equipment to make brightness measures. Consequently, this chapter lays the foundations of variable star observing and deals with eye-estimate techniques. The following chapter is devoted to methods of photometry.

Variable star nomenclature

One system for the naming of variable stars was originated by Friedrich Argelander in the nineteenth century. Using this scheme the first variable star discovered in a given constellation is denoted by R, followed by the name of the constellation. For instance the first variable star to be discovered in the constellation Cygnus was (and still is) named R Cygni. The second and third variable stars to be found in that constellation are named S Cygni and T Cygni, respectively. However, those letters soon ran out. The ninth discovered variable star in Cygnus is Z Cygni. To continue the scheme double letters were used. The tenth Cygnus variable is thus RR Cygni, the eleventh is RS Cygni, and so on.

After RZ Cygni the system continues with SS Cygni (**not** SR Cygni) and on up to SZ Cygni. After that we have TT Cygni (**not** TR or TS), and so on. As the years passed, so further variables were discovered taking the required designations beyond ZZ. Astronomers then started using the double letters in the first part of the alphabet – AA up to AZ, then BB up to BZ, CC up to CZ, etc. The letter J was left out because of its similarity to I.

Argelander's system is widely used today but the *Harvard Designations* are also very popular. In this scheme each variable star is given a six-digit number which represents its coordinates for epoch 1900. The first two digits give the number of hours right ascension (00 to 24) and the second two give the odd number of minutes (00 to 59). The third pair of digits give the declination (in italics for negative declinations). Thus the Harvard Designation of R Cygni is 193408, because its coordinates were $\alpha = 19^h\ 34^m$, $\delta = +08°$, on 1 January 1900. The Harvard designation of R Centauri, in the southern hemisphere of the sky, is 1409*59*.

It is always preferable to use **both** the designations to describe a given variable star when reporting observations. For instance 050953 R Aurigae.

This will minimise the risk of mis-identifying the star in your report. It is all too easy to make mistakes when writing down a set of numbers.

Stellar brightnesses and distance modulus

Most readers will already be familiar with the stellar magnitude scale used by astronomers. The star α Lyrae (Vega) is of visual magnitude (m_v) $0^m.0$. Brighter stars have negative magnitudes and fainter stars have positive magnitudes. A difference of one magnitude represents a brightness difference of 2.5 times. This is given a mathematical footing by the expression:

$$m_v = -2.5 \text{ Log } I,$$

where I is the apparent brightness of the star, in relative units. The difference in magnitude, Δm_v, then follows:

$$\Delta m_v = -2.5 \text{ Log } (I_1/I_2),$$

where I_1 and I_2 are the brightnesses of the compared stars. When using this formula, do remember that the **fainter** star has the more positive magnitude value.

Of course, the term *apparent magnitude* is used to distinguish a star's apparent brightness in the sky from its *absolute magnitude*, the brightness it would appear to be if placed at the standard distance of 10 parsecs (32.6 light years) from the Earth. For instance, the Sun's apparent magnitude is $-26^m.7$ but its absolute magnitude is only $+4^m.8$. The absolute magnitude values allow us to compare the real luminosities of stars, after correcting for distance.

Absolute visual magnitude is denoted by M_v. The quantity $(m-M)$, known as the *distance modulus*, is useful (at least to the theorist) because it fixes the distance of a given star. The equation is:

$$(m_v - M_v) = (5 \text{ Log } d) - 5,$$

where d is the distance of the star in parsecs.

Equipment

The most accurate method of measuring stellar brightnesses is the same as that used by professional astronomers. Namely, *photoelectric photometry*. *Photographic photometry*, though less precise than the photoelectric technique, is still objective. Both these procedures are described in the next chapter.

With care the eye of the observer can be used to make brightness estimates accurate to the nearest $0^{m}.1$. The techniques necessary will be discussed later in this chapter, though suffice it to say here that the apparent magnitude of the variable is compared with other stars of known magnitudes. For some variable stars the naked eye is best. Others require optical aid. One formula (widely published in various forms) which can be used to estimate a telescope's limiting magnitude, Lim. m_v, is the following:

$$\text{Lim. } m_v = 2 + 5 \text{ Log } D,$$

where D is the aperture of the telescope in millimetres.

This equation cannot provide any more than a rough guide. The transmission efficiency of the telescope optics and the atmospheric transparency, the quality of the optics and the seeing at the time of the observation (if the star's seeing disc is expanded its apparent brightness falls), the brightness of the sky background and the magnification (up to a point, increasing the magnification darkens the sky-background allowing fainter stars to be seen) – all will affect the magnitude limit of the faintest stars seen through the telescope. Last, but certainly not least, the observer's eyesight will have a large bearing on the faintest magnitudes seen.

Bradley E. Schaefer, of the NASA–Goddard Space Fight Centre, conducted a survey of telescope users and he published the results in the November 1989 issue of *Sky & Telescope* magazine. His results show that, for telescopes of 4-inch (102 mm) to 16-inch (408 mm) aperture one could see stars fainter than those predicted by the aforementioned formula by increasing the magnification to above ×100. In fact, there were still gains to be made by increasing the magnification to above ×200.

Schaefer's results indicate that one can expect to better the brightness limit suggested by the formula, though the advantage decreases somewhat with increasing aperture.

I would suggest the following practical formula for the limiting magnitudes obtainable with amateur-sized telescopes:

$$\text{Practical Lim. } m_v = 4.5 + 4.4 \text{ Log } D.$$

I have based this formula on Schaefer's published results. Once again, this can only provide a guide, though it should be more reliable than the previous formula. The magnification should be as high as possible, while still showing fairly sharp star images. This will be about ×200 for a 6-inch telescope, increasing to about ×300 for a 16-inch telescope. Note that the steadiness of the seeing will have an important effect on the faintest stellar magnitudes visible, particularly when using large apertures. Table 12.1

Table 12.1. *The limiting magnitudes of telescopes of differing apertures*

From the author's formula, which was based upon the published results of a study conducted by Bradley Schaefer. The figures are only intended as a guide. Unsteady seeing, as well as other factors, will reduce the magnitude limits obtained. Bad seeing has its greatest effect on large apertures.

Telescope aperture (inches)	Telescope aperture (mm)	Limiting magnitude
6	152	14.1
8	203	14.7
10	254	15.1
12	305	15.4
14	356	15.7
16	408	16.0
18	457	16.2
20	508	16.4

provides some values of limiting magnitude, using this formula. Binoculars cannot do as well as this equation predicts because of their inherent low magnifications.

As will be discussed later, the ideal aperture for a telescope is that needed to show stars to two or three magnitudes fainter than the variable. Thus, for the brightest variable stars, the naked eye is best. 7×50 binoculars should show stars of the tenth magnitude on a good night and so can be used for estimating stellar magnitudes down to about 7^m, or 8^m. Binoculars and the naked eye have one other advantage. They both have large fields of view. The *comparison stars* used to make the estimate can then be seen together in the same field.

Provide any telescope you use for variable star observing with an eyepiece giving the largest possible field of view (discussed in Chapter 2). However, there is one major pitfall that can catch the unwary. **Do make sure that the field of view of the eyepiece you use is fully illuminated by the object glass/primary mirror**. Using the equations given in Chapters 1, 2 and 3, you can calculate the size of the fully illuminated field at the focus of your telescope – taking into account any restrictions applied by the size of baffles and drawtubes (Chapter 3), secondary mirrors (Chapter 1) and/or Barlow lenses (Chapter 2).

If you wish to avoid measuring parts of your telescope and making the calculations, you can still check for vignetting using a simple practical method. Cut a cardboard disc to the same diameter as your telescope drawtube. Then cut a circular hole in the middle of it that is the same size as the field-stop aperture of the eyepiece you intend using. Offer the cardboard diaphragm up to the open drawtube. Adjust the focuser so that the diaphragm is as close as you can estimate to the focal plane of the telescope. Then bring your eye up to the diaphragm. You should be able to move your eye across the hole cut in the cardboard and not see any obstruction restricting your view of the telescope's objective lens or primary mirror.

If, for instance, you get your view of the primary mirror of a Newtonian reflector partially cut off by the edge of the secondary mirror when your eye is positioned close to the edge of your cardboard diaphragm, then the field of view of your eyepiece will not be fully illuminated. Unless you can make some change to the telescope to rectify whatever is causing the vignetting (in our example replacing the secondary mirror with a bigger one), you will have to use an eyepiece of smaller field of view for making magnitude estimates. Otherwise the apparent brightnesses of the comparison stars will be altered by their varying positions across the field of view.

It is true that visually estimating the brightnesses of stars is less exacting on the requirements of telescope quality than for other types of observation. However, I would caution against using the cheapest 'department store' telescopes for variable star work. For one thing, the telescope must be firmly mounted. Otherwise observing with it will prove a trial and your brightness estimates won't be anything like accurate.

Also, many 'department store' refractors have object glasses which are poorly corrected for chromatic aberration. The problem here is that the apparent brightnesses of stars of different colours will be artificially distorted. For instance, if the telescope sharply focuses a star in green light but has its red image out of focus, then any red star which is viewed through the telescope will be apparently dimmed as a result. In addition, the Huygens eyepieces that are usually fitted to, or supplied with, these telescopes have small fields of view, despite their low magnifications. To summarise, if you can't use a reasonable quality telescope for variable star observation, don't use any. Stick to using binoculars, instead.

Maps, finder-charts, and 'star-hopping'

The variable star observing section of the British Astronomical Association (BAA) and the American Association of Variable Star Observers (AAVSO) have already been mentioned. Other national groups exist for coordinating and collating variable star observations. Each of these groups can supply you with finder-charts.

The named variable star is positioned close to the centre of each chart.

Brief notes at the head of the chart list some of the pertinent details, such as the scale of the chart, the star's range of magnitude, its period (or a note on its irregularity), the 1950 and 2000 coordinates, etc. The AAVSO issues sets of charts suitable for both binocular viewing (large area of sky covered and the same orientation as the naked eye view) and with an inverting telescope (a field of view of a degree or so and oriented with south uppermost and west to the left). These charts also have the magnitudes of suitable comparison stars marked on them. A number like 87 means that the magnitude of the comparison star so marked is $8^{m}.7$. The decimal point is always left off to avoid confusion with the star images.

If your telescope has setting circles, you should be able to point it accurately enough to pick out the pattern of field stars and then centre on the variable. If not, you will have to use a technique known as *star-hopping*. For this use a star atlas (I can recommend *Sky Atlas 2000.0*, or *Uranometria 2000.0*) and a clear plastic overlay sheet with an inscribed circle of size equivalent to the field of view of the eyepiece. The telescope is set on the nearest prominent (usually naked eye) star. The circle on the overlay is then placed over the bright star. With the atlas correctly oriented you should be able to recognize the pattern of stars seen in the eyepiece.

The overlay is then moved a little (roughly in the direction of the variable) to cover an adjacent field of stars. Note the pattern. You might even be lucky enough to have a well recognised 'landmark' such as a nebula or galaxy to act as a 'stepping stone'. If so, centre on it. Then move the telescope a small amount to pick up the next new area of sky. Once again you should be able to recognise the stars. Keep doing this until you are centred on the variable. The finder-chart will enable you to make the final identification.

Try not to get thrown if your variable turns out to be unexpectedly faint, or even invisible! It could just be that you have found it at a time of deep minimum brightness. Your first attempt at setting on a given variable star may take you a long time. On the next occasion it will be a little easier. Eventually that particular star will settle into your repertoire and finding it will take less than a minute.

Making the magnitude estimate

Having found our chosen variable star, we now have to determine its brightness on the magnitude scale. There are several methods which observers use to do this and the three main techniques are discussed in the following notes.

Comparison star sequence method

This is the method most amateurs use to estimate variable star brightnesses. After first looking at the variable star in order to gauge its brilliance, the field of view is searched to try and find a comparison star of identical brightness. If successful, the variable is, of course, assigned the magnitude value of the comparison. Normally no such identical comparison will exist. In that case, the field is searched to find two comparison stars. One should be just a little brighter than the variable and one just a little fainter. If the comparison stars have magnitudes $7^m.3$ and $7^m.7$ and the brightness of our variable is exactly halfway in between then its magnitude is obviously $7^m.5$.

Of course, that was an easy example. If the variable is closer to the brightness of the magnitude $7^m.3$ star, then its magnitude is probably $7^m.4$. Often the comparison star brightnesses won't bracket that of the variable so conveniently. You will then have to estimate fractional differences of brightness. With practice you should be able to achieve estimates accurate to $0^m.1$.

Fractional method

This is really just the same as the foregoing, though the brightnesses of the comparison stars are looked up after the observation. Let us say that our variable's brightness is between that of two comparison stars A and B. After comparing we decide that it's magnitude is closer to A, say two fifths of the way from A to B. The observation is recorded as 'A 2 V 3 B'. If A's magnitude is $8^m.1$ and B's is $8^m.6$, then the difference between the two comparison stars is $0^m.5$. One fifth of this difference is $0^m.10$.

Hence, using star A:

$$\text{magnitude of variable} = 8^m.1 + (2 \times 0^m.10) = 8^m.3$$

As a check, we also use star B:

$$\text{magnitude of variable} = 8^m.6 - (3 \times 0^m.10) = 8^m.3$$

Note that I have used two decimal places in the value for the magnitude difference. Of course, this is well beyond the possible difference in magnitude detectable by the human eye. The reason is to avoid extra errors caused by rounding figures. For instance, suppose that the two comparison stars were $0^m.7$ different in brightness, with the brighter at $9^m.0$, and the variable was two fifths of the way from the brightest to the faintest. One fifth of this difference $0^m.14$. If we round this figure to $0^m.1$ then the magnitude of our variable star comes out as $9^m.2$. Leaving our rounding to the end of the calculation we get the more accurate figure of $9^m.3$.

Pogson's step method

In essence (though not in practice), just one star is used as a comparison for the variable. It's brightness should as closely match that of the variable as possible. The observer simply estimates the number of 'steps' of brightness separating the comparison star and the variable. The size of the steps should be as close to $0^m.1$ as possible. Obviously the observer must have experience in judging $0^m.1$ steps. One could get used to this technique by trying it out on non-variable stars before embarking on observing the variables. If your comparison star, let us call it A, is of magnitude $7^m.8$ and you judge the variable to be two steps brighter, then it's magnitude must be $7^m.6$ You could record your observation as 'A+2'.

In practice you should check your estimate using another comparison star. Let us call this one B. If you think that the variable is three steps fainter than B, then you would record 'B−3'. On checking, if you find that the magnitude of B is $7^m.9$, then that places the magnitude of the variable as $7^m.6$, in agreement with your first estimate. If the figures do not agree, then do the observation again. If possible, use another star to confirm your final estimate.

Difficulties

Just discerning, let alone accurately estimating, stellar brightness differences of $0^m.1$ is no easy thing. As an example look at the Belt of Orion. The westernmost of the bright three stars, Mintaka (δ Orionis), is obviously fainter than the other two. Of the other two stars, Alnitak (the easternmost) and Alnilam, one is $0^m.1$ brighter than the other. Can you tell which is the brighter? Try deciding for yourself before checking in a catalogue.

Another problem is the colour of the stars. Comparing the brightnesses of two stars of different colour is not easy. Many variables are red. There is an insidious phenomenon called the *Purkinje effect* which can badly distort the brightness estimates of red stars. What happens is that the human eye reacts to red stars in such a way that the apparent brightness of the star increases while looking at it. One way of countering this effect is to avoid taking lingering looks at the star. Instead, just use quick glances. Another way is to defocus the telescope and compare the expanded discs, in the same way as for estimating the brightness of a comet's nucleus (see

Chapter 10). The Purkinje effect is reduced when the light from the star is not concentrated into a small point.

The brightness of the sky-background can also affect the apparent brightness of a red star. Haze or bright moonlight tend to make a red star appear brighter than it really is, when compared with others which are not so red. This is another manifestation of the Purkinje effect.

Occasionally a variable will be too faint to observe with a pair of binoculars and yet it has no suitable comparison stars near enough to be seen in the smaller field of the telescope. The best that can be done is to rapidly move the telescope between the comparison and the variable, while trying to 'remember' the brightnesses. This is far from satisfactory and estimates made this way are not likely to be accurate to $0^{m}.1$.

When making naked eye brightness estimates, suitable comparison stars may lie large distances away. They are then likely to be affected by variations in atmospheric transparency (particularly for those stars at widely differing altitudes). Try to use stars which are close by the variable for comparison purposes, even if they differ from it in brightness by a little more than you would normally like.

Experience shows that for a reasonably accurate magnitude determination, the star ought not appear too bright, nor too faint. As already mentioned, whatever optical system is used the star should ideally be two or three magnitudes brighter than the faintest magnitude visible through the system. A telescope can be easily diaphragmed down to whatever aperture required and I recommend that a series of masks be made up in readiness.

Reporting variable star observations

As is the case for other observing projects, a variable star observer's results are most valuable when backed up, and supplemented, by those of other people. The coordinator of the group you belong to will supply you with that group's standard report forms. I can say little about these forms except that they must be filled out with meticulous care. Do make sure that the dates, times and variable star designations and magnitude determinations are correct. It is only too easy to slip up when writing down lists of figures. Do make sure you indicate any magnitude values of which you are less than certain. The normal way is by the use of a question mark following the figure.

As a check on the date and time, most groups report forms give you the option of including the *Julian date* of the observation. This is found from an ephemeris. On this artificial system Julian day 1 began at noon on 1 January 4713 BC. Thus each new Julian day begins at noon. Noon on 1 January 1991 marks Julian day number 2448258.0. Midnight on the same day is Julian day 2448258.5. As well as providing a check, using the Julian day numbers does make the analysis of the behaviour of a variable star easier. It avoids the need to reckon with the different numbers of days in various months.

Types of variable star

Basic knowledge is assumed here. A general description of the main types of variable stars are given in my book *Mastering Astronomy* (Macmillan Education Ltd., 1988 – 2nd edition to be released by Springer-Verlag under the new title *Astronomy Explained* in 1997). Other books specialising in variable stars are listed in Chapter 17. I would especially recommend *Observing Variable Stars* by David Levy (Cambridge University Press, 1989). This book contains the details, and in some cases finder-charts, of over 300 variables. The standard source for variable star data is the *General Catalogue of Variable Stars* published by the Astronomical Council of the Academy of Sciences in the USSR (Moscow). The fourth edition was published in 1985. However, I think that you will have difficulties in obtaining one of the rare copies of this work. Unfortunately there is only room enough here to give an extremely terse overview of the main types of variable stars, as they relate to the observer.

Eclipsing binary stars

In most cases useful work on these *extrinsic* variables requires a greater precision than is obtainable by making eye-estimates of brightness. However, they are handy 'training stars' for those beginning this sort of observation.

Mira-type variables

The largest main group of *intrinsic* variables, they are bloated red giant stars. They are semi-regular variables with absolute magnitudes around -1^m, radii about 100 times as large as the Sun and photospheric temperatures of about 3000K. Their periods range from a month to several years. The prototype is Mira (omicron ceti) which has a range of $3^m.4$–$9^m.3$ and a period of typically around 332 days. Estimates of these are useful.

Other semi-regular variables

These (eg. the R Lyrae-type) are also worth observing. Some have their cycles interrupted, becoming steady for a while. Others modify their behaviour from cycle to cycle.

Irregular variables

An especially valuable target for the amateur. They are fascinating because of their unpredictability.

Eruptive variable stars

The premier target for amateur variable star observation. There are many different types of eruptive variable. SS Cygni and U Geminorum stars undergo nova-like outbursts. These generally recur every few months, though the outbursts are far from predictable. T Tauri stars are young and hot, hiccuping before they settle down to a life as a main sequence star. Z Camelopardalis-type stars undergo nova-like outbursts, except that every so often one of these might cease its variability and settle down to a period of sedate quiescence. *Flare* stars can undergo rises of many magnitudes in just a few minutes. The subsequent return to normal brightness then takes several hours. Most flare stars are red dwarfs.

RR Lyrae stars

Pulsating variables. They are ageing giant stars, with radii around 8 times that of the Sun and photospheric temperatures of around 7000K. Their absolute magnitudes range around 0^m. Their brightness variations are as regular as clockwork. Periods range from a few hours to about a day, some stars pulsating in the fundamental mode, others in the first harmonic. They make good training stars.

Cepheid variable stars

Pulsating variables. Very regular. They are very luminous (absolute magnitudes ranging from roughly -1^m to about -6^m. They generally have photospheric temperatures very similar to that of the Sun and radii ranging from a little larger than the Sun's to over 100 times as big. The fact that their pulsation periods are related to their mean absolute magnitudes makes them invaluable as distance indicators. They make good training stars.

Miscellaneous variable stars

Some stars do not sit happily in any particular classification scheme. One example is R Coronae Borealis. Most of the time this star shines at around 6^m but every once in a while it plunges to the fourteenth magnitude.

Astronomers think that it puffs off clouds of soot which obscures our view of this star. When the stellar winds clear away this celestial smog, the star returns to its usual brightness.

Novae

I have been observing the skies for over a quarter of a century. Every clear night I have always found plenty which is of interest in the heavens. Every so often the sky produced surprises. One of the most dramatic occurred on 29 August 1975. As the sky was darkening early that evening I glanced upwards. The electric blue Vega and warm red Arcturus were already prominent against the deep blue sky. Cygnus the swan was flying high overhead. A moment's confusion – something was wrong. Cygnus didn't look quite right. A second magnitude star shone from a position north-east of Deneb (α Cygni) where no second magnitude star had shone before.

I telephoned an astronomer friend and he confirmed my sighting. I next telephoned the Royal Greenwich Observatory. I am sure the security officer who answered my call thought I was stark staring mad. 'A what?'. However, he did put my call through to an astronomer. Yes, they already knew about the *nova*. It had been discovered just hours earlier by several amateurs in Japan, while the Sun still shone down on the UK. The precursor was not shown on any pre-discovery photographs and so must have been fainter than the twentieth magnitude. Hundreds of other observers saw the nova on that first night. By the following night its brightness peaked at magnitude $1^{m}.8$. It had brightened by several tens of millions of times and had become the brightest nova visible to the naked eye in over thirty years. Yet this particular nova's rise and fall was particularly rapid and within a fortnight it had once again become too faint to be visible to the unaided eye.

Astronomers reckon that about thirty nova occur in our Galaxy each year. These stellar eruptions are caused when matter is transferred from one member of a binary star system to the other. Every so often the matter falling onto the receiving star triggers an explosive nuclear reaction. It is this explosion that we see as the nova. *Nova Cygni 1975*, as it became known, was especially luminous, peaking at over a million times the brilliance of the Sun. Others can be much less bright, the star only increasing its power output to perhaps fifty times it's initial value. We call these *dwarf novae*. Some dwarf novae give repeat performances, perhaps every few months. These are termed *recurrent novae*.

Keeping watch on the sites of old novae is an especially valuable activity. T Coronae Borealis is one example. It normally resides around the tenth or

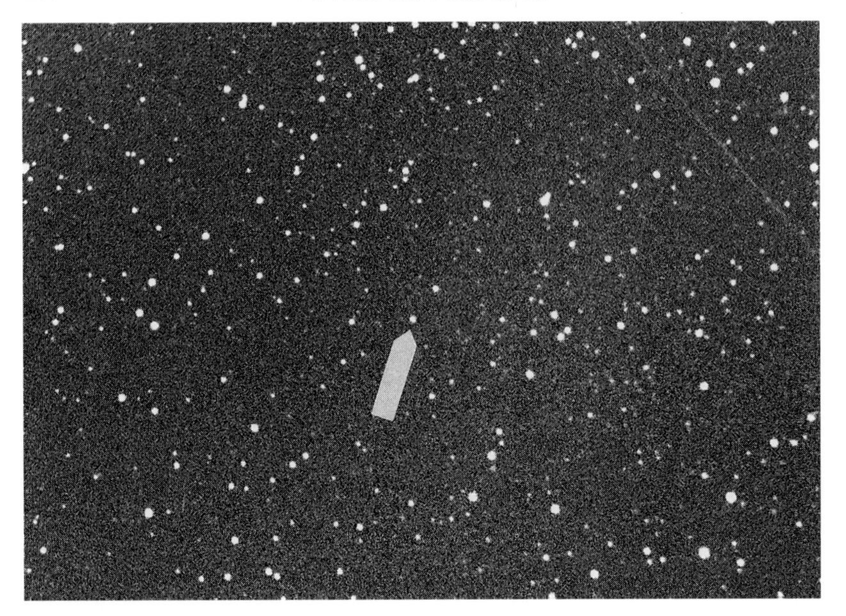

Figure 12.1. The arrowed star is the recurrent nova V404 in Cygnus, photographed by Martin Mobberley using his 14-inch (356 mm) reflector on 1989 May $30^{d}\ 23^{h}\ 06^{m}$ UT, when the object was at magnitude $12^{m}.5$. The 15 minute exposure was made on T-Max 400 film at the *f*/5 Newtonian focus of Martin's telescope.

eleventh magnitude but it suddenly rose to the second magnitude in 1866 and then gradually faded. Astronomers kept a watch on this star. It remained fitfully flickering around the tenth magnitude until one night in 1946 when it shot up to the second magnitude once more. Today it again unsteadily shines at the tenth magnitude. When will its next eruption occur?

In May 1989 an IAU circular reported the discovery of an X-ray nova, in Cygnus, by a scientist analysing data from the 'Ginga' satellite. Guy Hurst, coordinator of the UK nova/supernova patrol group, alerted the other members. Martin Mobberley used his 14-inch (356 mm) reflector to photograph the sky at the reported position. Measurements from Martin's photograph indicated that this object was none other than nova V404, which first flared up in 1938. One of Martin's subsequent photographs is shown in Figure 12.1. Other reccurrent novae are T Pyxidis, U Scorpii, V1017 Sagittarii and RS Ophiuchi.

As already mentioned, the bright nova that appeared in Cygnus at the end of August 1975 was named Nova Cygni 1975. Should a second bright nova have occurred later in 1975, and in the same constellation it would have been designated Nova Cygni 1975 No.2. Novae are also given Harvard designations. Thus Nova Cygni 1975 was 210847. Novae also are given an

IAU approved designation of V, followed by a number indicating the order of discovery. Nova Cygni 1975 is V1500.

Hunting for novae

Few novae ever become bright enough to be visible to the naked eye. One way of hunting for them is to use a pair of binoculars to methodically search the heavens for any new interlopers. Of course, one must memorise the star patterns visible. It is no good sweeping star fields unless you can be fairly sure of spotting any newcomer. Legendary in this respect is the British amateur George Alcock. Building on his knowledge of the naked eye stars, obtained during meteor-watching sessions, he began to memorise the patterns of stars visible in his binoculars in 1955. The long nights paid off in July 1967 when he discovered his first nova. Initially spotted when of the sixth magnitude, he caught it on the rise and this particularly slow nova did not reach maximum brightness until the following December. It then shone at $3^{m}.5$. During the 1960s and 1970s Alcock was to find four novae and several comets.

If you are prepared to memorise much of the binocular sky and then clock up several hundred hours of observing time before your first discovery, you might follow in Alcock's footsteps. A good pair of binoculars (7×50 or 8×50 would be ideal), a good star atlas (*Uranometria 2000.0* is probably the best) and a deck-chair is all that you need in the way of equipment.

When memorising the sky, don't attempt too much in one go. Choose no more than one new small area to learn each night. After a few nights you will begin to recognise that area well enough to spot any reasonably prominent interloper. Meanwhile the previously chosen areas will become like old friends and you will be becoming familiar with your newer targets. If you feel that the mental effort is becoming too much and you are losing grip, then stop taking on any new areas to learn until you are fully familiar with the heavens you have so far selected. Sooner than you expect you will be familiar with the appearance of a large, and ever-widening, patch of sky.

As happened for Alcock, comets and other celestial bodies may come your way. If you think you have discovered an interloper, do check first before issuing an alert. Check at least one star atlas to see if your suspect object really is new. Look up the positions of asteroids of similar brightness – is your object one of these? If you are satisfied of your discovery, then contact the coordinator of your observing group. You could also telegram the Central Bureau for Astronomical Telegrams (see Chapter 10 for details and the address).

Figure 12.2. Pre-discovery photograph of a nova in Vulpecula, taken by Nick James on 1987 November 14^{d} 19^{h} 04^{m} UT. He used a conventional camera fitted with a 55 mm focus lens, set at *f*/2.8, for this three minute exposure on Tri-X film. The nova (arrowed) was roughly of the eighth magnitude at the time of the exposure.

Perhaps a better way of hunting for novae is to use photography. A 35 mm camera fitted with a standard lens is sufficient, if fixed on a mount which is driven to track the stars (see Chapter 4 for details). The lens should be set at maximim opening and the exposure should be of several minutes duration. It could even be continued until the sky fog limit is reached. Hypered TP2415 film will give the best results, combining speed and high resolution. Failing that, any fast film such as T-Max 400 or HP5 Super will suffice, provided it's grain is not too obtrusive. Chapter 4 provides more details of this sort of photography. Interlopers can be searched using a blink, or 'problicom' technique, or overlaying negatives, as described for asteroid hunting in Chapter 10.

On the evening of 12 November 1987 Nick James was taking photographs to patrol the sky for novae. His normal routine is to process the film immediately after the session in order to quickly search the negatives for anything unusual. However, on the following day he was moving house and so on this one occasion he put aside the processing. I am certain that you can guess what comes next. Sure enough, on November 15 the nova, in Vulpecula, was discovered by Kenneth Beckman and Peter Collins in the

USA. Nick had photographed that very area of sky on the 12th (and 14th) and when he eventually processed the film, there it was. Figure 12.2 shows the later pre-discovery photograph.

Some observing groups have split the sky into small sections and allocated their members specific areas, either for photographic or visual hunting. This is the method which is most likely to produce the greatest number of discoveries. However, the rub is that you might well be unlucky and never discover a comet or nova in 'your patch'. On the other hand . . .

13
Methods of photometry

This chapter, building on the last, extends to methods of quantitatively measuring the brightnesses of celestial bodies, rather than relying on eye-estimates alone.

Artificial star photometer

The general principle and construction of this device is illustrated in Figure 13.1. The idea is that an artificial star is created in part of the field of view of the telescope eyepiece. The real star is brought to a position nearby and the brightness of the artificial star is adjusted until it matches the real star. The eye is better at matching brightnesses than it is at gauging brightness differences.

In practice the bulb changes its colour together with its brightness. In any case, this colour might well be different to that of the star and so cause difficulties when making the comparison. Using a strong yellow filter to unify both the artificial star and the real star overcomes this problem.

Another difficulty is that the artificial star may look significantly different in size and brightness distribution to the real star. Defocusing the eyepiece slightly may help. The two small discs should now look much more similar. I haven't included the mechanical arrangement for the provision for altering the focus of the eyepiece in the diagram, for the sake of simplicity. A small rackmount, or a helical focuser, or a simple sliding tube should do the job.

If you use a 6V 3W bulb it will draw a current of 0.5 A and have a resistance of 12Ω when run normally. Though not critical, the potentiometer ought to have a similar resistance to the bulb. A little experimentation will sort out the details of correct bulb, potentiometer and ammeter to use. Much will depend upon the size of the hole you make in the covering to the

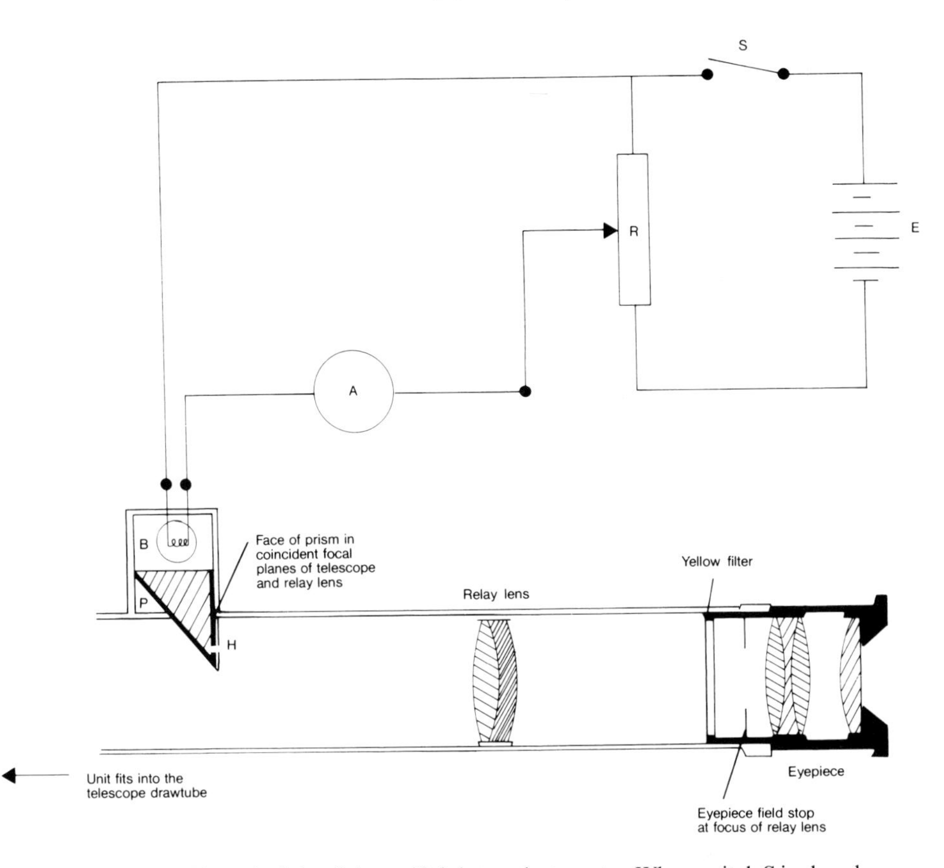

Figure 13.1. The principle of the artificial star photometer. When switch S is closed the battery, E, delivers a current to the bulb, B, via the wiper of the potentiometer, R. As the wiper is moved, so the current delivered to the bulb varies, causing a change in its brilliance. This current is monitored by the ammeter, A. P is a 45° prism. The heavy lines indicate where the faces of the prism are masked by heavy black paint. The hole, H, in the paint is positioned at the focal plane of the relay lens. This is also the position of the focal plane of the telescope. The eyepiece views the relayed image. To simplify the diagram the provision for focusing the eyepiece has been left out. The yellow filter is necessary to unify the colour of the bulb (the artificial star) and the star, in order to compare their brightnesses.

face of the prism. It ought to be as small as possible. Any sharp pointed object can be used to make this hole.

Just as for making unaided visual estimates of star brightnesses, you still need to record the brightnesses of comparison stars. The corresponding current readings obtained by matching the brightnesses of the comparison stars with the artificial star are noted, together with their given magnitudes. The comparison stars ought to have brightnesses ranging from a little less than that of the star you are interested in, to a little more. A graph is then plotted of current v. magnitude. Our star of unknown magnitude is then observed and the corresponding current reading obtained. Using the graph, the star's magnitude is then read off.

Before relying on the brightness measures produced by this device you should try it out on a number of stars of known brightness. Unless you can get results reliably accurate to $0^m.1$ you would be better off training yourself to make eye-estimates of stellar brightnesses in the manner described in the last chapter. In particular, the ammeter you use should be of good quality and have a large scale (a good quality digital meter is even better). Also, the potentiometer must allow you to control the current with precision.

Stellar brightnesses from photographs

Flick through the pages of this book and you will find many photographs showing stars of a range of brightnesses. How is it that you can tell that one star is brighter then another? Look closely and you will see that it is the diameters of the star images that cause a particular star to look faint, or otherwise, on the photographs. This effect is called *halation*. With care you can make eye-estimates of these photographed images in the same way as at the telescope.

Better still, you can project the negative onto a screen and then measure the diameters of the star images directly, using a clear plastic ruler. The enlargement factor should be as high as possible so that the faintest star images you measure should be at least several millimetres across. You should measure the diameters of a number of stars of known magnitude in the vicinity of the star of interest. From your results then plot a graph of magnitude v. image diameter. The magnitude of the subject star's image can then be translated into a magnitude reading using your graph.

If the comparison star readings are widely scattered, rather than following a smooth curve fairly closely, then you know that something is wrong and your results are unlikely to be accurate. Do beware of the effects of optical aberrations on the star images. Images away from the centre of the field of view, in particular, may well be distorted by coma, or their images enlarged by a change in focus. Hence the need to use comparison stars which are close to the one in question.

One other problem with using this technique arises from the spectral response of the film (see Chapter 4). Films vary, and in any case have a response which is quite different to that of the naked eye. Comparisons between stars of different colours are then rendered invalid. The problem is largely overcome by using a deep yellow filter (better still a standard photometric filter – dealt with later in this chapter). The required exposure time is, of course, lengthened but this is not a serious problem when using today's very sensitive photographic emulsions.

Photographic densitometry

Even in these times of electronic imaging, astronomers still commonly use *microdensitometers* for analysing images stored on photographic plates or film. Basically, a very thin beam of light is shone through the photographic image. The intensity of the transmitted beam is then measured by a suitable detector. The whole of the image can be scanned and the digitised information stored for further manipulation on a computer. Effectively, the microdensitometer and computer builds up an electronic version of the photographic image. This might be further manipulated. For instance, to bring out subtle details, or to enhance a particular feature. On the other hand, the object could be to make certain quantitative measures, such as brightness or position.

Unfortunately, you would have to be rather wealthy in order to purchase this equipment ready-made from the manufacturer. However, one or two rare individuals have built digitising equipment for themselves.

There is one 'low-tech' approach which will allow the average amateur to perform a kind of densitometry. That is to project the photographic image onto a screen in a darkened room. A sensitive light-meter is then held close to the screen, with its aperture facing the projector lens. Moving the light-meter about over the image will allow you to take readings of the image brightness at different positions. A standard photographic light-meter may not prove convenient in use but you can make your own if you are adept at simple electronics. A photodiode, or phototransistor would make an ideal detector. It then operates an ammeter, via an amplifier. The detector could be mounted in a small box, or even a small block of wood, with a small opening to allow the light to reach it.

Photoelectric photometry

The amateur astronomer equipped with a *photoelectric photometer* can make brightness measurements of an accuracy on a par with the professionals. Measures accurate to $0^m.01$ are attainable for stars brighter than about 8^m or 9^m, when the photometer is used in conjuction with a telescope of about 8-inches (203 mm) aperture. The photometric equipment necessary to do this will currently (1996) cost around £700 ($1100), for instance the Model SSP-3 stellar photometer marketed by 'Optec' in the U.S.A (see Chapter 17 for a full listing of manufacturers). One can achieve the same accuracy on fainter stars, even with the same size of telescope, but the cost then multiplies to several thousands of pounds or dollars.

For the amateur, it is probably better to go to a larger telescope if he/she wants to measure the brightnesses of fainter objects. This is especially true because the optical accuracy required is not so high as for other types of observation. Large, cheap, telescope mirrors can be obtained from several sources. The listing given in Chapter 17 will help. Suppliers of photometric equipment are also listed there. They will provide you with all the necessary information about their products.

One can make scientifically useful brightness measures of: intrinsic variable stars, eclipsing binaries, asteroids, the moons of the major planets, planetary surfaces and much more. I would strongly recommend any interested reader to pursue the list of references given in Chapter 17. The purpose of the following notes is to provide you with an overview of the techniques and the required equipment.

Photoelectric photometry: the photomultiplier tube

The heart of the photoelectric photometer is the *photomultiplier tube* (called a PMT, for short). Figure 13.2 illustrates the principle of a PMT. It consists of a series of electrodes, labelled a to g on the diagram, enclosed in an evacuated glass envelope. Electrode a is made of a material which liberates electrons when light falls on it. It is known as the *photocathode*. The alkali series of metals, particularly sodium, potassium, rubidium and caesium, are usually used, often in combinations with each other and with other materials.

The diagram schematically illustrates light falling upon the photocathode and liberating a single electron from it. The electrode b, or *dynode*, is made about 100 volts positive with respect to the photocathode. Thus the ejected electron (which is negatively charged) is accelerated towards it. The external electrical circuitry is not shown in the diagram.

When the electron strikes dynode b it causes, by a 'billiard-ball effect', more electrons to be emitted. On my diagram I have shown two electrons being knocked from the surface of b. Dynode c is also about 100V positive with respect to dynode b (and thus about 200V positive with respect to the photocathode). As shown in the diagram, more electrons are liberated in the same way. This process continues, each dynode being about

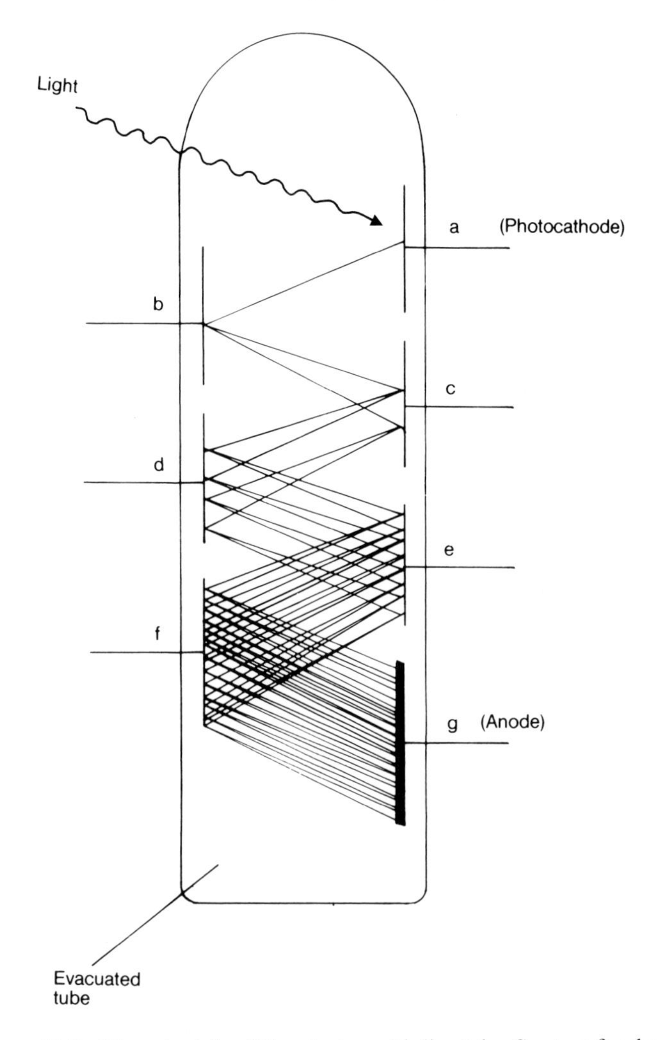

Figure 13.2. The principle of the photomultiplier tube. See text for details.

100V more positive than the last. Electrode g on my diagram is the final *collecting electrode*, or *anode*.

Note how a single electron emitted from a is multiplied to a virtual avalanche of electrons at g. Thus a very low light intensity falling on the photocathode is made to produce measurable pulses of current that can be counted by the external circuitry. The brighter the light falling on the photocathode the higher the number of current pulses counted in a given time. Put another way, the number of pulses counted in a given time tells the operator how bright the source of light is.

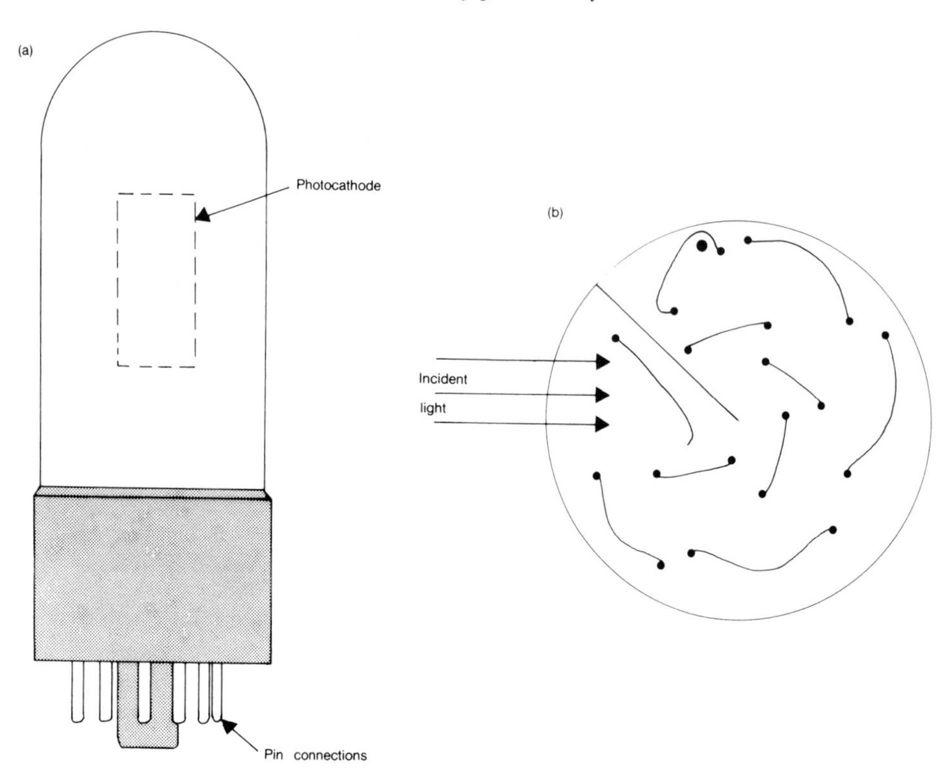

Figure 13.3. (a) Shows the physical appearance and (b) shows a plan view of the electrode configuration of a side-on and circular-cage type photomultiplier tube. The popular IP21, manufactured by Hamamatsu, is of this type. In the case of the IP21, the overall height (including pins) of the PMT is 94 mm and its maximum width is 32 mm. The receptive area of the photocathode is 8 mm×24 mm. These diagrams, and the data, are based upon those in Hamamatsu's catalogue of photomultiplier tubes and are used with their permission.

In order to explain the principle of the PMT clearly, I have made my diagram rather simplistic. In practice a given PMT may have any one of a number of different electrode configurations. The number of dynodes is typically 9 to 15. In addition a PMT might be a *side-window* type (also called a *side-on* type), or an *end-on* (also called a *head-on*, or *end-window*) type. The end-on type of PMT has a semi-transparent photocathode mounted at the flat top end of a cylindrical glass tube.

Figure 13.3 illustrates the general appearance and design of a side-on and *circular-cage* type of PMT. One example of this tube is the IP21 manufactured by Hamamatsu – a company with branches in most countries throughout the world. It is, probably, also true to say that the IP21 is the most popular choice of tube for the amateur astronomer building his own photoelectric equipment. Certainly this has been my impression talking to many photometrists.

Hamamatsu describe their IP21 as a medium gain tube with a very low *dark current*. Even if absolutely no light falls onto the photocathode of a PMT a small current is liberated from it because of thermally liberated electrons. It is, of course, desirable that this dark current be as small as possible and the IP21 is particularly good in this respect. After the tube has been switched on for 30 minutes, a typical dark current value (at room temperature) is about 1 nA (0.000001 mA). This tube, responds to light of wavelength between 300 nm and 650 nm. The operating voltage of the IP21 is fairly typical: 1 KV spread across its 9 dynodes and the anode. Undoubtably one major reason why the IP21 is so popular is that it is one of the cheaper photomultiplier tubes, at well under £100 (=$160) at mid-1990s prices.

Photoelectric photometry: the photomultiplier head

Figure 13.4 illustrates the basic layout of a photometer head. The unit plugs into the telescope and the telescope's focal plane is made coincident with a small pinhole aperture, S. If the telescope's focal ratio is less than about *f*/7, it may well be desirable to install a Barlow lens before S. The reason is simply to reduce the divergence of the beam after it passes through S, since the other optical parts are spread out along several centimetres and vignetting might otherwise result.

The size of S is chosen so that it spans about 1 arcminute (60 arcseconds) of sky. The formula given in Chapter 1, and repeated in later chapters, can be used to calculate the size of the hole necessary. As an example, if the telescope's effective focal length (taking into account the Barlow lens, if one is used) is 2 m, then 1 arcminute corresponds to 0.58 mm. The size of the pinhole is not critical but shouldn't be much more than 2 arcminutes across, otherwise too much sky background will be included when the star is centred for measuring.

M is a hinged mirror. When in the position shown in the diagram it directs the light passing through S upwards into a 'viewing-tube'. This consists of two lenses, R_1 and R_2, which relay the image from S to a crosswire eyepiece (not shown in the diagram). Focusing can be achieved either by having the whole of the 'viewing-tube' move, or just the eyepiece. R_1 and R_2 should preferably be two identical plano-convex achromatic lenses.

The distance (measured along the optical axis) of R_2 from S should equal

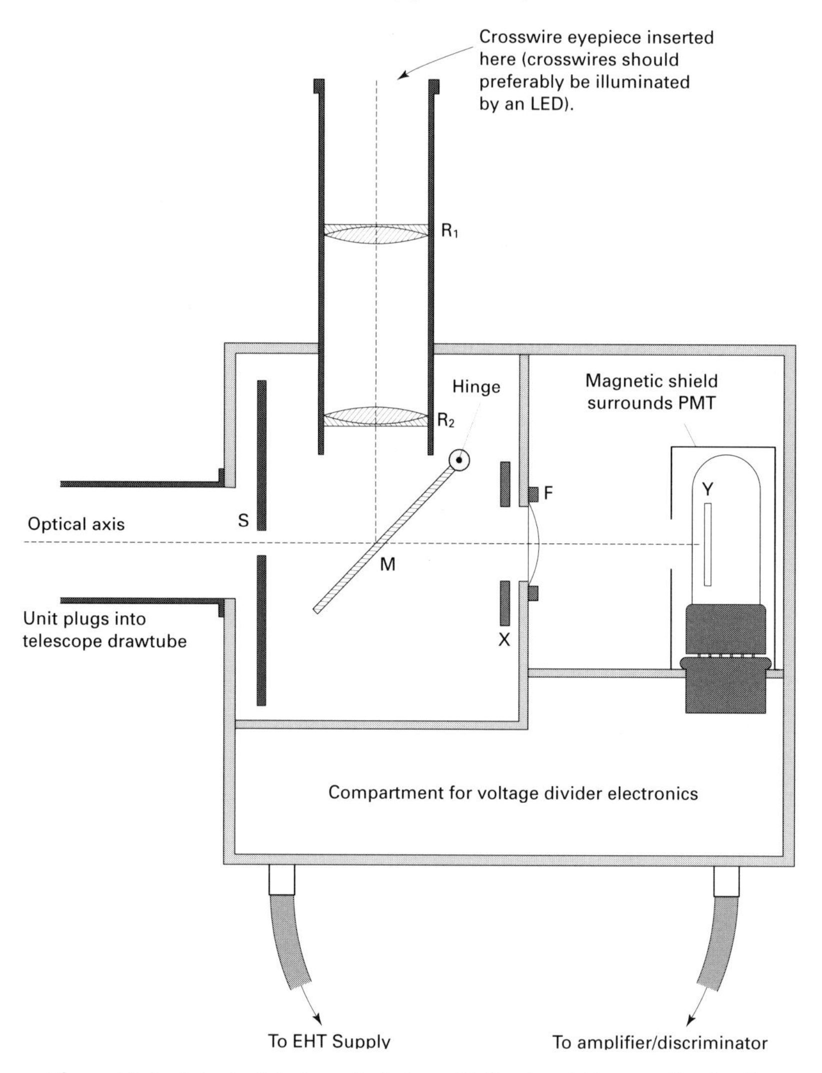

Figure 13.4. A typical design of photomultiplier head. See text for details.

its own principal focal length. The light rays passing between R_2 and R_1 will then be parallel. The distance separating them can then be as little or great as is convenient without affecting anything else. The distance of the final focus from R_1 is equal to its own principal focal length. It is at this position that the crosswires are placed when correctly focused. Hence, when the mirror is in the position shown in the diagram, the aperture S can be viewed through the eyepiece. If a star is centred in S, then that star can also be seen from the eyepiece.

In practice, S is mounted on a slide, together with a much larger aperture. The large aperture is used to initially centre the star (using the eyepiece crosswires). The aperture S is then selected for precise centring and for making the brightness measurement.

When the mirror, M, is swung out of the way it blocks off the end of the 'viewing-tube'. This is desirable to stop any stray light that may enter through the eyepiece affecting the PMT. The light from S is then free to pass through a slide, X, carrying a clear aperture and filters. Next, the light passes through the *Fabry lens*, F. A simple, non-achromatic, plano-convex lens will do to act as a Fabry lens.

The Fabry lens **does not** focus the image of the star on the photocathode, Y. Instead it focuses an image of the telescope's objective lens or primary mirror on the photocathode. The reason is that the photocathode will not be evenly sensitive across its surface. Even tiny movements of the star's image would then cause significant changes in the reading obtained. By imaging the telescope's primary mirror or objective lens, the system becomes relatively insensitive to small changes in position; at least it does, provided the star image stays well inside the aperture defined by S.

What diameter and focal length should the Fabry lens be? How does one calculate its correct position? The following formulae will provide the answers:

$$D = A/f_T,$$

$$\text{and } F_F = B\,(F_T + A)\;/\;(F_T + A + B),$$

$$\text{and } B = F_F\,(F_T + A)\;/\;(F_T + A) - F_F,$$

$$\text{and } d = (B \times \text{telescope aperture})\;/\;(F_t + A)$$

where,

F_T is the effective focal length of telescope,
f_T is the effective focal ratio of telescope,
F_F is the focal length of the Fabry lens,
D_F is the minimum required diameter of the Fabry lens,
A is the distance of the Fabry lens from S,
B is the distance of the Fabry lens from the PMT photocathode,
d is the diameter of the image of telescope objective/primary mirror formed on the photocathode.

As always, consistent units must be used throughout when using these equations. The image of the telescope objective formed on the PMT's

Figure 13.5. A photomultiplier head built by the late Jack Ells of Bexley Heath.

photocathode should ideally nearly cover it. The IP21 photocathode covers an area of 8 mm × 24 mm. Hence the ideal diameter of the image of the primary mirror is 8 mm. It could be a little less but any bigger and some of the precious light will be wasted.

The photomultiplier tube is shielded from the effects of the Earth's magnetic field (and that caused by electrical currents in the vicinity) by a metal screen. This is particularly important since the photomultiplier head rides on the telescope and changes its orientation with respect to the Earth's field. Otherwise, the flow of the electrons within the photomultiplier is disturbed and may well result in incorrect readings.

One feature of paramount importance is safety. If you are building a photoelectric photometer for yourself, do make sure that all electrical insulations and connections are absolutely safe. Remember the PMT has to be supplied with about 1KV and the unit will be used in the dark and dampness of night.

You must make sure that the casing of the photomultiplier head is absolutely light-tight. The only light that should fall on the PMT is that passing through the aperture S. Figure 13.5 shows an external view of one unit designed and built by the late Jack Ells, of Bexley Heath, England.

Photoelectric photometry: telescope requirements and ancillary equipment

The users of small telescopes often have problems balancing them when they try and use heavy pieces of auxiliary equipment. If the telescope tube can't be re-positioned and its existing counterweights won't cope, then there is nothing for it but to add extra, strap-on, counterweights. Try to ensure that the telescope remains balanced both in declination and right ascension, no matter how it is pointed. A little experimentation should solve this problem. The photometer head described in the foregoing notes may well weigh several pounds/kgf.

Norman Walker, formerly of the Royal Greenwich Observatory, has designed a 'Joint European Amateur Photometer' (JEAP) system which has the PMT totally separate to the telescope, instead of being carried on it. Only the aperture, relay-lens and eyepiece, and mirror are attached, in a light-tight box, to the telescope. A fibre-optic cable (or a fluid light-guide) conveys the light from the box to the photomultiplier tube, which is mounted in its own container. Details of Norman Walker's design may be found in a paper he wrote for the December 1986 issue of the *Journal of the British Astronomical Association*, 'The Joint European Amateur Photometer'.

The telescope must also track well. While the measurement is being made the star is not visible in the photometer eyepiece. Typical integration times may vary from 10 to 30 seconds. During this interval the star's image must stay well within the 1 arcminute aperture, S. If your telescope does not drive well enough to do this try to identify just what is wrong. I offer some advice on how to correct driving problems in Chapter 5. If the fault cannot be cured then there is nothing for it but to replace all, or at least the affected part, of the telescope mounting.

Figure 13.6 (a) to (d) illustrates the alternative arrangements of ancillary equipment necessary to turn your photometer head into a working photometer. In each case the voltage supply unit must be stable to better than 0.1 per cent at 1 kV. The voltages to the various dynodes of the PMT are usually divided off using a standard series resistor voltage divider circuit, with the anode earthed and the photocathode at -1000 V.

(a) Shows the simplest arrangement. Here the output from the amplifier is simply read by an analogue meter (either an ammeter, or a voltmeter depending on the output of the amplifier). The problem with this arangement is that the output is likely to constantly vary because of the flickering starlight. A degree of damping should be built into the meter. A chart recorder could be used as an alternative to the analogue meter but will be expensive if bought new. The same thing applies for high quality digital meters.

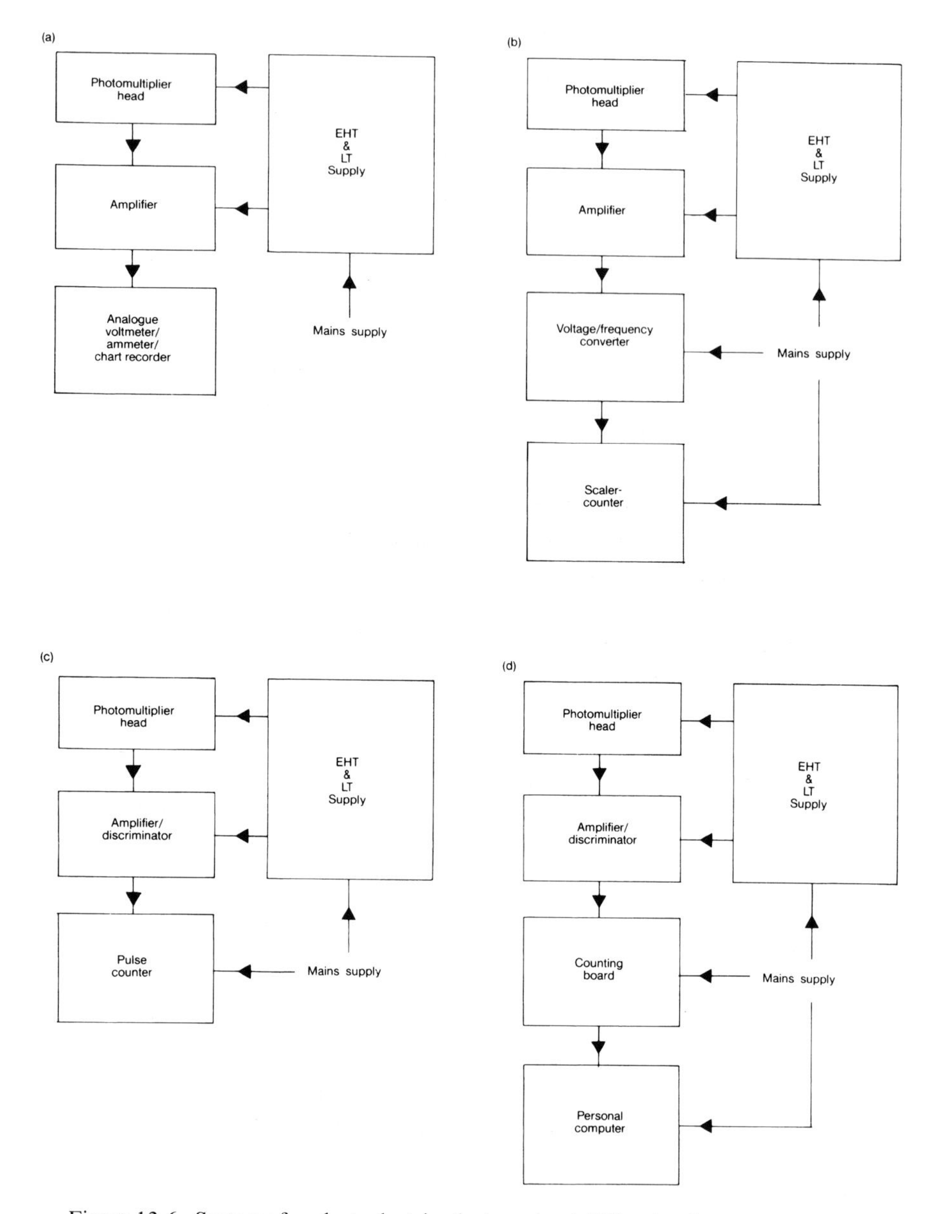

Figure 13.6. Systems for photoelectric photometry. (a) The simplest system uses an analogue meter at the output of the amplifier. Another alternative is to use a chart recorder. (b) A better solution is to convert the voltage amplifier output to a stream of pulses. The pulse frequency is proportional to the output voltage from the amplifier. The pulses are then counted over a given time interval by a scalar-counter. (c) Provided high quality electronics are used, the best accuracy is obtained by photon counting. (d) A further refinement is to use a computer for data-logging. It could also be used to automate the process of taking the readings, or it could even take over the control of the telescope.

(b) Shows a better design. Here the output from the amplifier (the output signal being in the form of a voltage) is converted to a series of pulses by a voltage/frequency converter. The output frequency is proportional to the voltage. The scalar-counter unit counts the number of pulses in a pre-set time and then displays the result.

(c) Is a refinement which is possible if the highest quality electronics are used. This is *photon counting*. In effect, individual photons of light cause pulses within the photomultiplier and these are amplified and counted. The PMT will not register every photon of light that falls on the photocathode (the detector quantum efficiency, DQE, falls short of 100 per cent) but a constant fraction of them will be recorded. The amplifier should be of particularly high quality and it and the pulse-counter must allow for the possibility of recording very high count rates. The amplifier contains a built in descriminator which 'weeds out' most of the spurious signals which would otherwise cause the counter to register.

(d) One could use a computer to record and analyse all the data, or even partially automate the taking of the readings. Figure 13.7 shows a photograph of John Watson's 10-inch (254 mm) Newtonian reflector equipped with a photometer head and ancillary and computer data-logging equipment. Going one step further, one could put the whole operation of the telescope and the photometer under the control of a computer. The late Jack Ells was one of the few amateurs so far that have done this. Figure 13.8 (a) to (d) shows four views of the 'automatic photoelectric telescope' (APT) that he, together with his sons, had built.

The APT is built around an 8.3-inch (212 mm) *f*/4 Newtonian reflector and utilises a JEAP type photometer, equipped with an EMI 9924 end-window PMT. The photometer is used in the 'photon counting' mode. The ancillary equipment, including computer, is carried on a wooden trolley. This is stored inside the run-off shed during the observing runs. Meanwhile Mr Ells was free to pursue other interests while the telescope busied itself selecting and measuring variable stars. As an example of the results that Mr Ells obtained with his equipment, Figure 13.9 shows the light-curve of the eclipsing binary star V1143 Cygni.

The automation of telescopes is briefly described in Chapter 3 of this book and the construction and operation of the APT is described in three successive articles beginning in the December 1989 issue of the *Journal of the British Astronomical Association*.

A local astronomical society has re-sited the APT and taken over the running of it. It continues to be a highly successful and productive monument to the skill of its originator.

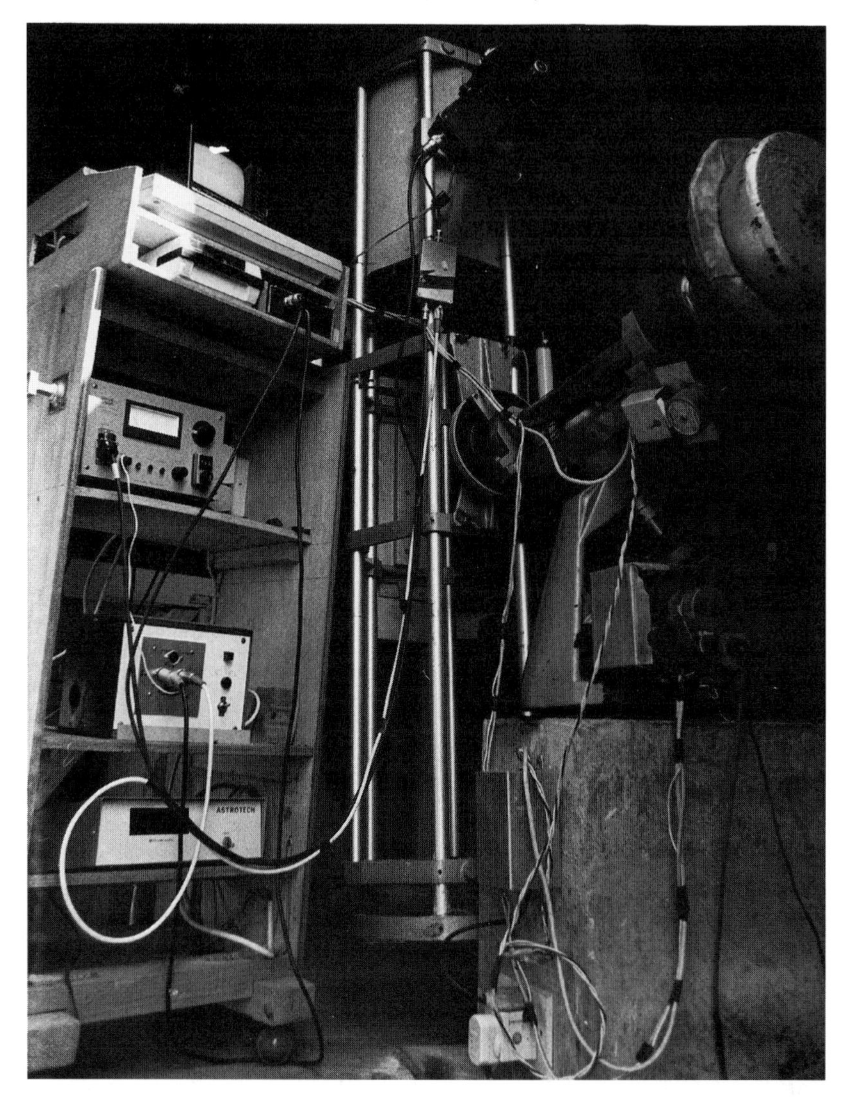

Figure 13.7. John Watson's 10-inch (254 mm) Newtonian reflector, equipped for photoelectric photometry. The exterior of the dome covering this telescope is shown in Chapter 3 (Figure 3.15). The interior is painted matt black to reduce stray light. It is also for the same reason that the top end of the telescope's skeleton tube is closed in with a baffle. The cabling attaches the photometer head to the anciliary equipment and the data-logging computer which are stacked in the movable trolley.

Photoelectric photometry: taking the readings

There are two basic methods of photometry: *all-sky* and *differential*. All-sky photometry involves taking direct readings of star brightnesses over a large portion of the sky. It requires observing conditions of a standard that the backyard telescope user will almost never encounter. So, having mentioned all-sky photometry, we will now discard it and concentrate on the alternative.

Differential photometry involves measuring the differences in brightness of the star in question and nearby comparison stars. All the methods of photometry described so far use this technique.

A photomultiplier tube must be allowed to settle down after first switching on (see also the section on precautions). Allow at least half an hour. Next locate the area of sky containing the variable star, first using the finder and then the photometer eyepiece.

With no light falling on the PMT (i.e. with the mirror down and any necessary baffles tightly closed), take a reading. This is the *dark current reading, D*. Next move the telescope so that the photometer aperture covers a blank area of sky. Take a reading. This is the *sky reading, S*. Next centre the comparison star in the photometer eyepiece and take a reading. This is the *comparison star reading, C*. Now take another sky reading. Now centre the variable star and take a reading. This is the *variable star reading, V*. Once again repeat the sky reading and then, once again, the comparison star reading. Finally note the date and time of the observation. To summarise, the sequence goes:

$$D\ S\ C\ S\ V\ S\ C$$

This sequence is not sacrosanct but every star reading should be bracketed by sky readings. The sequence should then be repeated with a different comparison star. If all is well, the two sets of figures will give identical values for the magnitude of the variable star.

Beginning each sequence with a dark current reading will allow you to keep a check on the dark current throughout the observing run. It should remain constant, or at least gradually change at an even rate. If the dark current reading is a large fraction of the star readings, and it changes unpredictably, then your results won't be very reliable or accurate.

If the readings are displayed on an integrating system such as a scalar-counter, then an integration period of 10 to 30 seconds will normally be found adequate for each of the readings. Longer integration times may not be possible owing to inaccuracies in the telescope drive rate.

So far I have made little mention of filters. This is treated more fully later

(a)

(b)

Figure 13.8. Four views of the late Jack Ells' 'Automatic Photoelectric Telescope' (APT), formerly at Bexley Heath, England. Note the swivelling head section of the telescope tube. Another feature is the baffle covering the end of the tube (best seen in (d)) which restricts the amount of off-axis light that can enter the telescope.

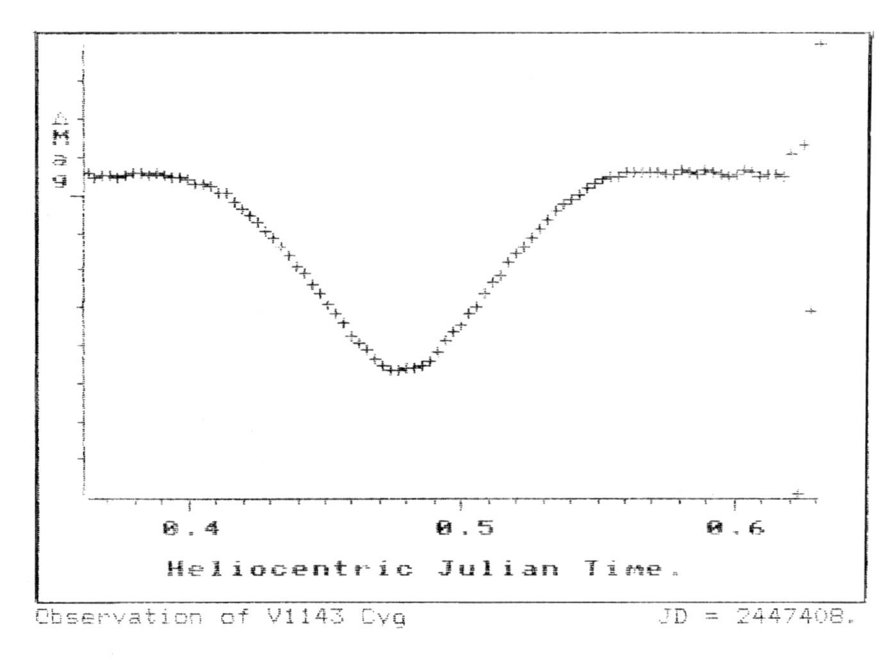

Figure 13.9. The minima of the eclipsing binary star V1143 Cygni, obtained by the late Jack Ells using his APT, on 1988 September $3^{d}/4^{d}$ (Julian Day 2447408). The scale along the bottom shows the time elapsed in fractions of a Julian day. The scale up the side expresses the change in magnitude of the star. The marks on this scale are separated by $0^{m}.1$.

in this chapter. Suffice it to say here that certain tasks, such as following the light-curve of an eclipsing binary star in order to determine its period, do not demand the use of a filter. However, one should use a standard photometric filter when making the more usual sort of brightness measures on stars if your results are to be scientifically comparable to those of others.

Photoelectric photometry: analysing the readings

To find the magnitude of the variable star proceed as follows:

1. Subtract the dark current reading from all of the others (interpolating for the value at the time of the observation if it is found that the dark current reading slowly changes throughout the observing run).
2. Subtract the sky-background readings from the comparison star and variable star readings (interpolating between the measures taken before and after each star reading).

3. Determine the difference in magnitude, Δm, between the variable star and the comparison star using the formula:

$$\Delta m = -2.5 \text{ Log } (V'/C'),$$

where V' and C' are respectively the variable star and comparison star readings **after** subtraction of the dark current and sky-background readings.
4. Calculate the magnitude of the variable from:

$$m_{\text{var}} = m_{\text{comp}} + \Delta m,$$

where m_{var} and m_{comp} are the magnitudes of the variable and comparison stars, respectively.

How accurate is your answer? In theory the uncertainty of a reading, N, is equal to $1/\sqrt{N}$. Hence if 100 pulses are counted in a given integration time then the uncertainty amounts to 0.1, or 10 per cent. This is equivalent to $0^{\text{m}}.1$. The uncertainty of a count of 10000 is then only 0.01 or 1 per cent, corresponding to $0^{\text{m}}.01$.

In practice instrumental defects may cause an error several times larger than this value. The best way of determining the uncertainty of a brightness determination is to take several readings. Accept the average as the most likely true reading. The difference between the highest and the lowest readings divided by the number of readings then gives the probable uncertainty in the result.

For more about the methods of photoelectric photometry and the analysis of the results refer to the listing of books and articles given in Chapter 17.

Photoelectric photometry: precautions

(a) The PMT is a very delicate device. Avoid subjecting it to sharp knocks.
(b) If you have to touch the PMT clean it in alcohol (not methylated spirits which will leave a deposit).
(c) Under no circumstance subject the PMT to bright light when the E.H.T. supply is switched on. The PMT will be permanently damaged (the major effect noticeable will be a large increase in the dark current). You should make absolutely sure that the photometer head is absolutely light-tight and that all light baffles are adequate. Make sure that the mirror is in the correct position before switching on any bright lights.
(d) Try to prevent the inside of the photometer head, particularly the PMT itself, from becoming damp. Stray currents may significantly distort the readings. An enclosed heater may be undesirable because it will increase

the dark current. A better solution is to enclose a packet of silica gel but do ensure that it is removed and dried out frequently.

(e) **Above all – do make sure that all the electrical apparatus is well earthed and is properly insulated. Don't ever take risks with electricity.**

Filters and star colours

In order to classify star colours, astronomers use filters in combination with a photometric device. The UBV system was originated by Johnson and Morgan. The U filter has a passband which is centred on a wavelength of 365 nm in the ultra-violet part of the spectrum. Similarly the B filter is centred on the blue, at 440 nm, and the V (for 'visual') filter has a passband centred on the yellow, at 550 nm. The system has been extended to include red (R, 700 nm) as well as a number of infrared wavebands. If one observer's photometric measures are to be strictly comparable with those of another, then a standard photometric filter (usually V) should always be used. These can be supplied by equipment manufacturers (listed in Chapter 17). If you buy a ready-made photoelectric photometer you will normally find B and V filters included as standard.

The *colour-index* of a star is its difference in magnitude as recorded by two different wavebands. The value usually quoted is the difference between the blue and the yellow magnitudes, (B−V). This can also be written as ($m_B - m_V$). The star Vega (α Lyrae) has a blue magnitude of $0^m.0$, the same as its visual magnitude. Hence Vega also has a colour index of 0.0. Bluer stars than Vega have negative colour indices. Redder stars have positive values.

As far as I know, not many amateurs follow the colour variations of stars, together with their V-band alterations in brightness. Yet this is a fascinating field. Since the colour of a star is linked to its photospheric temperature, measurements of colour-index add an extra dimension of insight to its variations. I heartily commend this study to those with the necessary equipment.

CCDs and photometry

On the face of it, the CCD camera is the ideal photometer. It has great sensitivity and has a large dynamic range. It is also relatively easy to use and it can image enough of the sky to perhaps contain suitable comparison stars, as well as the star to be measured, all in the one frame. All this is true. With software already in existence all the operator has to do is move a cursor around on a monitor screen and 'click' to read off the brightnesses recorded at chosen locations in the image.

However, there is one main problem that has not been overcome at the time of writing these words: no suitable filter set has yet been settled on that will work with given CCDs to generate the same responses as the standard UBV system adopted with photomultiplier tubes. The standard system has been around for so long and is so deeply rooted that abandoning it is virtually unthinkable. This seems like a problem that should fairly easily be overcome and I think that it will be. When that happy day arrives an ever larger number of amateur astronomers will be enthusiastically 'getting numbers from the sky'.

14

Double stars, star clusters, nebulae, galaxies and supernovae

Double star measures provided the astrophysicist with essential raw data. It is from these data that binary star orbits and stellar masses can be determined. Combining these with the results of other types of observation, the astrophysicist can build up a portrait of the component stars and gain insights into the structure, behaviour and evolution of stars in general. Sadly, though, the number of amateur astronomers carrying out this work is vanishingly small. In part this is due to it being perceived as far less exciting than, say, hunting novae or planetary observing. Many are also put off by the exacting demands made upon the observer and the telescope.

By contrast, observing *star clusters, nebulae* and *galaxies* is a highly popular pastime among backyard observers. The results may not advance astronomy but this is more than made up for by the knowledge that one is observing the largest structures in the Universe, situated in the depths of space and time.

Are the vast stellar systems, even the external galaxies, totally devoid of life and wheeling in cold loneliness in the void? Perhaps. On the other hand, when we view a galaxy as a hazy smudge, are we really looking down on to myriads of worlds teeming with almost unimaginable varieties of life? Maybe some alien 'backyard observer' is, perchance, pointing his/her/its telescope in our direction and wondering the same thing!

Supernovae, collosal stellar explosions where a star can shine as brightly as a galaxy, are particularly exciting. It is thanks to some star destroyed aeons ago that the Sun, the Earth, you and I exist. Supernova hunting is a very valuable pastime. No two supernovae are ever quite the same and there are always lessons to be learned from their study, particularly if they are spotted when the star is still rising towards maximum brilliance.

Observing double stars: telescope and observer

Countless numbers of double stars are within reach of amateur sized telescopes. Some, such as those with strongly colour-contrasting components, are particularly beautiful. Albireo (β Cygni) and Almaak (γ Andromedae) are popular examples but there are many more. As previously indicated, the amateur observer can do particularly valuable work in measuring the separations of double stars **provided** he/she can do it accurately. Inaccurate measures are worse than useless, as they are actively misleading. Note that I use the term *double stars*, interchangeably with *binary stars*. Strictly, a binary star system is a double star where the components are truly gravitationally associated with one another. There are still many double stars in the sky which may or may not be actual binary star systems. Our knowledge is very far from being complete. Measures made over a long period can resolve this issue in each case.

The few professional astronomers are involved in this work mainly concentrate on the binary stars of the closest separations, since these tend to orbit each other in relatively short periods. Some others work on *spectroscopic binary stars*, which are far too close together to be resolved by ordinary techniques. They use a high dispersion spectrum of the binary pair in order to sort out the spectra of the two stars. The cyclical relative motions of the spectral lines then reveal the radial velocity component of the motions of the stars. A very few others work with other techniques, such as speckle interferometry, capable of producing results from stars too close for 'straightforward' resolution. At the other extreme the widest pairs are scrutinised by astrometrists, particularly in connection with proper motion studies.

The amateur astronomer will not have the resources necessary to compete with the professionals making measures of very close or very wide double stars. However, he/she can contribute by measuring some of the large numbers of double stars with separations ranging between about 1 and 15 seconds of arc. Some of these have only ever been measured once or twice since their discovery! Many more have not been measured for decades.

Telescope resolution is discussed in Chapter 1. The empirical formula given by Dawes, for refractors, is that usually quoted in connection with resolving double stars. This is:

$$R = 4.56/D,$$

where D is the telescope aperture, measured in inches and R is the minimum resolvable separation, in arcseconds, of two sixth magnitude stars. In metric units this becomes:

$$R = 116/D,$$

where D is measured in millimetres. This formula assumes that the optical quality of the telescope and the seeing conditions are good enough to produce fairly well-defined stellar diffraction patterns. The obstruction due to the secondary mirror and its support vanes inherent in a compound telescope will, it is true, cause some re-distribution of light within the diffraction pattern. However, the actual dimensions of the Airy disc and rings are only marginally altered provided the central obstruction is smaller than about one third the diameter of the primary. If that is so then the potential resolving power of a compound instrument is virtually equivalent to that of the refractor of equal clear aperture. More troublesome are tube currents. On this score a totally enclosed optical system, such as a refractor or a catadioptric telescope, might prove superior.

As discussed in Chapter 2, the minimum magnification to see diffraction-limited detail through a telescope is about ×20 per inch (×8 per centimetre) aperture, for a normal-sighted observer. Magnifications up to about twice this value are useful for observing double stars near the diffraction limit in order to avoid eye-fatigue. However, it is one thing merely to **see** a double star. It is quite another to actually **measure** it. For the latter, the highest possible powers are recommended; perhaps even up to ×80 or ×100 per inch (×32 to ×40 per centimetre) of aperture if the conditions allow.

Double star lists and nomenclature

One primary source of double star data is R. G. Aitken's *New General Catalogue of Double Stars*, which was published in 1932. In this catalogue, the doubles are given 'ADS' numbers. For instance ADS 6175 is Castor (α Geminorum) and ADS 10087 is λ Ophiuchi. Doubles are often also listed by a code which indicates their discoverers. Alphabetically, the key letters of the code covering most doubles of interest to the amateur are: Aitken (A), Burnham (β), Couteau (C), Espin (Es), J. Herschel (h), W. Herschel (H), Hough (Ho), Hussey (Hu), Innes (I), Kuiper (Kpr), Milburn (Mlb), Muller (Mlr), Rossiter (Rst), See (λ), O. Struve (OΣ), W. Struve (Σ), and Van den Bos (B). For instance, Castor is also known as Σ1110 and λ Ophiuchi is also known as Σ2055.

The brighter component of a binary star is termed the *primary* and is denoted by the letter A. The companion is called the *secondary*, or the *comes*, and is denoted by the letter B. The letters C, D, etc. are used for the additional component stars if the system is multiple.

Obviously, Aitken's catalogue can only be obtained through a good technical library, or by inter-library loan. National astronomical society libraries are also likely to carry copies of this volume. A number of current works do provide listings of double stars. One such is *Webb Society Deep-sky Observer's Handbook*, Vol 1, *Double Stars*, edited by Kenneth Glyn Jones, Enslow publishers, 1986. See Chapter 17 of this book for a further listing of books and articles for the double star observer.

Measuring double stars: using an eyepiece Micrometer

Figure 14.1(a) shows a crude eyepiece micrometer. This item was included in a package of equipment I purchased many years ago. It was home-made by its original owner. The micrometer is plugged into the telescope draw-tube. An eyepiece is plugged into the device in order to view the pointers. The position of the eyepiece is adjusted until the pointers are seen in sharp focus. The telescope focuser is then adjusted until the stars, or other celestial bodies, are also in focus.

The gap between the pointers is adjusted by means of the large wheel on the right. A 'tachometer' is also operated by this adjustment and so registers the size of the gap as a proportionate number. For convenience both the pointers can be moved, as one unit, by adjusting the small control to the left. In use the double star, planetary disc, or whatever else is to be measured, is positioned between the pointers and the gap adjusted. In the case of a planetary disc, the pointers should both just 'kiss' opposite limbs of the planet. For measuring the separation of double stars, each pointer should just bisect each star image. The tachometer reading is then converted to an actual measurement in arcseconds by means of the *micrometer constant*, described shortly.

I have tried out this micrometer by measuring double stars of known separation. With this device used at the 102-inch (2.59 m) Newtonian focus of my 18¼-inch (464 mm) reflector I can measure separations to within an accuracy of about 3 arcseconds (almost irrespective of the actual separation). This is nowhere near good enough for useful results on double stars. The addition of a Barlow lens, or amplifying relay-lens, would help and this should be standard on any micrometer. However, an amplification factor of ×10 would produce an effective focal length of 1020 inches used with my 18¼-inch telescope (and therefore a magnification of ×1020 when a 1-inch focus eyepiece is plugged into the micrometer). Even so, the measures would still only be reliably accurate to within about 0.3 arcsecond – still not good enough. 0.1 arcsecond accuracy is required.

Much better is a proper *filar micrometer*. Figure 14.1(b) shows a cut-away view of the cross-slide arrangement of a twin drum filar micrometer.

(a)

(b)

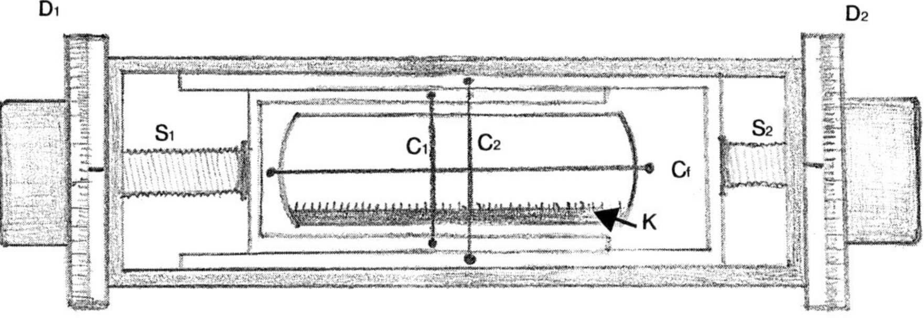

Figure 14.1. Eyepiece micrometers. (a) shows a crude, home-made, version, suitable only for low accuracy measures. In this view the eyepiece is removed to show the pointers. The gap between the pointers is altered by adjusting the large control on the right. This also 'dials up' a proportionate number on the tachometer. Turning the small control on the left causes both the pointers to move sideways as one unit. Note the springs which keep the cross-slide firmly in contact with the adjuster. (b) is a cut-away diagram of the cross-slide arrangement of a proper two-drum filar micrometer. The tensioning springs are not shown on this diagram, nor the eyepiece and drawtube assembly. Turning the calibrated drum D_1 advances the lead screw S_1 and so moves the cross-slide containing the crosswire C_1. Similarly drum D_2 adjusts the position of crosswire C_2. The comb, K, has teeth which are at the same pitch as the threads on the lead screws S_1 and S_2.

In other respects the mechanical arrangement is very similar to the crude eyepiece micrometer shown in Figure 14.1(a). The crosswires are traditionally made of spider's web (that of the Black Widow being favourite), though the very finest drawn wires could also be used.

On Figure 14.1(b), C_1 and C_2 are two crosswires movable by adjusting the calibrated drums D_1 and D_2, respectively. The scales on the drums are each divided into hundredths of a revolution. C_f is a fixed crosswire, set perpendicular to C_1 and C_2. One turn of each drum advances the corresponding crosswire a distance equal to the pitch of the thread of the lead-screws S_1 and S_2. The comb, K, has teeth of the same pitch as the threads on the lead-screws. Hence the separation of the crosswires can be determined by counting the number of teeth on the comb between the wires and noting the drum readings. For instance, if the crosswires are separated by slightly over four teeth and the drum readings are 43 and 56 (meaning 0.43 turn and 0.56 turn) then the separation is recorded as 413. This number is converted into arcseconds using the known value micrometer constant. How this is found is discussed shortly.

A slightly simpler version is the single-drum micrometer. In this type one of the vertical crosswires is fixed and the two can be moved together as one unit by an adjuster. Making a truly accurate filar micrometer will prove challenging to the amateur machinist. Certain manufacturers do supply filar micrometers but make sure that what you buy really is suitable for making accurate measures.

It is highly desirable that the micrometer has illuminated crosswires. Failing that you can arrange your own illumination by means of a strategically placed bulb to throw a glow over the sky-background, as seen through the eyepiece. However, this remedy does have the disadvantage that the fainter stars will tend to be swamped and so be much more difficult to measure. The following notes detail the procedure for making measures with the micrometer, the instrument already being set into the telescope and the eyepiece and telescope both focused.

Determining the value of the micrometer constant

The first step is to close the gap between the crosswires completely. The reading should be zero. Is it? If not, then the zero reading will have to be subtracted from all readings obtained with the micrometer.

Next, separate the crosswires by a known amount, say 1000 divisions. Then choose a suitable star (reasonably bright so that it is easily recognisable through the micrometer's eyepiece) and bring it to the fixed crosswire.

With the telescope drive switched off the star's image should exactly track along the crosswire. If it does not then rotate the micrometer and try again. Keep trying until the star image remains in contact with the crosswire as it drifts across the field. This orientates the fixed crosswire exactly east–west. Then, without disturbing the micrometer further, time the passage of the star between the two movable crosswires. Record this time, t, in seconds.

The angular separation, r, of the crosswires, in arcseconds, is then given by:

$$r = 15t \cos \delta,$$

where δ is the declination of the star. Ideally, a star should be chosen of declination which causes the time to traverse the gap between the crosswires to be around 30 seconds. If the star only takes a few seconds, then the timing will suffer from a large uncertainty (the start and stop times are each not likely to be accurate to better than 0.1 or 0.2 seconds). Finally the value of r is divided by the micrometer reading (1000 in our example) in order to find the number of arcseconds per division. This is the value of the micrometer constant.

Provided the micrometer is always plugged into the same telescope, the value of the micrometer constant need only be determined once. However, if anything changes, such as the position of a Barlow lens, or of a Cassegrain reflector's secondary mirror, etc., then you will need to make a fresh determination. It is a good idea to run a check by making separate determinations of the micrometer constant for a range of crosswire separations. Naturally all the figures should agree, though do bear in mind that the closest spacings will produce the greatest percentage uncertainty in the values obtained. You could plot a graph and note any general trend.

Measuring a double star's position angle

The value of the micrometer constant having been determined, we can now get down to the business of measuring double stars. The first measure to make is that of the star's *position angle* (PA). Position angles, in connection with the Moon, are mentioned in Chapter 7 of this book. The position angle of a binary pair is always that of the secondary, B, component measured with the primary, A, component as the origin. Due north defines the zero of position angle. The value of position angle increases eastwards, being 90° at due east, 180° at due south, 270° at due west and further increasing to 360° (identical to 0°) at due north.

The first step is to render the fixed crosswire's orientation exactly east–west as previously described. If the micrometer (or the telescope) has a

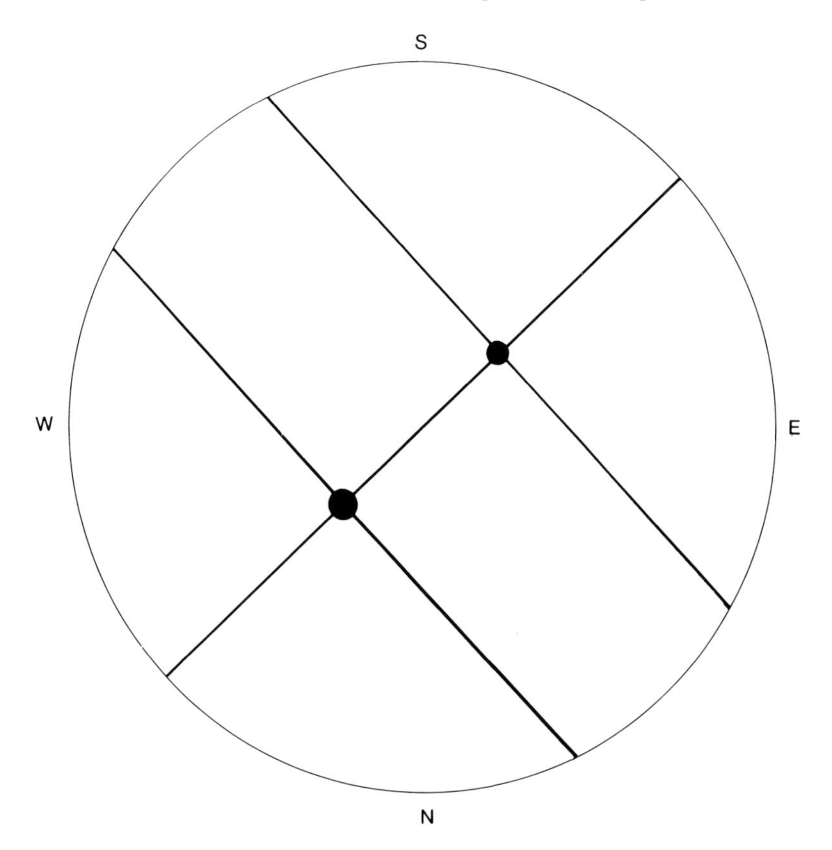

Figure 14.2. The field as seen through an eyepiece micrometer, with the crosswires set on a double star. The position angle of this double is roughly 135°.

movable position angle scale fitted then set it to 270°/90°. If the scale is fixed then take a note of this east–west reading.

Carefully move the telescope and rotate the micrometer as necessary to get both the A and B components of the double star on the fixed crosswire (see Figure 14.2). Now take the new reading. If the PA scale was first set to 270°/90°, then the new reading is the correct PA of the star. If not then use the previous reading (adding or subtracting as necessary) to calculate the correct PA.

Remember, if the B star is anywhere in the semicircle east of the primary, then the PA of the pair is between 0° and 180°. If the B star lies anywhere to the west of the primary then the PA is between 180° and 360°. The position angle of the binary illustrated in Figure 14.2 is roughly 135°.

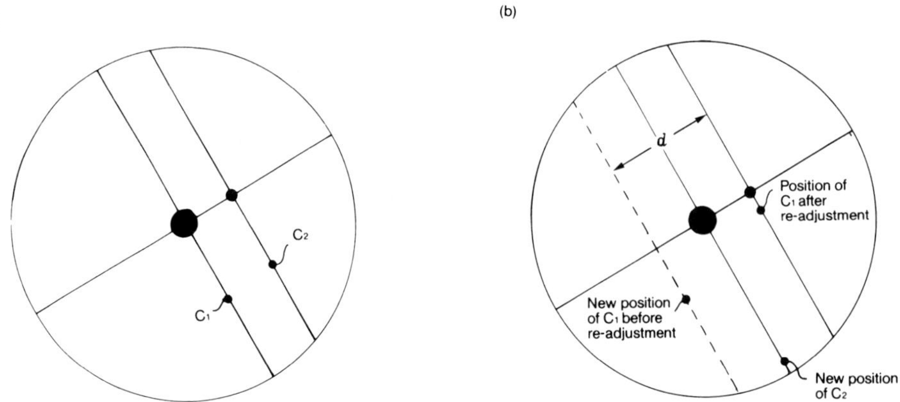

Figure 14.3. Measuring the separation of the components of a double star by finding the *double distance*. The first step is to set the crosswires on the star as shown in (a). In this example the crosswire C_1 is set on the primary. Then rotate the micrometer by 180°. Next move the telescope so that the crosswire C_2 is now on the primary. The dotted line shows the position of the crosswire C_1. C_1 is then aligned on the secondary by adjusting the micrometer drum D_1. The shift, d, in C_1 corresponds to twice the separation of the double star components.

Measuring a double star's separation

In principle at least, all the hard work has already been done. With the fixed wire still aligned along the pair of stars, the movable wires are adjusted until they bisect the stars. The drum readings and number of teeth on the comb between the wires then give a number which is multiplied by the value of the micrometer constant (remember to adjust for the zero error if applicable). This is the separation of the star in arcseconds.

A better method is to set the crosswires on the star this way, **then rotate the micrometer by exactly 180°**. If crosswire C_1 was on the primary star, then adjust the telescope slow motions so that crosswire C_2 is now centred on it – **but don't touch the micrometer drums at this stage**. The crosswire C_1 will now lie on empty space (see Figure 14.3(a)). **Now** note the drum reading D_1. Then, keeping crosswire C_2 aligned on the primary, adjust the position of crosswire C_2, passing it across C_1 until it is aligned with the secondary star (see Figure 14.3(b).

Now note the new reading of drum D_1. Don't touch drum D_2 at any stage during this operation. Treat crosswire C_2 as fixed and make any necessary re-alignment of the primary star with it by means of the telescope slow motions only. The difference between the two readings of drum D_1 represents the *double distance* of the primary from the secondary.

Measuring the double distance may seem very involved but it has the twin benefit of being more accurate, since the micrometer is being used to effectively measure twice the separation of the star, and any zero error is effectively cancelled out, owing to the 180° reversal of the micrometer.

Naturally one should repeat each measure a number of times. The average is taken to be the likely true reading and the range in values divided by the number of measures gives the value of the uncertainty.

Filar micrometers do vary in design. Some have only one movable crosswire. Some have no horizontal fixed crosswire. In the latter case the instrument is rotated 90° in order to measure position angles. The reader should be able to adapt the foregoing notes to suit the instrument used without too much difficulty.

Measuring double stars: other techniques

Other types of eyepiece micrometer, not necessarily incorporating crosswires, are used by professional astronomers. As far as I know, few, if any, amateurs use these more elaborate devices. However, an account of the different types can be found in the *Webb Society Deep-Sky Observer's Handbook*, already mentioned. In addition, an account of a 'grating-type' micrometer consisting of a rotatable grid, with vernier scale, placed at the sky end of the telescope tube, is given in the June 1980 issue of *Sky & Telescope* in an article called 'Measuring stars with a grating micrometer'.

Photography is potentially a very accurate method of measuring widely spaced double stars. The telescope-camera is set up to record high-resolution images in the same way as for photographing the planets (see Chapter 5). Multiple exposures on the same piece of film will be used. So, the camera's wind-on mechanism should be overriden, or the camera exposure control set to the 'B' setting and a separate shutter used for making the exposures.

One star image is trailed across the frame to provide an east–west orientation. This can be achieved by making the exposure with the telescope drive switched off. The double star image, or preferably several images of the double star, are then exposed on the same film frame (the telescope is moved a little in R.A. between the exposures). Knowing the image-scale, the separation and position angle of the double star can subsequently be found by measurement. In essence it is as simple as that.

In practice there are difficulties. One is that the image-scale may not be constant over the film frame. One can test this by recording images of the double over a large portion of the frame. Measurement will then reveal any variations. The image-scale can be found by measuring the image of a double star of known separation. Use the shortest exposures that just record the star images. Pinpoint star images are measurable, enormous great blobs are not! Obviously, this technique puts high demands on the accuracy of the telescope tracking. This is another good reason for keeping the exposures short!

Hypersensitised Kodak TP2415 film may well be the best for this type of work because of its fine grain and high-resolution. The exposure times will likely be just a few seconds, to several tens of seconds, depending upon the telescope aperture, the atmospheric conditions and the brightness of the stars. A little prior experimenting will establish the correct exposures needed.

The final problem is actually measuring the negative. To measure positions directly requires the use of a two-dimensional plate measuring machine capable of 1 μm accuracy. Failing that, one must project the image onto a flat screen (such as a white painted wall). This will inevitably cause the image-scale to vary over the projected frame. In addition the negative should not be mounted into the projector in the usual way. It should be sandwiched between two thin pieces of glass in order to keep it flat. Provided the enlargement factor is sufficient (an arcsecond being at least 2 cm in the projected image) and great care is taken when measuring, then a millimetre ruler ought to be sufficient to enable one to achieve 0.1 arcsecond accuracy.

So, the photographic method while potentially very accurate is, in practice, more than a little difficult. However, some observers have achieved success using this technique. Two such people are Denis Buczynski and Rob Moseley. They have described their work in a short paper entitled 'An experiment in double-star photography' in the October 1989 issue of the *Journal of the British Astronomical Association*.

Perhaps the CCD allied to the computer will soon supercede all previous methods. I don't know of any amateurs actually measuring double stars this way at the time of writing these words but surely some will soon do so. Very high EFRs should be used so that each 'blob' formed by the stars' seeing discs cover several pixels. Using suitable software, the positions of the exact centres of each blob are recorded. Knowing the image-scale and true E–W orientation (found in the same way as for the photographic method) the software can calculate the value of the separation, in arcseconds, and the PA directly. The procedure should be repeated several times for each double star because of the motions of the star images caused by atmospheric turbulence. As always, the average value should be accepted and the uncertainty estimated by dividing the range in values obtained by the number of determinations made.

Observing star clusters and nebulae

First find your object. A listing of suitable atlases and books is given in Chapter 17. One work I would give special mention to is *The Universe from your Backyard* by David Eicher (Cambridge University Press, 1988). This book is arranged in a largely alphabetical listing of the constellations. Each constellation is surveyed and the author picks out a great many deep-sky objects, by their NGC and Messier (if applicable) numbers, and gives the reader hints on successfully viewing them. Star maps are provided and these show the locations of just about every deep-sky object accessible to the amateur's telescope.

I have covered the use of maps and atlases, together with the construction of finder-charts and 'star-hopping' techniques earlier in this book (Chapter's 10 and 12). These same techniques can be used for locating deep-sky objects.

What instrument is best for observing deep-sky objects? The answer to that question depends very much on the objects themselves. Light grasp is always important but so is field of view. For viewing large, sprawling, *open star clusters*, such as M44 (Figure 14.4 and Figure 10.1), binoculars are best. The typically 1° field of view of a medium sized telescope fitted with a low power eyepiece can only show a small portion of this type of object in one go. This is especially true for the largest forms, such as the star clouds of the Milky Way (Figure 14.5). Even a tighter star cluster such as the Pleiades (Figure 14.6 and Figure 4.9), or the famous 'Double Cluster' of Perseus (NGC 869/884, shown in Figure 4.12) still demands a field of view of over a degree across if all the main stars are to be seen together.

For viewing visually smaller deep-sky objects I would say use the largest aperture possible. However, it is also true that the magnification cannot be too high, otherwise the field of view will be too small. Yet, a low magnification could limit the effective aperture of the telescope. This is discussed in Chapter 1. As an example, if the apparent field diameter of the eyepiece used is 80° and a real field of 1° is required, then the maximum magnification that can be used is ×80. The resulting size of the exit pupil would be larger than the pupil size of the observer's eye for any telescope greater than about 50 cm (about 20-inch) aperture. Hence, however large the primary mirror, the telescope is effectively stopped down to about 20-inch aperture when a power of ×80 is used.

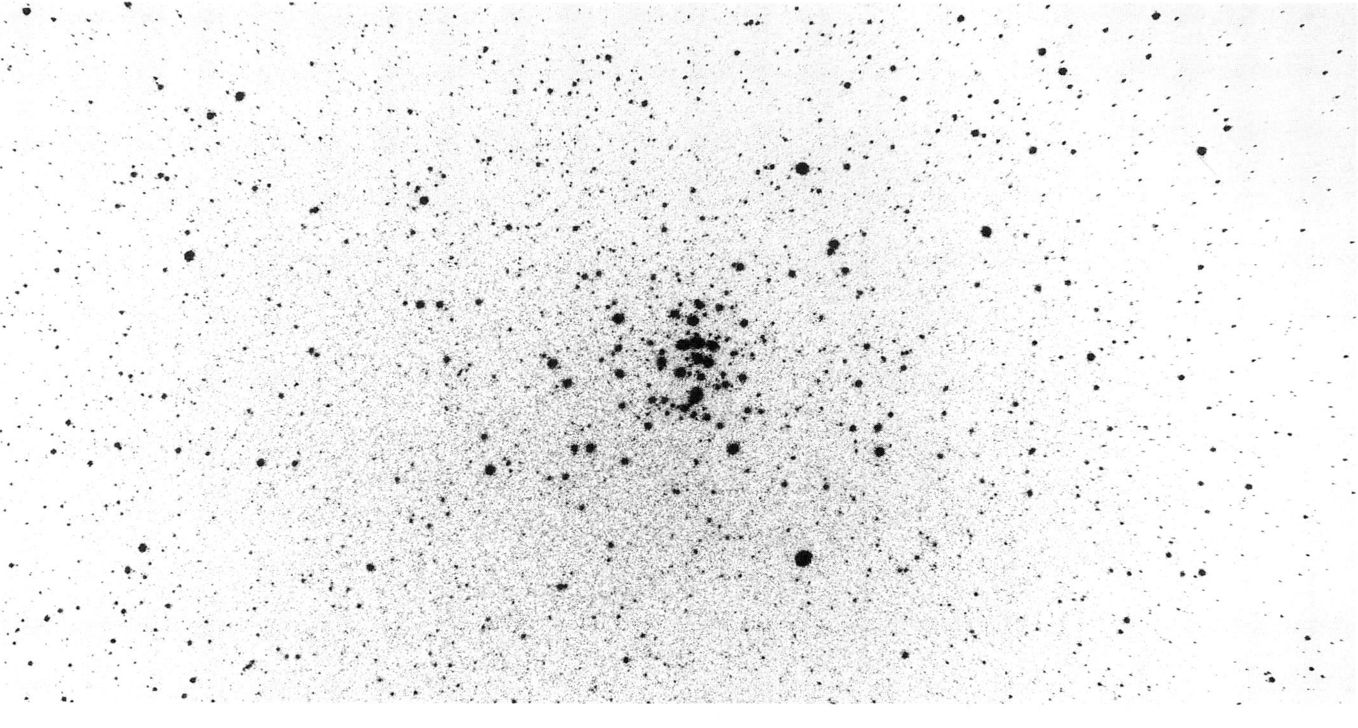

Figure 14.4. Negative reproduction of the Beehive star cluster, M44 in Cancer. The original photograph was taken on 3M Colourslide 1000 film by the author on 1988 February $12^{d}\ 21^{h}\ 57^{m}$ UT. The camera was fitted with a 135 mm telephoto lens, set to *f*/2.8, and fixed to a guiding telescope for the 5 minute exposure. North is uppermost.

Figure 14.5. The Milky Way in Cygnus photographed by Nick James. Nick used an ordinary camera fitted with a 55 mm lens set at *f*/2 for this 3 minute exposure made on Tri-X film on 1987 October 17^{d} 19^{h} 02^{m} UT. The camera was driven on Nick's 'barn door' camera mount. North is approximately uppermost and the bright star near the top of the frame is Deneb.

Figure 14.6. The Pleiades, photographed by Ron Arbour using his 16-inch (406 mm) Newtonian reflector stopped down to 11.4-inch (290 mm) *f*/7. (a) is a 10 minute exposure made on 1989 November 5^{d} 00^{h} 04^{m} UT on 120 format T-Max 400 film. (b) shows the result of superimposing two identical negatives when making the print, the second negative copied from the original. Note how the visibility of the nebulosity is enhanced. Ron stopped his *f*/5 telescope down to *f*/7 in order to sharpen the outfield star images, which would otherwise suffer from coma.

In addition, many deep-sky objects are better seen using higher powers than about ×4 per inch (×1.6 per centimetre) aperture. Owing to a reciprocity failure effect in the human eye, the apparent contrast between many objects and the sky-background glow is markedly increased by using higher powers. Just how well this works depends very much on the object. Planetary nebulae, in particular, respond well to higher magnifications. Even a fairly diffuse example, such as the Dumbbell Nebula (M27) in Vulpecula (see Figure 5.2) looks better at a higher power. I find that M27 looks best at a power of ×144, through my 18¼-inch reflector and the view of the more compact Ring Nebula, M57, is improved by going still higher. *Globular star clusters* also look at their most spectacular at high powers.

There are plenty of examples of gaseous nebulae, big and small, dim and bright. One of the brightest in the sky is the Great Nebula in Orion, M42 (shown in Figure 4.8 and in Figure 5.10). This brings me to a controversial subject: colours. *Reflection nebulae*, such as that surrounding the Pleiades (Figure 14.6) shine with a bluish colour caused by scattered starlight. *Emission nebulae*, as their name suggests, shine because their gas is excited into emission by the stars within the nebulosity. The chief colour of this emission is red, due to the Hα spectral line of hydrogen gas but the cyan light of Hβ is also present (together with still weaker emissions at other wavelengths). Since the nebula is mostly hydrogen the emissions of this gas dominate. However, other elements do contribute their own colours in small amounts.

Some authorities state that no colours, whatsoever, can be seen in any deep-sky object. They insist that the light levels are too low to activate the colour receptors in the eye. Others state that all planetary nebulae and gaseous nebulae look greenish. They say that the light levels are bright enough to activate the receptors of green light, since the eye is most sensitive to that colour. It is true that planetary nebulae, in particular, emit strongly in the green part of the spectrum owing to the oxygen gas they contain.

A few people state that colours, including faint reds and blues, are visible in the brightest nebulae. Clearly the detection of colours depends as much upon the individual observer as the equipment he/she uses. My own experience is that I can see no colours, green or otherwise, in most planetary nebulae (NGC 7009, the Saturn Nebula, is an exception. It seems to have a strong cyan colour to me). However, I **can** see colours in gaseous nebulae of sufficient surface brightness, when using a sufficiently large telescope. M42 is a good example. Close to the Trapezium stars, the nebula, to my eye, looks predominantly blue and green, with faint streaks of yellow. The arms of thick nebulosity (which remind me of the wings of some great bird) extending east and west from the bright region and arching to the south look blue.

Figure 14.7. The Horsehead Nebula photographed by Ron Arbour, using his 16-inch (406 mm) Newtonian reflector, on 1989 January 12^d 21^h 00^m UT. The 20 minute exposure was made on Hypersensitised Kodak 2415 film. The over exposed star to the left is Alnitak (the easternmost of Orion's 'Belt stars'). West is uppermost in this view.

The enclosed 'clam-shell' backdrop seems a lovely rose pink tint to my eyes.

Am I seeing things? No, I don't think so. Let me stress that the colours I see are only pastel shades. Yet they are easily strong enough to be definite and not brought on by wishful thinking. Over the years I have had plenty of experiences which have shown me that my eyes are more colour sensitive than most peoples and I am quite sure that this is the explanation. It really is a case of 'seeing is believing'. Some people will be able to see colours in nebulae if they use a sufficiently large telescope. Others, probably the majority, will not.

Not many years ago objects such as the Horsehead Nebula in Orion (Figure 14.7) were considered entirely beyond the range of visual detection with backyard telescopes. So were many other deep-sky objects. However, when amateur observers really started working at seeing these objects the popular conception changed. The Horsehead Nebula still is considered extremely difficult – but no longer impossible. Other, formerly 'difficult' objects have now become 'easy'.

Of course, you would not set up your telescope, complete with tarnished and dirty mirrors, under the light of a halogen car-park lamp, plug in a high power eyepiece and expect to get a good view of some dim nebulosity! Nonetheless, it is worth maximising your chances of success if you wish to try for the most difficult objects.

The first factor is the most obvious: a dark sky. The glow from town lights is ruinous. Not only is the sky-background bright, swamping the object, but your eye will never reach its maximum possible sensitivity.

If you can't move the telescope to a better site (mounting telescopes on trailers is very popular in the U.S.A.) then you might try one of the 'anti-light pollution' filters on offer from companies such as Lumicon, Orion, Meade and Celestron (see the listing of manufacturers given in Chapter 17). These work by selectively blocking out most of the light at wavelengths between about 550nm and about 620nm, at which most of the common sources of light pollution occur. These companies also offer various narrow-band filters which are suitable for different deep-sky objects. For instance, an OIII (oxygen) filter is particularly useful when observing planetary nebulae. Any of the companies listed will provide you with the literature about their available filters on request.

The second factor is the cleanliness of the telescope optics and the efficiency of the mirror coatings. If either is poor then much light will be scattered, so reducing image contrast. Faint, low surface brightness, objects may well be rendered invisible due to this. Especially so if there is a bright star nearby to swamp the field with scattered light. The third factor is technique. Make sure that nothing disturbs your dark-adaption. It will take nearly an hour in virtually complete darkness before your eyes will be at their maximum sensitivity. Try to look for the difficult object when it is highest in a dark sky. Finally, try a range of magnifications. Generally the lowest powers will work best on low surface brightness objects but also try higher powers to see if they can help.

See Chapters 4 and 5 for details of how to photograph deep-sky objects. See also Chapter 6 for information on the powerful technique of CCD imaging. As far as drawing these objects goes, some of the advice given in Chapter 8 (on drawing the planets) will apply – and as for the techniques you can use to make the rendition, well, the sky's the limit!

Observing galaxies and hunting supernovae

Much of the foregoing applies equally well to observing galaxies. Some examples are bright, such as M31 in Andromeda – the most distant object clearly visible to the naked eye (Figure 14.8). Others, if visible at all, are extremely small and faint. A few can show quite a lot of structure, such as dark dust lanes, knots of nebulosity in the disc and indications of spiral arms, even for the visual observer using a moderate-sized telescope.

On average a supernova occurs in a galaxy every fifty years, or so. On that

Figure 14.8. The Galaxy M31 photographed by the author using a 13-inch (330 mm) *f*/10.4 astrographic refractor on 1989 December 5^{d} 21^{h} 05^{m} UT. The image was recorded on a 10 cm × 15 cm sheet of T-Max 400 film. This shows that a high focal ratio instrument can be used for photographing faint extended objects, though the exposure time was a rather long 45 minutes. The original negative clearly shows details in the disc of the galaxy, as well as the satellite galaxies M32 (NGC 221) and NGC 205.

reckoning, if you keep fifty galaxies under surveillance for a year, you will quite likely see a supernova in one of them. **You** might even be the first to see it! Premier among visual hunters of supernovae is Rev. Robert Evans, of New South Wales, Australia. In the period 1981 to 1988 he discovered fourteen supernovae and co-discovered three others, mostly using a 16-inch (406 mm) *f*/4.5 Newtonian reflector. His procedure is simple – he looks at as many galaxies as he can every clear night. He has memorised the appearance of hundreds of galaxies and can spot any starlike interlopers.

Others conduct photographic patrols, though it does seem that the visual technique is the most productive. Certainly it is the cheapest! However, photographs are useful for confirming, or otherwise, any suspicious candidates. A list of useful reference works is given in Chapter 17.

Supernovae are generally denoted by the letters SN, then a space, followed by the year of discovery, followed by a capital letter which indicates the order of discovery in that year. For instance, the supernova that erupted in M58, shown in Figure 14.9, is known as SN 1988A. Another example is SN 1994I. That particular stellar explosion happened near the heart of the spiral galaxy M51 and is shown in the specially processed CCD image in Figure 14.10.

Figure 14.9. M58 and the supernova SN 1988A, photographed by Martin Mobberley, at the *f*/5 Newtonian focus of his 14-inch (356 mm) reflector, on 1988 February $12^{d}\ 23^{h}\ 49^{m}$ UT. The 16 minute exposure was made on Kodak Tri-X film.

Figure 14.10. Supernova SN 1994I imaged by Martin Mobberley on 1994 April 11^{d} $20^{h}\ 39^{m}$ UT, using a 'Starlight Xpress' FSX camera on his 19-inch (490 mm) *f*/4.5 Newtonian reflector. The supernova can be seen very close to the nucleus of the galaxy in this 160 second integration. In order to avoid it being lost in the 'burn't out' nuclear region, Martin subsequently processed the image with a logarithmic stretch.

15
Spectroscopy

Spectroscopy is surely the most neglected of all the possible methods of observing by the amateur astronomer. This is not really very surprising. For one thing few amateurs have the necessary grounding in physics to make much headway with this analytical technique. Even if they have, a large telescope is required to obtain high dispersion spectra (explained later) of even quite bright celestial bodies and there is little that one can do to advance modern astronomy by taking low dispersion spectra of bright objects (with, perhaps, just one or two notable exceptions).

Nonetheless, I thought that I would include a few notes on spectroscopy in this book. Not every activity of the amateur astronomer necessarily has to advance science. There is also the enjoyment factor and the kinship created by pursuing a technique much used by his/her professional colleagues. Consequently I have restricted my treatment of this subject to just the areas likely to be of most interest and use to the backyard observer.

Fundamentals of electromagnetic radiation

Visible light is just one form of electromagnetic radiation. Figure 15.1(a) shows the classic 'radio tuning dial' representation of the electromagnetic spectrum. In some ways this radiation behaves as a stream of particles, or *photons*, but in other situations it behaves more like a stream of waves. In reality it is both. Figure 15.1(b) illustrates the concept of wavelength. Some people like to think of a photon as a 'packet of wave energy'. The energy of a photon, E, is related to its wavelength, λ, by:

$$E = hc/\lambda,$$

where c is the speed of light (3×10^8 ms^{-1}) and h is a constant of proportionality, known as Planck's constant. $h = 6.63\times10^{-34}$ Js (note the unit of h

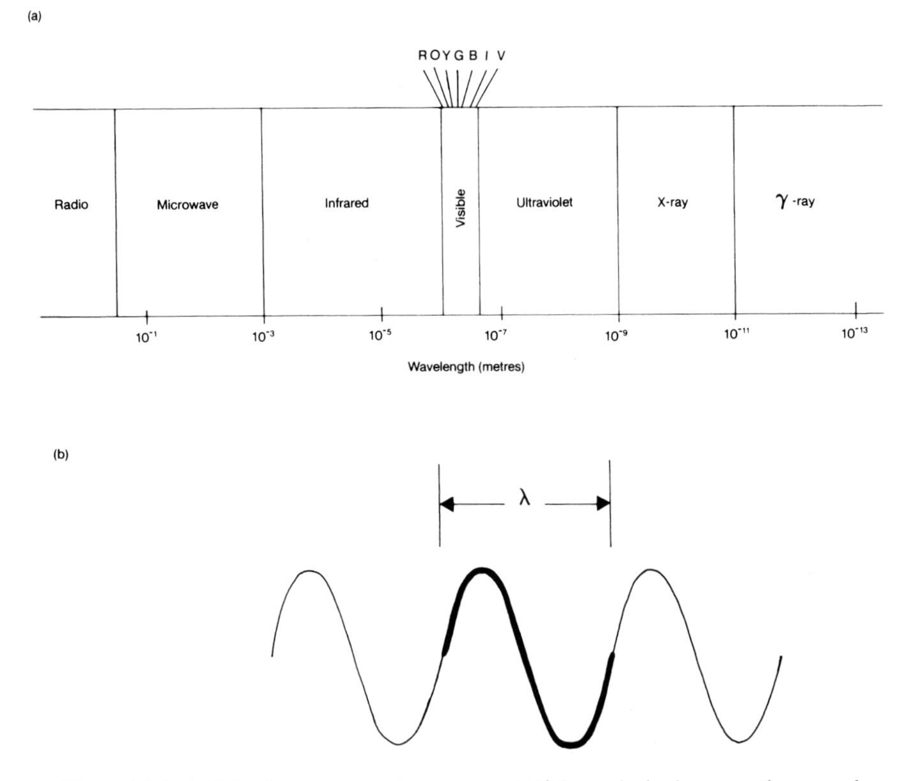

Figure 15.1. (a) The electromagnetic spectrum. The boundaries between the named regions are not hard and fast. Radiations are as much classified by their methods of production and detection, so producing some overlap. (b) The concept of wavelength. The length λ is one wavelength. The wavelength to which the eye is most sensitive (that of yellow-green light) is about 5.5×10^{-7} m. This is also conveniently represented as 550 nm, or 5500Å.

is the joule second, **not** the joule per second). Thus a photon of short wavelength has a higher energy than a photon of long wavelength.

How spectra are produced

Spectra have their origin in atoms. The nucleus of an atom consists of a cluster of *protons* and *neutrons*. The protons are positively charged particles, the charge on each being $+1.6\times10^{-19}$ Coulomb. It is the number of protons in the nucleus of one atom of that element, known as the *atomic number*, that determines which element it is. If the nucleus contains one proton, then the element is hydrogen. If it contains two, it is helium. If it contains three it is lithium, and so on.

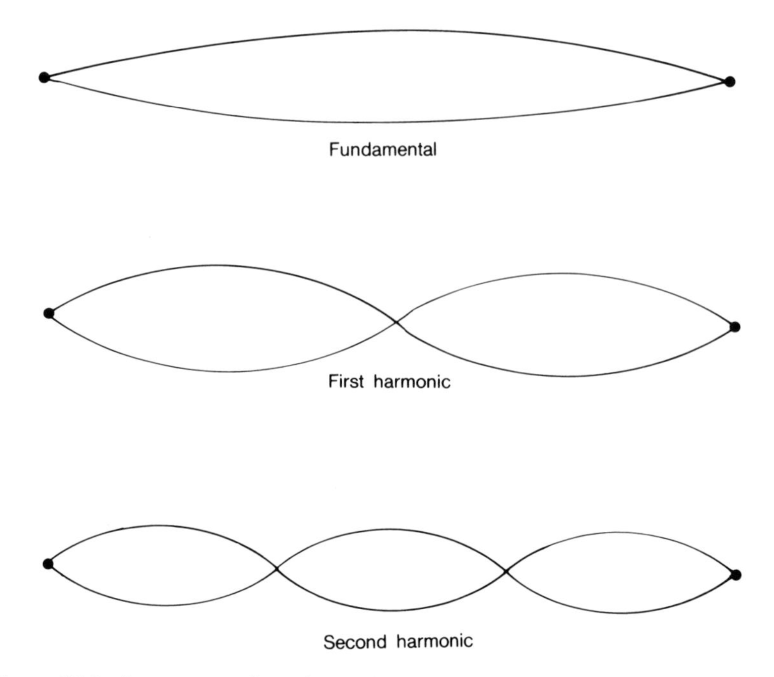

Figure 15.2. A representation of standing waves, such as those on a bowed violin string. The electrons surrounding the nuclei of atoms exist as clouds of charge in complicated three-dimensional standing wave patterns. Each pattern corresponds to a certain amount of energy.

The neutrons are, as their name suggests, electrically neutral. Their main function is to dilute the mutual repulsive forces of the protons and, in effect, glue the nucleus together. The positive charges in the nucleus of the atom are balanced by negatively charged *electrons* 'orbiting' outside the nucleus. The charge on each electron is -1.6×10^{-19} Coulomb, the same size as, but opposite sign to, that of the proton. Hence, an electrically neutral atom has the same number of electrons as protons.

Elementary physics texts compare the electrons surrounding the nucleus to planets orbiting the Sun but this is very far from accurate. Really these electrons don't exist as solid particles at all. They are 'smeared out' in a *standing wave* pattern that wraps around the nucleus. A simple example of a standing wave is that on a violin string as it is being bowed (see Figure 15.2). The ends of the string remain fixed. When one 'loop' can be fitted onto the string it is said to be vibrating in its 'fundamental mode'. The violinist can bow the string in such a way as to excite two 'loops' onto the string, the ends still remaining fixed and now one stationary point existing in the middle. This is the 'first harmonic' and the frequency is twice as high as the fundamental.

The string can be exited to higher frequencies but **only those that produce a whole number of loops on the string**. Since the ends of the violin string are always fixed, it cannot vibrate in such a way that, say, two and a third loops can be fitted along the string.

The three-dimensional electron standing waves that surround atoms are rather more complicated than their violin string counterparts. However, the principle of fitting whole numbers of 'loops' around the nucleus still applies. Now, each electron standing wave corresponds to a certain amount of energy. The lowest energy corresponds to the 'fundamental' standing wave and larger energies correspond to the higher 'harmonics'. The atom can accept energy in order to drive its electrons to higher energies in two basic ways. Either another particle of matter could collide with it, or it could absorb a photon of electromagnetic radiation. However, do remember the example of the violin string – it could only be excited to specific frequencies (the fundamental and the higher harmonics). **The atom can only absorb energy in specific quantities and the electrons will then rise to specific allowable energy levels**.

Assume the atom has absorbed some specific amount of energy. Let us say that all this energy has gone to raise one of its electrons from one energy level to the next. So, what happens next? What is likely is that that electron will hop back down to its original, lower, energy level. In the process the energy which the atom absorbed is given back out, **as a photon of electromagnetic radiation. Also, the wavelength of that photon is determined by the energy difference between the two electron energy levels**.

In practice any given atom will have a number of electrons and a more or less complex arrangement of possible energy levels for them to occupy **but each of the atoms of any given element are identical**. This means that the atoms of one specific element will give out electromagnetic radiation in a unique and recognisable pattern of wavelengths. Recognise the pattern and the element can be identified.

Types of spectra

Matter in a state of relatively low density, such as a gas, will produce photons of light of specific wavelengths, as previously described. Viewed through a *spectroscope* the emission will be seen as a pattern of coloured lines. Each line is in a position according to its wavelength (and hence colour). This sorting out of the wavelengths is the function of the spectroscope. The spectrum the gas produces is known as a *line spectrum*.

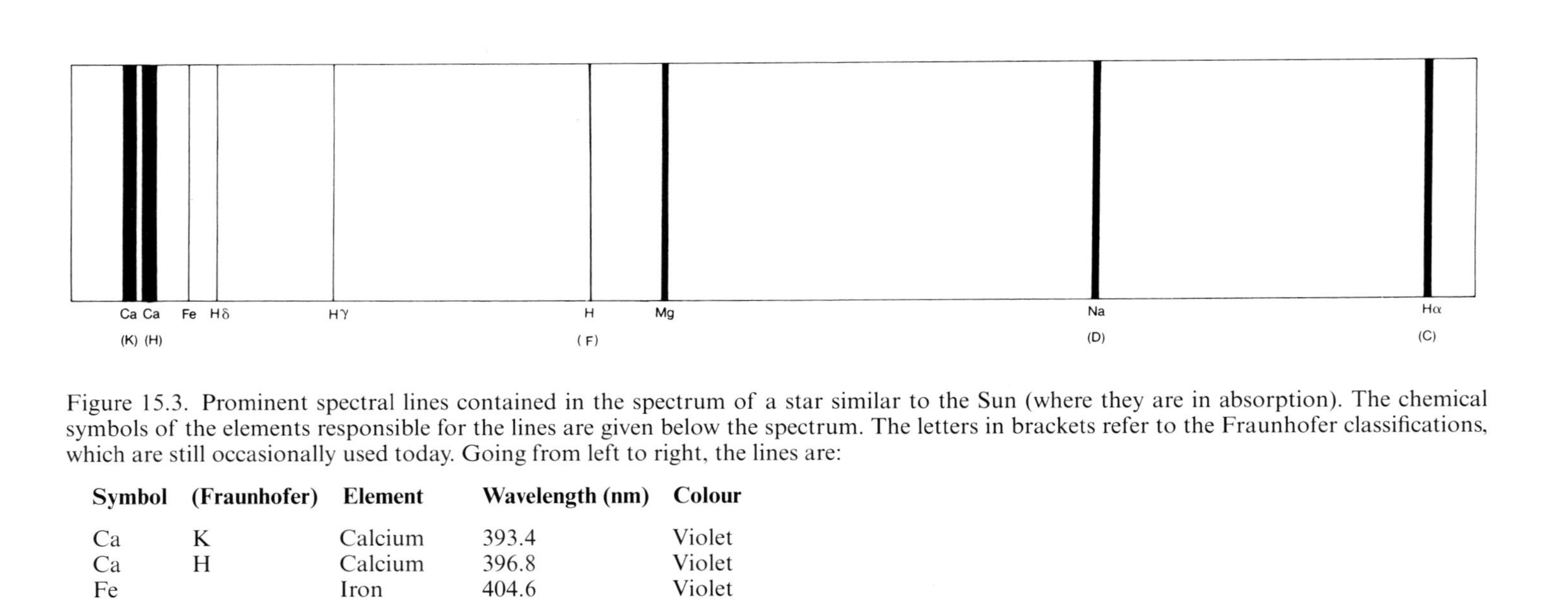

Figure 15.3. Prominent spectral lines contained in the spectrum of a star similar to the Sun (where they are in absorption). The chemical symbols of the elements responsible for the lines are given below the spectrum. The letters in brackets refer to the Fraunhofer classifications, which are still occasionally used today. Going from left to right, the lines are:

Symbol	**(Fraunhofer)**	**Element**	**Wavelength (nm)**	**Colour**
Ca	K	Calcium	393.4	Violet
Ca	H	Calcium	396.8	Violet
Fe		Iron	404.6	Violet
Hδ		Hydrogen	410.2	Violet
Hγ		Hydrogen	434.0	Blue
Hβ	F	Hydrogen	486.1	Cyan
Mg		Magnesium	517.5 (average)	Green
Na	D	Sodium	589.3 (average)	Yellow
Hα	C	Hydrogen	656.3	Red

Some people can see the H and K lines of calcium, others, in common with the author, cannot. The wavelength of the magnesium line is an average, since under moderate resolution this feature is seen to consist of three separate lines spread over 1.7 nm. Similarly the sodium D feature is really two lines separated by 0.6 nm.

Figure 15.3 illustrates some of the most prominent spectral lines in sunlight (and in the light of similar stars). Note that the lines are not equally intense. This is because some of the electron energy levels are more 'popular' than others. Therefore certain transitions occur more frequently and so the emission contains a greater amount of those particular wavelengths. In the spectrum of hydrogen the Hα line is most intense, etc.

Some materials, particularly those containing molecules rather than single atoms, can exhibit *band spectra*. These, as their name suggests, consist of broad bands rather than sharp lines. However, seen at very high spectral resolutions these bands are broken down into series of extremely close-spaced lines. The reason is that molecules have very much more complex arrangements of energy levels than the simpler atoms.

Materials at higher densities, such as liquids or solids, don't produce band or line spectra. Instead one gets a *continuous spectrum*. This is because the atoms in a liquid or a solid are much closer to each other and they each affect each other's energy levels, causing the levels to merge. The electrons are then all free to hop up and down through any energies. Overall, the sample then emits radiation smoothly distributed over a wide range of wavelengths. The way the energy varies with wavelength depends upon the effective temperature of the sample. Figure 15.4 illustrates this energy distribution for a *black body*. The term black body is used in physics for any object which is a perfect emitter and absorber of radiation and has certain well-defined properties. To a large degree the Sun and other stars behave as black bodies.

Two effects happen when the temperature of a black body is increased. Its total power output increases and the wavelength at which the maximum of that emission occurs decreases. The relation is:

$$\lambda_{max} = \frac{2.9\times10^{-3}}{T},$$

where T is the absolute temperature of the black body in kelvins and λ_{max} is the peak wavelength in metres. The temperature of the Sun's photosphere averages 5800 K and so the peak wavelength at which the Sun emits is 5×10^{-7}m, or 500nm. This is the main reason why adaptive evolution has caused our eyes to be most sensitive to light of wavelengths close to this value. This is also the reason why stars cooler than the Sun appear reddish and hotter stars appear bluish. Note, though, that this is the peak wavelength of emission and **not** the only wavelength. This is why stars of different temperatures show only pastel shades, and not vivid colours.

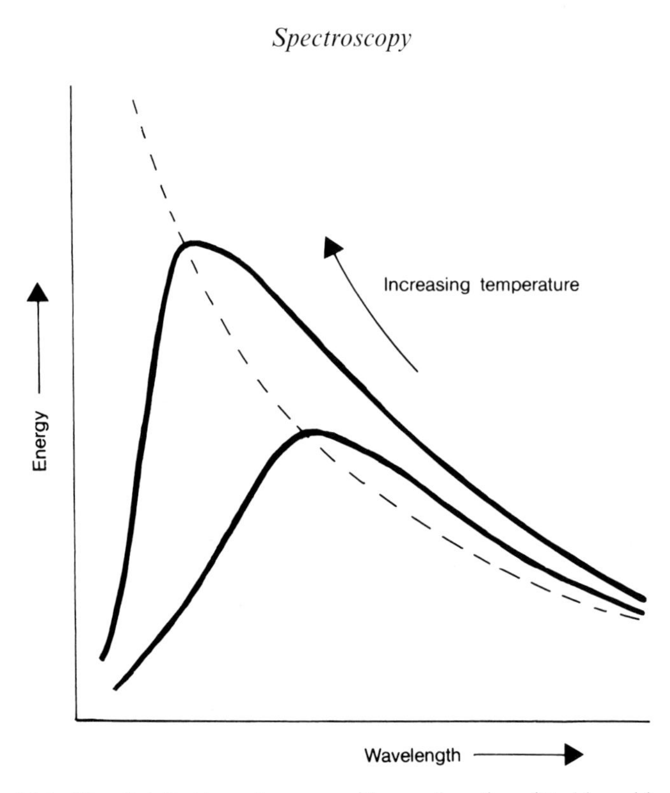

Figure 15.4. The distribution of energy with wavelength emitted by a black body. Two curves are shown, representing a black body at two different temperatures. The dotted line indicates the variation in the wavelength of peak emission with temperature.

In the case of the isolated gas emitting light the appearance of the spectrum is of bright lines, or bands, on a dark background. A spectrum of this type is known as an *emission spectrum*. However, if the gas is seen against a much brighter continuum background then an *absorption spectrum* is formed instead. The reason is that the electrons in the atoms of the gas can be excited into higher states by absorbing photons of just the right energy, and hence wavelength. **If a gas can emit light at a given wavelength, due to an electron de-exciting, it can also absorb light at precisely that same wavelength, the electron then being excited as a result**.

So, much of the light at the specific wavelength that was heading in the direction of the spectroscope is removed by the gas. However, when the electrons de-excite this energy is radiated, as photons of the given wavelength, in *all* directions. Hence the light reaching the spectroscope is deficient in the light of that specific wavelength. The line or band is then seen as dark against the bright background continuum.

The Sun and most stars display absorption spectra (see Figure 15.5). The tenuous chromosphere, on its own, would produce an emission spectrum. However, the denser layers below produce a brilliant continuum emission and this, shining through the chromosphere, produces the absorption spectrum.

The action of a glass prism

Figure 1.7(a) shows the manner in which a triangular glass prism bends light, while also splitting it up, or *dispersing* it, into its component colours. Figure 15.6 shows the path of a monochromatic ray of light through such a prism. The angle A shown on the diagram is known as the *apex angle* of the prism and the angle D is the *angle of deviation* of the light ray. Maximum dispersion happens when the light passes symmetrically through the prism (as shown in Figure 15.6). However, the deviation of a ray of any one given wavelength is then at a minimum. To summarise, in order to produce maximum dispersion, the prism is set to *minimum deviation* – the light then passing symmetrically through it.

The *refractive index, n*, of the glass (taking that of the surrounding air to be 1.00), and A and D are related, for the condition of minimum deviation, thus:

$$n = \frac{\sin\left(\frac{A+D}{2}\right)}{\sin\frac{A}{2}}$$

However, the value of n changes with wavelength. As an example, for a crown glass prism with a 60° apex angle n for red light ($\lambda = 6.5\times10^{-7}$m) is 1.51. The corresponding value of D is 38°.0. For blue light ($\lambda = 4.5\times10^{-7}$m) n is 1.52 and D is 38°.9. Hence the blue ray is deviated 0°.9 more than the red ray. This value is termed the *angular dispersion* produced by the prism.

An identical shaped prism made of flint glass ($n = 1.65$ for red light and $n = 1.67$ for blue light) will deviate red and blue rays (at minimum deviation) by 51°.2 and 53°.2. The dispersion this time is 2°.0, demonstrating that flint glass prisms are better at producing higher dispersions, than are crown glass prisms.

Objective prism spectroscoscopy

A prism can be mounted in front of a camera lens, or a telescope objective, in the manner shown in Figure 15.7. The prism should be oriented for minimum deviation. In that case the camera/telescope should be aimed at

H_β

Fe

Ca Cr Fe H_γ

Ca Fe H_δ Fe

Ca Ca

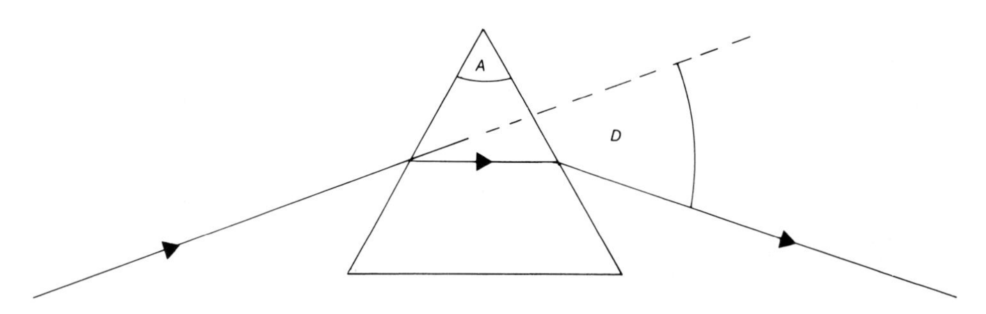

Figure 15.6. A light ray passing symmetrically through a triangular glass prism. See text for details.

an angle D from that of the subject. The spectra of stars will be short coloured lines. The length, L, of these spectra in the focal plane of the camera lens/telescope can be found from the approximation:

$$L = F \tan \theta,$$

where θ is the angular dispersion of the prism, measured in degrees and F is the focal length of the telescope/camera lens. L and F are measured in the same units, say both in millimetres, or both in metres, etc. For instance, with a 60° flint glass prism the length of the spectra (extending from red to blue) about 1.8 mm at the focus of a 50 mm camera lens.

Figure 15.5. [left] High-resolution spectrum of sunlight reflected from the surface of the Moon, taken by the author, using the 30-inch (0.76 m) coudé reflector and high-dispersion spectrograph of the Royal Greenwich Observatory, Herstmonceux. The wavelength range covered is aproximately 355 nm (left-hand side of bottom strip) to 504 nm (right-hand side of top strip), wavelength increasing going from left to right along each strip. The central stripe is the solar spectrum. The lines above and below are a copper–argon comparison spectrum exposed on the photographic plate at the same time as the solar spectrum. In order of increasing wavelength, the identified lines are:

Symbol	**(Fraunhofer)**	**Element**	**Wavelength (nm)**
Ca	K	Calcium	393.4
Ca	H	Calcium	396.8
Fe		Iron	404.6
Hδ		Hydrogen	410.2
Fe		Iron	414.4
Ca		Calcium	422.7
Cr		Chromium	427.5
Fe	G	Iron	430.8
Hγ		Hydrogen	434.0
Fe		Iron	438.4
Hβ	F	Hydrogen	486.1

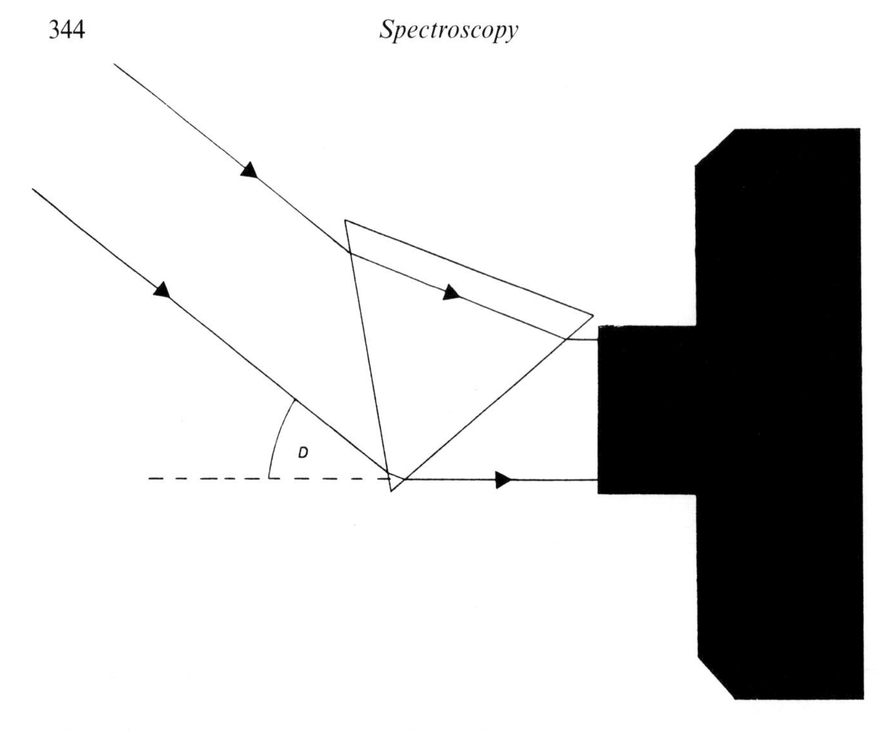

Figure 15.7. A prism positioned in front of a camera lens. In practice the prism would be enclosed so that the only light entering the camera is that which has passed through the prism. The prism could be mounted in front of a telescope objective in the same way. See text for details.

The linear spread of wavelength at the focal plane is known as the *linear dispersion*. It is conveniently measured in nanometres per millimetre (nm mm^{-1}), or Ångstroms per millimetre (Å mm^{-1}). In the previous example, a wavelength range of 2×10^{-7}m (200 nm, 2000Å) is spread over 1.8 mm. Hence the dispersion is 111nm mm^{-1}, or 1110Å mm^{-1}.

This brings us to one other distinction. An instrument used for purely visual scrutiny of spectra is called a *spectroscope*. One which is used with a movable crosswire and has a graduated scale for wavelength measurement is termed a *spectrometer*, while an instrument which can photographically record spectra is called a *spectrograph*.

Spectral resolving power, R, is defined in terms of the minimum difference in wavelengths, $\Delta\lambda$, of two spectral lines which can be resolved as separate and the mean wavelength, λ, of the two lines:

$$R = \lambda/\Delta\lambda,$$

λ and $\Delta\lambda$ both being measured in the same units. The potential resolving power of a prism is governed by its refractive index and the difference, x, in the longest and shortest path lengths of the rays passing through it. Hence for the maximum potential spectral resolution all of the prism should be used to disperse the light. The difference in the path lengths is then virtually equal to the length of the base of the prism. In practice (for most types of glass):

$$R \simeq 100x,$$

where x is measured in millimetres. The smallest resolvable wavelength difference is then given by:

$$\Delta\lambda = \lambda/100x,$$

As an example, a fully illuminated prism with a 30 mm long base could potentially just resolve two spectral lines, close to a wavelength of 500 nm, as separate if they were separated by no less than 0.17 nm. Whether this spectral resolution is achieved in practice depends upon other factors.

For instance in our example, the 60° flint prism mounted in front of the 50 mm focus camera lens produces a linear dispersion of 111 nm mm^{-1}. In order to photographically resolve 0.17 nm the film resolution (see Chapter 4) would have to be an impossible 653 lines mm^{-1}. Even if such a film existed, what about the problems in focusing the camera that precisely, let alone the quality of its lens? In practice you would be doing well to achieve a resolution of 2 nm using a 50 mm lens. Of course, with longer focal lengths the theoretical resolution can be approached – at least for the stars.

One useful area of research involves recording the spectra of bright meteors. For this a prism fitted to the front of a 35 mm camera is ideal. The procedure is then exactly the same as for conventional meteor photography (see Chapter 10). Of course the best spectra will be those obtained with the apparent motion of the meteor perpendicular to the dispersion produced by the prism. In practice most meteors brighter than about zero magnitude will produce good spectra with a fast film in the camera. Be careful to select a film which will allow you to record a spectrum over the full range of wavelengths you desire. For instance, many black and white films respond very poorly to red light. The spectral response of photographic emulsions is discussed in Chapter 4. Suffice it to say here that hypersensitised TP 2415 would be a good choice for spectrography.

Any extended body would produce a series of overlapping spectra – basically a mess! To get good quality spectra of extended objects, and to get the best quality spectra in general, one must select out just one portion of the image for dispersion by means of a slit. This is the basis of the spectroscope/spectrograph proper.

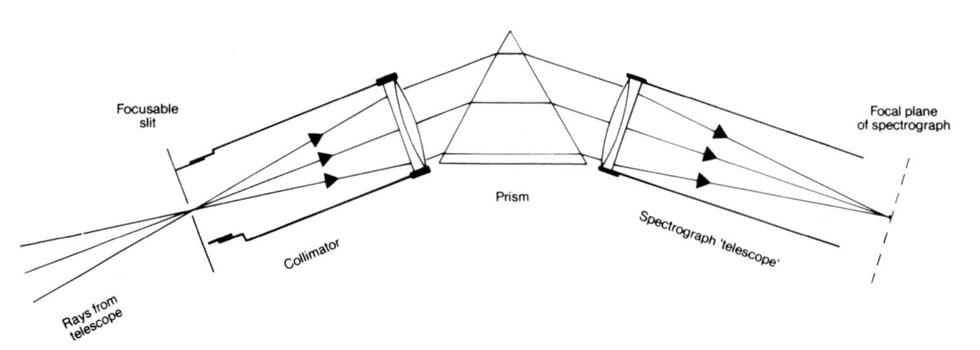

Figure 15.8. A single prism spectroscope/spectrograph. See text for details.

The prism spectroscope/spectrograph

Figure 15.8 shows the basic layout of the instrument. The slit is set into the focal plane of the telescope. After passing through the slit the light rays diverge. However, a lens renders these rays parallel (*collimate* the rays) before they enter the prism. Thus the slits also have to be in the focal plane of the collimator lens and provision is made for the two adjustments. The collimated rays then pass through the prism symmetrically and are subsequently converged to a focus by the spectrograph 'telescope' objective. A photographic film could be positioned at the focal plane, turning the instrument into a spectrograph, or an eyepiece could be used to examine the focal plane image, the instrument then being a spectroscope.

The spectrum formed at the instrument's focal plane consists of a series of images of the slit, each formed in a position dependent on its wavelength. So the slit should be set to the minimum possible width in order to achieve the greatest possible spectral resolution. The width, s' of the slit image is related to the actual width, s, of the slit and F_c and F_t, the focal lengths of the collimator and spectrograph 'telescope' lenses by:

$$s' = \frac{sF_c}{F_t}.$$

The linear dispersion at the focal plane is dependent on the focal length F_t and is calculated as described earlier for the objective prism.

There are some conditions to be met if the spectrograph is to have the best spectral resolution and brightness of spectrum:

1. The focal ratio of the collimator lens must be no greater than that of the telescope feeding light to the slit. Otherwise, not all of the light passing through the slit will be used in forming the spectrum.

2. The focal length of the collimator should be that required to produce a collimated beam which **just** fills the prism with light. Any less and the potential spectral resolution will be limited. Any more and some of the precious light will be wasted.

3. The aperture of the spectrograph 'telescope' should be such as to accept all of the light passing through the prism.

4. The width of the slit should be equivalent to the diameter of the spurious disc imposed by seeing (and/or the quality of the optics) at the focus of the telescope. Any wider and the spectral resolution might be unnecessarily reduced. Any less and some of the light in the seeing disc will not pass between the slit jaws and enter the instrument.

5. The width of a single slit image (see the previous equation) should match the limiting resolution of the film if the instrument is to be used as a spectrograph.

The conflicts between these separate requirements are lessened if the telescope feeding light to the spectrograph/spectroscope has a large focal ratio. The high dispersion spectrographs on professional telescopes are most often used with effective focal ratios larger than $f/30$.

Razor blades make good slit jaws and the rest of the optical components can be obtained from an optical-goods supplier. Anyone who can make the mechanical parts of a telescope should have little difficulty making a spectroscope or spectrograph. The problems associated with attaching a spectroscope/spectrograph to the telescope are the same as those of attaching any other heavy equipment. These problems, and their remedies are discussed earlier in this book.

One other problem is keeping the star, or other, image centred over the spectrograph slit during the exposure. This is solved if the slit jaws are polished and set at 45° to the optical axis. A long-focus eyepiece can then be used to focus on the slit, which is coincident with the image at the focal plane of the telescope. The telescope is adjusted to keep the required part of the image over the slit. For taking stellar spectra the telescope slow motions should also be used to trail the star along the slit. Otherwise the spectrum will just consist of a fine line running through the wavelengths, rather than a wide stripe.

Probably the most scientifically valuable work (particularly because this area is so controversial) that an amateur spectroscopist could perform is to monitor the Moon for Transient Lunar Phenomena (see Chapter 7) and take spectra of any suspect appearances.

A little prior experimentation will establish the range of exposure times needed. As a typical example, a 10-inch (254 mm) reflector feeding light through a 2 arcsecond slit to an efficient 10 nm mm^{-1} (100 Å mm^{-1}) dispersion spectrograph, should record a good lunar spectrum, on a 1000 ISO film, in around 20 minutes. This value is for a region close to the terminator around first quarter Moon. The exposure needed at full Moon is only about 4 minutes. The same set up and a half-hour exposure time will allow the spectra of stars brighter than about the second magnitude to be recorded. Going to lower dispersions allows one to record proportionately fainter spectra. Of course, the CCD can do much better than the photographic emulsion (see the section at the end of this chapter).

How a diffraction grating works

In many ways a *diffraction grating* is superior to a glass prism for producing spectra. Diffraction gratings come in two basic types. *Reflection gratings* consist of fine rulings etched onto a mirror surface and *transmission gratings* consist of a grid of fine lines. The former type are expensive and not so easily obtainable and so only transmission gratings will be considered here.

Imagine monochromatic light falling onto a mask with a single fine slit cut in it. A natural consequence of light passing through a narrow gap is that it spreads out (this is called *diffraction*). It happens because light behaves like a wave and diffraction occurs whenever waves are made to pass through a narrow gap.

Now imagine monochromatic light falling onto a mask which has two parallel slits cut in it. Some light passes through, and is diffracted by, each slit. If we put a white screen some way from the slits, in order to intercept the light, we would see something quite interesting. Instead of a general 'blur' of light on the screen, we would see an array of bright and dark bars, or fringes.

What is now happening is that the rays from each slit are *interfering*. In some places the rays meet together crest to crest. The rays then add up to produce a bright fringe. This process is called *constructive interference*. In between these positions a crest of one wave meets with a trough from the other and the rays cancel out. Hence in these positions there is no light and a dark fringe is seen. This is called *destructive interference*.

Interference is another property that is common to all waves. Accoustic dead spots in rooms are caused by the destructive interference of sound waves after the waves have been reflected off walls and ceilings.

If we now go one stage further and imagine monochromatic light falling onto a grill made up of very fine parallel bars, with small gaps between each bar, we will again see the processes of diffraction and interference at work. This time, though, the fringes are very narrow and sharp and are widely spaced. This is because the large number of slits admitting light causes the angles at which light can constructively interfere to be very precisely defined. At slightly different angles some or other of the slits cause destructive interference. So, for a given wavelength of light there are only specific angles to which the light can be directed as it passes through the grill, or grating.

A mathematical analysis of the situation would give a formula for calculating these angles. The relationship is:

$$N\lambda = d \sin\theta.$$

In this formula λ is the wavelength of the light measured in metres. d is the spacing of the slits in the grating (again measured in metres). θ is the angle for constructive interference, measured in degrees.

N is known as the *order* of the bright line formed. Some of the rays constructively interfere at an angle of 0°. This is where the rays from all pairs of slits (for instance the two outermost slits) travel along the same lengths of paths and so all rays arrive in phase (crest meeting crest). This bright line on the screen is known as the *zeroth order* line. To either side of this line are the *first order* lines. These occur where the paths lengths of the arriving rays are one whole wavelength different. For instance the millionth crest of one ray might meet with the millionth and one crest in the other ray. N has the numerical value 1 for the first order lines.

Out at greater angles are the *second order* ($N = 2$) lines where the rays from corresponding pairs of slits are two wavelengths out of phase. At still greater angles are the *third order* ($N = 3$) lines, and so on.

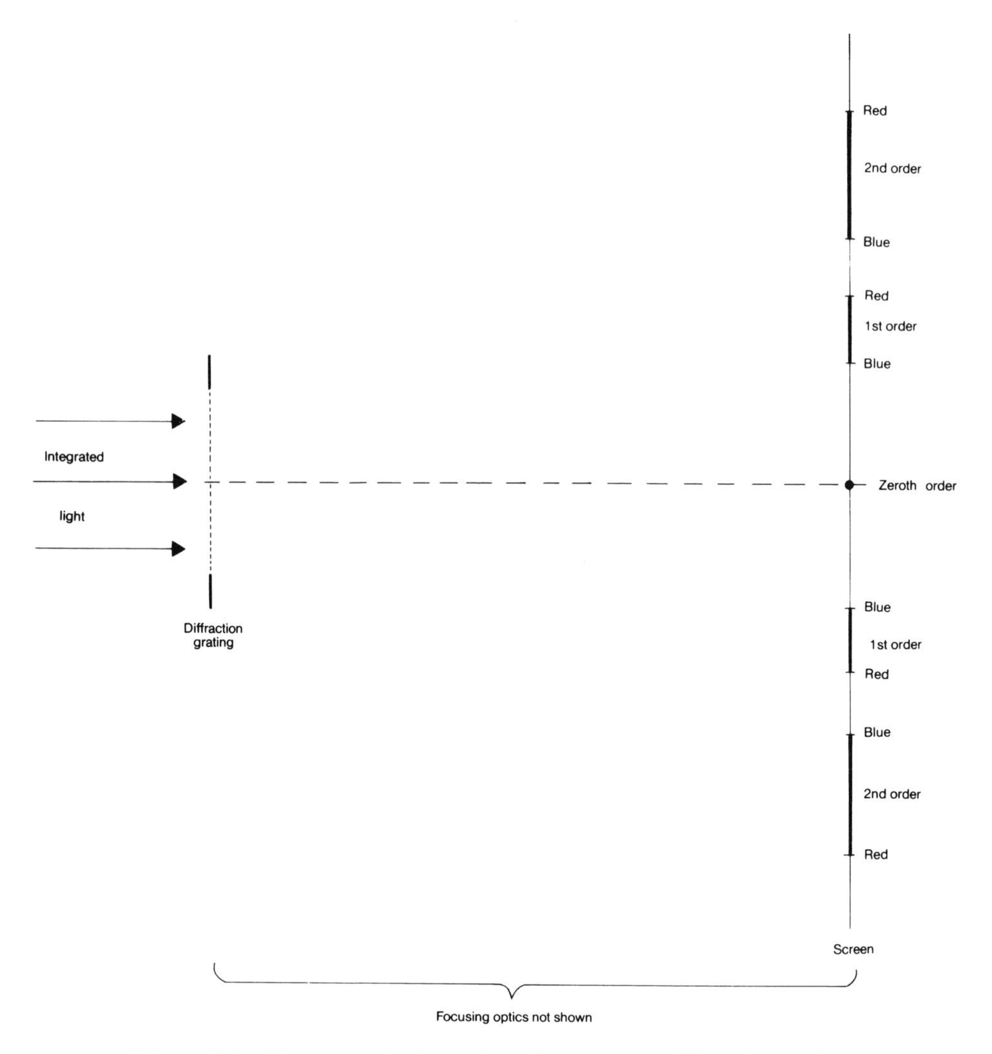

Figure 15.9. Illustrating the formation of spectra by a diffraction grating.

We have considered how monochromatic light behaves when it passes through a fine grill, or diffraction grating, but what happens to light that is composed of a mixture of different wavelengths, as is usually the case? The formula gives the answer. For a given order, each value of wavelength, λ, produces a different value of θ (the angle for constructive interference). The end result is that the grating produces two complete spectra for each order of diffraction (see Figure 15.9). The spectra are spread out in a direction which is at right angles to the slits in the grating. If a fine grating is used

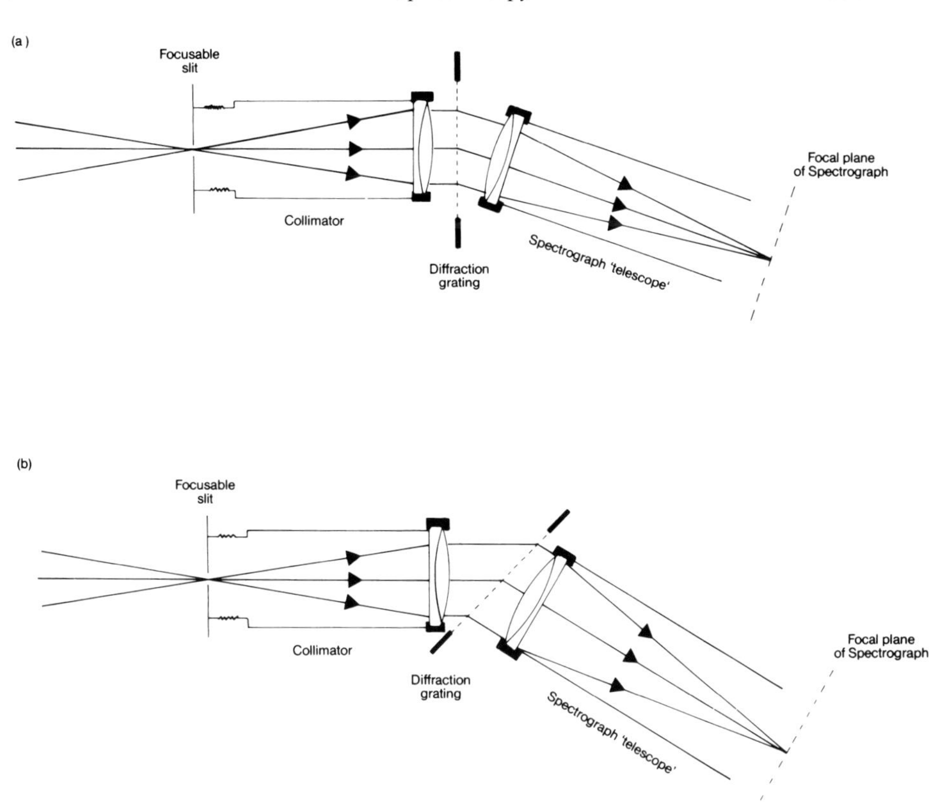

Figure 15.10. (a) A diffraction grating spectrograph with the grating oriented normally to the collimator. (b) A more efficient arrangement has the diffraction grating oriented at an angle (most conveniently 45°) to the collimator beam.

(small value of d) the angles (θ) for the various lines are all increased, giving a higher dispersion.

A diffraction grating spectrograph

Transmission-type diffraction gratings can be bought from optical suppliers for around £10 (\$16) at mid-1990s prices. They are usually mounted in 2-inch (5cm) square photographic transparency mounts with 23 mm × 35 mm clear apertures. The lines in the grating run parallel to the 23 mm side. The finest gratings available have around 600 lines mm^{-1} ($d = 1.7 \times 10^{-6}$ m), though gratings of 300 lines mm^{-1} are more easily available and are a little cheaper.

The diffraction grating can simply replace the prism in a spectrograph as shown in Figure 15.10(a). Unfortunately the simple diffraction grating is not so efficient at using all the collected light to form a spectrum as is a prism. This is because the light is shared between two sets of spectra, each with several orders. The situation is improved by angling the grating to the incident beam, as shown in Figure 15.10(b). More of the light is then preferentially thrown into the spectra on one side of the zeroth order, at the expense of those on the other . It also has the spin-off benefit of increasing the effective numbers of lines per millimetre presented to the beam from the collimator, so increasing the dispersion.

As an example, consider a 600 lines mm^{-1} grating angled at 45° to the collimator beam. The effective number of lines per millimetre is then 848 ($d = 1.18 \times 10^{-6}$ m). The diffracted angles (measured from the zeroth order line in a direction towards the normal of the grating) are then 19°.8 for violet light ($\lambda = 4$ X10^{-7} m) and 33°.4 for red light ($\lambda = 6.5 \times 10^{-7}$ m), with the other colours ranging in between. The first order spectrum on the other side of the zeroth order makes a large angle to the normal (the perpendicular direction) of the diffraction grating and the amount of light directed into it is consequently reduced. The second and higher order spectra ($N = 2$, 3, etc.) on the same side cannot exist as they would be diffracted by more than 90° to the normal of the grating.

Note how much greater the angular dispersion is for a diffraction grating than for a prism. This is the chief advantage of a grating. However, also bear in mind that only fairly low linear dispersion spectra are photographable with amateur sized telescopes, except for the brightest subjects or when using a CCD to do the recording of the spectrum. The potential spectral resolution of a diffraction grating is determined by the number of grating lines used to form the spectrum. The relation is:

$$R = \lambda/\Delta\lambda \simeq \text{Number of lines.}$$

Consequently the requirements for an efficient diffraction grating spectrograph are identical to those for a prism spectrograph. The diffraction grating, like the prism, should **just** be filled with light from the collimator. The grating used in the foregoing example could potentially deliver a spectral resolving power of 21000. This is enough to split two spectral lines, close to 500 nm wavelength, separated by only 0.02 nm! In practice, the other parameters in the spectrograph will limit the actual spectral resolution achieved.

The solar spectroscope

Unlike other celestial bodies, the Sun is more than bright enough to allow high dispersion spectroscopy/spectrography with even a small telescope feeding light to the slit. I recommend using a diffraction grating to achieve the high dispersion, though older solar spectroscopes do employ a train of prisms instead. If the spectroscope has a fairly long focal length 'telescope', fitted with an eyepiece, then the instrument can be used for observing prominences on the Sun's limb. The procedure is described in Chapter 11.

Spectroscopy with a net curtain

How many telescope owners would like to do spectroscopy but are put off by thoughts of elaborate and expensive equipment? In fact, a telescope can be made to produce spectra by the addition of only one very inexpensive item: a fine gauze screen attached to the 'sky' end of the telescope tube!

The gauze should be as fine as possible but a piece of unpatterned net curtain will do for moderate or large aperture telescopes. It forms an objective diffraction grating, splitting the incoming light into its component wavelengths.

A grating with a spacing of, say, 1 millimetre produces rather small angles of diffraction. The angular spread between the first order violet ($\lambda = 4 \times 10^{-7}$ m) and deep red ($\lambda = 7 \times 10^{-7}$ m) rays is only 0°.017, or 61 arcseconds.

However, a telescope enables one to appreciate these small angles. In the above case the spread of the first order spectrum covers a greater length in the image than the diameter of the planet Jupiter seen through the same telescope. The second order spectra are spread over twice the angle, the third are spread over three times the angle, and so on.

The grating material may be mounted in a light frame which can be neatly fixed to the front of the telescope. It can even be directly held using clothes pegs, though care should be taken to get the grating reasonably flat if the resulting spectra are not to be confused and of low contrast. For the same reason the grating should have a simple meshwork pattern and should not carry fancy designs.

Seen through the telescope eyepiece the results are spectacular. The view of a diffracted star cluster, such as the Pleiades, is very beautiful. A fine gauze applied to a moderate aperture allows some individual spectral lines to be resolved. The effect of the grating on emission nebulae is especially interesting: discrete multiple images rather than smeared out rainbows.

The spectra can be photographed through the telescope in the same way as any celestial body. The effective focal length of the telescope's optical

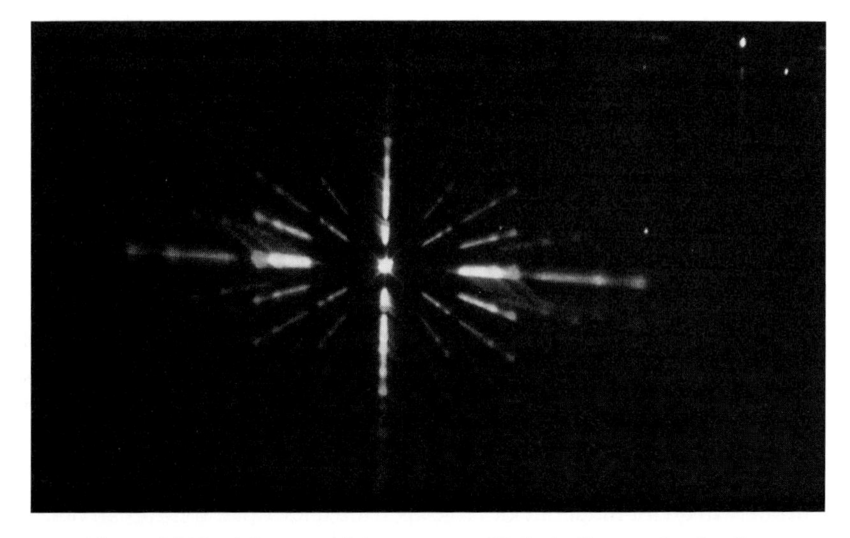

Figure 15.11. A 'net curtain' spectrum of Polaris. See text for details.

system (taking account of a Barlow lens or eyepiece projection, if used) determines the image-scale, as given by:

image-scale = 206265 / focal length.

As an example, an effective focal length of 3000 millimetres produces an image-scale of 69 arcseconds mm^{-1}. The first order spectrum obtained using a gauze with 1 mm spacings between the lines will be nearly 1 mm long in this case. The second order spectrum will be nearly 2 mm long, etc. A finer gauze will produce longer spectra and better spectral resolution. For this reason it is important to use the finest gauze possible with small aperture telescopes. A piece of black chiffon would be ideal – but please do ask the wife before cutting a hole in her favourite nightdress!

Nonetheless, as a 'worst-test case' I offer the photograph of the spectrum of Polaris (Figure 15.11), recorded in just 8 seconds on 3M Colourslide 1000 film, using an old scrap of net curtain (spacing very roughly 1.3 mm in one direction, 0.4 mm in the perpendicular direction) pegged to the end of my 18¼-inch reflector. The image was recorded at the 2.59 m Newtonian focus with no additional optics.

The author's parsimonious spectrograph

I make much of my own equipment, as many amateur astronomers still do, though nowadays the trend is towards buying everything ready made. If you want a spectrograph built onto your telescope then, currently at least, you are going to have to make it yourself, unless you are wealthy enough to have an optical specialist do the work for you.

Figure 15.12 shows the unit I built onto my own 18¼-inch telescope in the summer of 1990. By extensively using scrap materials I kept the total cost of the project down to under £50 ($80). In fact, nearly half of that cost was for the lead I needed to buy to make the counterweighting necessary to re-balance the telescope tube!

Figure 15.13 shows the diagrammatic layout and Figure 15.14(a) to (c) shows views of some of the component parts of the spectrograph. All the principles discussed in this chapter were brought to bear. The framework construction of the telescope tube was a help to start with, as was the fact that the secondary mirror is rotatable (it locks into any one of four positions – directing the light from the primary mirror towards either of the four faces of the tube).

There isn't room enough here to include full contructional details. However, I hope that the brief notes following will, taken together with the photographs and diagram and the principles outlined earlier in this chapter, help anyone wishing to make such a unit for themselves.

Figure 15.14(a) shows how I allied a star diagonal to the lens and focuser unit of a discarded slide projector. The metal plates were cut to shape from a sheet of scrap aluminium and drilled as appropriate to take the threaded rods which act as spacers (and in addition allow for the provision of collimation) and smaller holes for fixing to the projection lens focuser casting. I drilled matching holes in the casting, sized to take screw threads which I cut using a hand wrench and tap.

The star diagonal having re-directed the optical axis of the telescope to run parallel with the side of the telescope tube (and 50 mm from it), the telescope focal plane is formed between the two plates. I used a 'tank cutter' to make the larger central holes in each plate which have to pass the light. Keeping those to a couple of centimetres diameter each helps with the baffling of any stray light that might get into the system (of course, allowing the rays I do wish to pass to get through unvignetted).

The projection lens is an $f/2.8$ of 100 mm principal focal length, adequate to avoid vignetting, and the image is projected 75cm down to the slit. I used the *'lens maker's formula'* ($1/f = 1/u + 1/v$) to calculate the positions (u = distance of telescope focal plane to lens nodal plane, v = distance of lens nodal plane to spectrograph slit and f = principal focal length of the lens) and the amplification factor produced by the projector lens ($m = v/u$). In my case, the amplification factor is 6.5. If the 'nodal planes' I have mentioned worry

Figure 15.12. The components of the author's spectrograph can be seen mounted on the north face of the telescope's eight feet (2.45m) long tube. The Newtonian secondary mirror can rotate to direct the beam from the primary through a hole in the underside of the spectrograph casing, near the top of the tube.

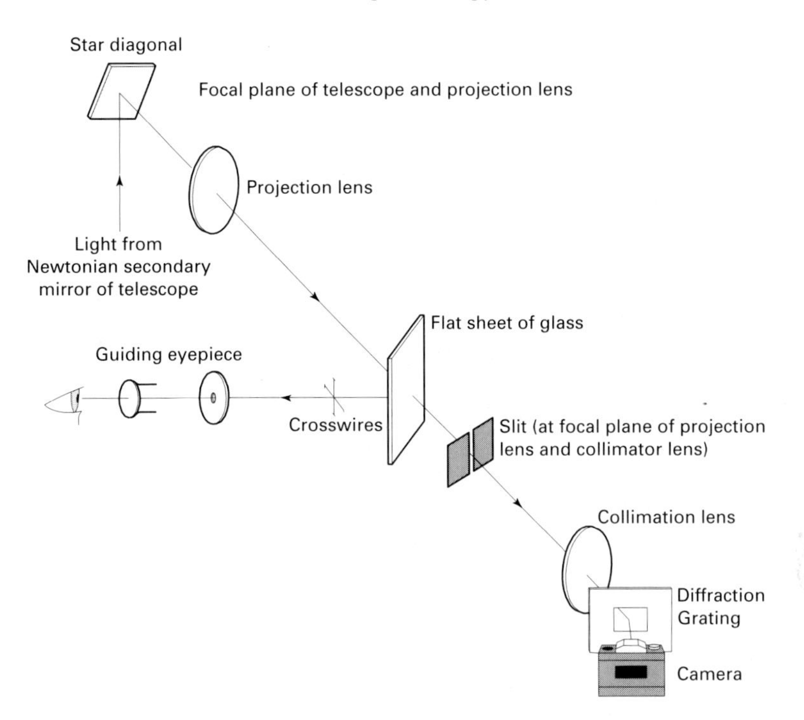

Figure 15.13. The schematic layout of the author's spectrograph.

you, then just take the measurements from what you judge to be the mid-point of the lens assembly and you won't be far wrong, if you intend following my design.

The 6.5 times amplification factor produces an effective focal length of 16.8 metres with my telescope and an effective focal ratio of *f*/36.

I had an old slit unit from a piece of discarded school optical equipment and so pressed that into service (see Figure 15.14(b)). I mounted it at the correct height (the centre of it 50 mm above the side of the telescope) on a small aluminium tray, together with a mounted sheet of glass at 45° to the incident beam set a little before the slit. Most of the light from the projector reaches the slit, where it is focused. The image-scale at the slit is about 12 arcseconds mm^{-1}. The slit is 1 mm wide (and 25 mm high) and so picks off a slice of image 12 arcseconds wide.

Conveniently forgetting about the function of the glass sheet for a moment, the now diverging light that has passed through the slit passes along a piece of plastic drainpipe at the end of which is mounted a 50 mm diameter achromatic lens of 900 mm focal length (part of an otherwise useless homemade telescope I was once given). The lens is set its own focal

(a)

Figure 15.14. (a) View of part of the spectrograph during construction. When on the telescope, the light from the Newtonian secondary mirror passes up through the hole in the base (lower right in this photograph) and is directed by the star diagonal into the projection lens. (b) [opposite, top] Another view during construction. The light from the projection lens is focused onto the slit (on the right in this photograph). Before reaching a focus, a little of the light is re-directed by the plane sheet of glass to form an image in a crosswire eyepiece of long focal length (not shown). (c) [opposite, bottom] This shows the relationship of the parts of the spectrograph shown in (a) and (b). After the converging light beam passes through the slit (lower right) it goes on to diverge until it encounters the collimating lens (not shown).

length from the slit and so the diverging rays from the slit are rendered parallel after passing through it. Given that the beam from the slit is an *f*/36 one before the lens, the lens intercepts it when it has a width of approximately 26 mm (900/36 + the width of the slit) and so the parallel rays entering the grating (see Figure 15.13) also have this width.

I bought the 600 lines mm^{-1} transmission grating from a laboratory supplier for about £15 (about $24) and received a surprise when I examined it. It appeared to preferentially throw the light to one side of the zeroth order. The spectrum on that side was rather brighter than on the other. This is a property normally manufactured into reflection gratings, called *blazing*. I didn't expect to find it with a transmission grating. I obtained the manufacturer's telphone number from the supplier and gave them a call. It turned out that what I had bought was a ruled grating (the lines being cut, as grooves, into the glass), rather than the photographic type, and that it was

(b)

(c)

indeed blazed. When I expressed surprise that this could be done for the price, I was told that the grating is considered to be only of low quality!

Of course having the grating blazed is an advantage because less light is wasted in producing unused spectra. Using a little more metalwork I set the grating in the position to intercept the light from the collimator. I angled the grating at 33° and arranged for my old SLR camera, complete with a

58 mm *f*/2 lens, to view the light from the grating at normal incidence. By this I mean that the camera lens looks directly at the grating, the optical axis of the camera lens being perpendicular to the plane of the grating. This bring two advantages: the camera lens can be set as close as possible to the grating, to ensure that it intercepts as much light as possible, given the large angle of dispersion produced by it; and the maximum brightness of spectrum results when the grating is used in this way.

Of course, I made sure that the grating was orientated the correct way so that full advantage was taken of the blazing. I used it to brighten up the second order spectrum.

As well as getting the camera lens as close as possible to the grating, it is important that it does not vignette the light from the grating. Having the largest possible aperture helps. In the case of my *f*/2 lens of 58 mm focus, it is 29 mm.

It is true that the diameter of the beam from the collimater is a couple of millimetres wider than the height of the grating aperture but the amount of light clipped off is truly minimal – I worked with what I had available, rather than what would have been ideal.

Using the formulae given earlier in this chapter (remember that the grating presents 715 lines mm^{-1} to the collimator beam), I was able to calculate the angles. This was why I decided to orientate the grating and the camera at 33° to the beam from the collimator. I also planned to have the camera image the entire first order spectrum and the second order from near ultraviolet through to yellow (beyond which the spectrum starts to become confused with the overlapping third order spectrum, anyway) all on the same film frame. The approximate linear dispersions in the first and second orders are 22 nm mm^{-1} and 9.7 nm mm^{-1}, respectively.

The slit width is 1 mm and so a monochromatic image of the slit on the film has a width of 1 mm × 58/900 (58 and 900 are the camera and collimator focal lengths, in millimetres), or 0.064 mm. Given the values of the linear dispersions this implies potential spectral resolutions of 1.4nm (14Å) and 0.62nm (6.2Å) in the first and second orders. This is the limiting factor in my spectrograph. The potential resolution provided for by the number of lines in the grating and by the film resolution are each several times better than this.

Returning to the function of the inclined glass plate which is set in the converging beam a little before the slit: its function is to pick off a little light (about 6 per cent) and direct it towards a guiding eyepiece. I mounted a spare achromatic lens of about 3cm diameter and 15cm focal length in a plastic tube and set crosswires (made from copper wire) at the end of the tube in the coincident focal planes of the lens and that relayed by the glass. I fashioned an eyeguard several centimetres beyond the lens for comfortable viewing.

I provided some degree of lateral adjustment of the lens–crosswire assembly for initially lining everything up. For this I used the limb of the Moon: when I could just begin to see a spectrum through the camera viewfinder I set the crosswires to the position of the limb, as seen through the guiding eyepiece. Locking everything up, I know that whenever an object is brought to the guiding eyepiece crosswires then most of the light from it is passing through the spectrograph slit.

In truth, two overlapping images are actually seen: one is a reflection from the front surface of the glass and one from the rear surface. However, it is a simple matter to ignore one and set on the other (say, the right-hand image).

The casings for the parts of the spectrograph (Figure 15.12) were also made from scrap materials. The section housing the projection lens and slit used to be an automobile dealer's advertising sign. I fitted doors in it to give access to to projection lens (for focusing) and the glass sheet (for cleaning) and made holes in it and fitted shaped pieces of wood for mounting the guiding eyepiece and collimator assemblies, etc. Note that the cover only functions to keep the light and weather out. All the optical parts are bolted through to the framework tube of my telescope, otherwise problems would arise with flexture.

The photograph also shows the manner in which the colimator and diffraction grating assemblies are mounted on the telescope; folded sheet metal being used for the lower casing, and foam rubber being used at the junctions of the collimator and camera to keep out stray light. The camera is removable, only being on the telescope when taking spectra.

All the interior surfaces, and the outside surfaces that faced into the telescope tube, were painted matt black.

As Figures 15.15(a), (b), and (c) show, the spectrograph works despite being very crude and cheap in its contruction. When I rebuild this instrument, I shall replace the 1 mm slit with one about half the width. This will double the spectral resolution and increase the contrast of the lines in an absorption spectrum, which at the moment is rather poor. There are many other improvements that could be made: using reflective slit jaws set at 45° to provide for guiding rather than the method I used, provision for recording comparison spectra, a CCD camera rather than the photographic type . . .

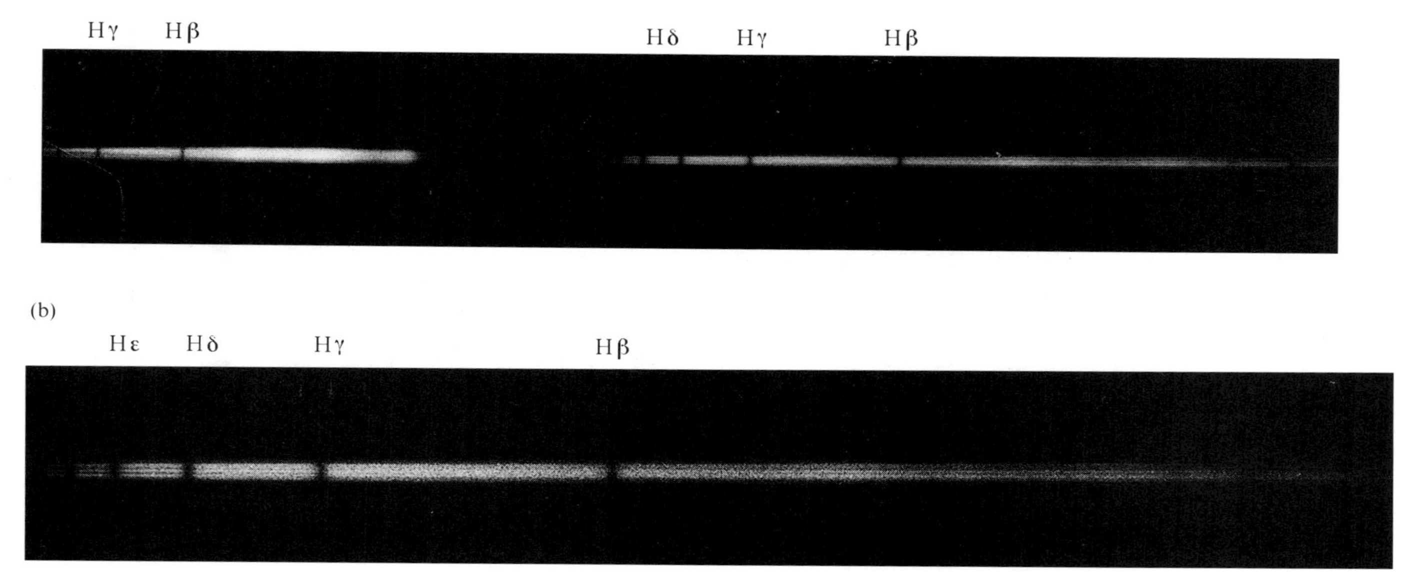
(a)
Hγ
Hβ
Hδ
Hγ
Hβ
(b)
Hε
Hδ
Hγ
Hβ

(c)

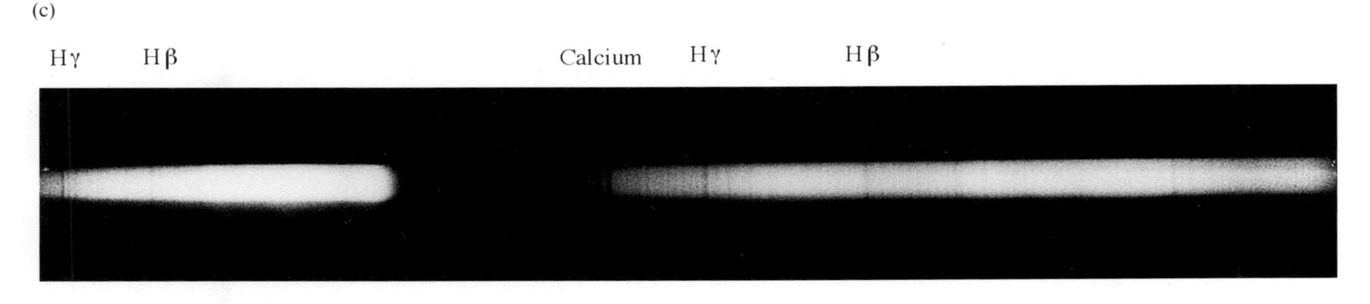

Figure 15.15. (a) Spectrum of Vega (α Lyrae) obtained by the author using his home-made spectrograph on 1990 September 19^{d} 18^{h} 55^{m} UT. The 2 minute exposure was made on T-Max 400 film, processed in D19 developer for 8 minutes at 23°C. Note the prominent Balmer series of lines typical of an A0 star. The short stripe, on the left, is the first order spectrum, covering from the near ultraviolet (at left) to red (at right). The longer stripe is the second order spectrum, covering the near ultraviolet (at left) to the yellow (at right). (b) Close-up of the second order spectrum shown in (a). (c) Spectrum of sunlight reflected from within the lunar crater Copernicus obtained by the author on 1990 October 4^{d} 22^{h} 28^{m} UT. The high throughput of the spectrograph allowed for a very short exposure, a mere 2 seconds in this case! Other details the same as for (a).

Spectrography and the CCD

CCDs and their use are discussed in detail in Chapter 6. Replacing the photographic emulsion with a cooled CCD produces something like a fifty-fold increase in the sensitivity of the spectrograph. Consequently the spectra of objects about four magnitudes fainter can be recorded, or the spectra of objects of the same brightness can be recorded in one fiftieth of the exposure time.

Even better, the appropriate processing software will allow a variety of manipulations, such as subtracting one spectrum from another to reveal differences and the measurements of the wavelengths of unknown spectral features given a prior calibration of the CCD-spectrograph on certain known lines in the spectrum, or even from a spectrum of a completely different source. As is the case for those who directly image celestial bodies, CCDs are most definitely the way forward for the spectroscopist.

16

Radio astronomy

Currently there are several hundred amateur radio astronomers active over the world. The leading international organisation for coordinating this work is SARA (the Society of Amateur Radio Astronomers). A contact address is: 37, Crater Lake Dr., Coram, N.Y. 11727, U.S.A.

Advances in modern electronics have enabled small radio telescopes to be constructed which are sensitive enough to detect scores of deep-sky objects as well as the much more powerful old favourites: the Sun, meteors, and the Jupiter–Io system. Nonetheless considerable expertise in practical electronics, as well as a good grounding in the theoretical aspects, is necessary to contruct such an instrument **and get it working successfully**. The final cost is likely to be at least several hundred pounds. Some of the necessary items of equipment can be purchased but the likely cost of a halfway decent radio telescope will then definitely rise to several thousand pounds. Also, this does not include the purchasing of the various items of test equipment, such an oscilloscope (capable of handling u.h.f.), a high quality multimeter, an r.f. signal generator, etc.

So, the most likely potential amateur radio astronomer will already be both an electronics 'buff' and a radio ham. Consequently I offer just a brief overview of this subject in this chapter and I strongly recommend that anyone interested join SARA and pursue the list of references given in the final chapter of this book.

Wavelengths, signal strengths, and observing conditions

The powers received from radio sources in the heavens are almost all extremely low. The Sun is the most powerful source and the *radio flux* received on Earth from it normally varies between 10^{-23} and 10^{-18} W m^{-2} Hz^{-1}. The only other strong natural source in the Solar System is the planet

Jupiter. Its interactions with its satellite, Io, can produce a flux of about 10^{-20} W m^{-2} Hz^{-1}. The most 'radio bright' of the deep-sky objects we can pick up is Cass A, the received flux being around 10^{-22} W m^{-2} Hz^{-1}. Others are very much less intense. They are listed in the *Astronomical Almanac* (published annually in the UK by HMSO).

Optical astronomers are used to dealing with wavelengths but radio astronomers tend to work with frequencies. Wavelength, λ, (in metres) and frequency, f, (in Hertz) are related by:

$$\lambda = c/f,$$

where c is the speed of light (3×10^8 ms^{-1}). The wavelength (and frequency) range practical for amateur work is 0.5 m (600 MHz) to 30 m (10 MHz). Longer waves cannot as easily penetrate the Earth's ionosphere and building equipment to operate in a predictable manner at higher frequencies is very difficult.

Even within this frequency range, one will only find a few narrow bands not cluttered with television, radio and an extraordinary variety of other terrestrial emissions. Even the precious 'quiet bands' will be disturbed by blasts of radio noise ripping through the frequencies caused by motor car ignitions, railway locomotives and other electrical machinery. Consequently, unless you live in a rural area, you would probably be best advised to make your equipment portable and follow the example of the many who cart their optical telescopes to suitable sites away from light-pollution.

Types of antennae

The simplest possible radio antenna is the *dipole*, illustrated in Figure 16.1(a). It works because the electric part of an electromagnetic wave causes a tiny induced voltage across the two halves. It responds only to the electric component parallel to the dipole. The transmission line then conveys this tiny voltage to the receiving and amplification equipment. To receive the incoming radio waves efficiently the overall length of the dipole should be a simple fraction of the wavelength of the radiation. The usual length chosen is $\lambda/2$.

The *antenna pattern* of the dipole is shown in Figure 16.1(b). This pattern represents the variation of sensitivity around the dipole. In this case the 90° and −90° directions are in the same direction as the dipole rods. The sensitivity in these directions is effectively zero, while it is at a maximum in the perpendicular directions.

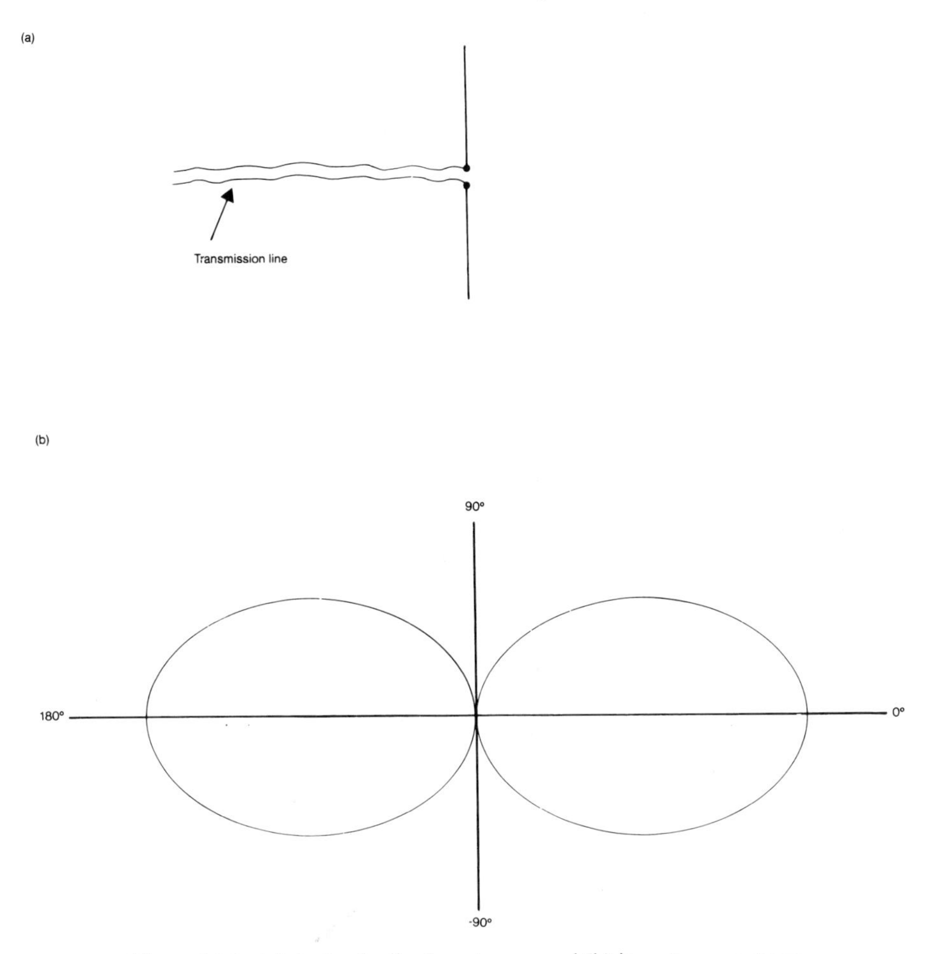

Figure 16.1. (a) A simple dipole antenna, and (b) its antenna pattern.

A considerable improvement on the dipole is the *Yagi antenna*. The Yagi is the basis of the common u.h.f. television aerial. As Figure 16.2(a) shows, it consists of one dipole, the *feeder*, or *driven dipole*, fixed to a rod along which other metal rods are positioned. The dipole is usually very slightly under $\lambda/2$ in total length, with the reflector about 5 per cent longer. On the other side of the feeder are the *directors*, each successive one being about 4 per cent shorter than the previous.

The reflector and director rods are not electrically connected to anything but their combined effect on the incoming radio waves is such as to produce an antenna diagram like the one shown in Figure 16.2(b). Notice the improved directionality (and hence angular resolution) of the Yagi

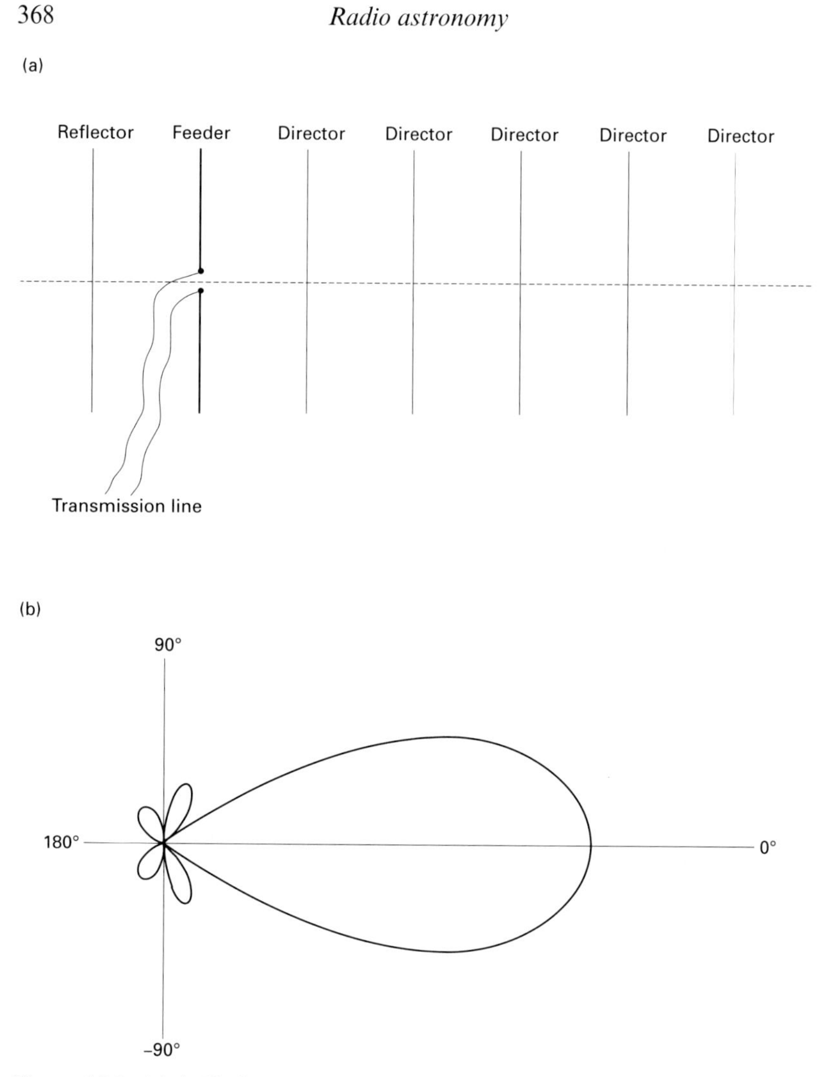

Figure 16.2. (a) A Yagi antenna. The dotted line represents the cross-staff along which the elements (antenna rods) are fixed. (b) A typical Yagi antenna pattern.

antenna, compared to the simple dipole. Like the simple dipole, the Yagi antenna works best at its designed frequency but the useable bandwidth is increased by increasing the number of directors.

Other designs of radio antenna are in current use but the Yagi has much to commend itself to the needs of the amateur. Those living at sites good enough to set up a permanent radio observatory can indulge in one of the more elaborate and specialised designs, such as the fully steerable parabolic dish (Figure 16.3), which is most people's conception of a radio telescope.

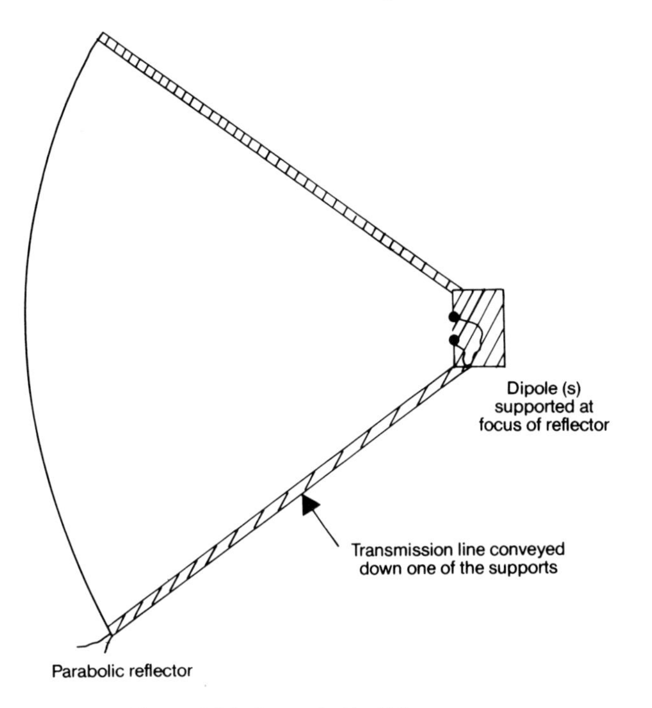

Figure 16.3 A parabolic dish antenna.

Of course the parabolic dish does have one obvious advantage over the dipole and dipole arrays. It has a bigger flux collecting area. The effective collecting area, A, of a half-wave dipole is very roughly given by:

$$A = 0.1\lambda^2,$$

A being measured in square metres and λ, the wavelength of the radio waves, being measured in metres. Thus a half-wave dipole working at 300 MHz has an effective collecting area of only about 0.1 m^2. A Yagi antenna does rather better. Again very roughly:

$$A = 0.3n\lambda^2,$$

where n is the number of directors. The effective collecting area of a parabolic dish, of diameter at least several times its working wavelength, is simply equal to its physical area.

In terms of effective resolution, the single dipole is very poor, as its polar diagram illustrates. Its effective beamwidth is about 90° which means that is cannot distinguish between sources much closer than that angular distance apart. Moreover, it cannot distiguish between sources 'in front of' or 'behind' it (shown on the diagram as 0° and 180°, respectively). A two-director Yagi antenna has a beamwidth almost as large but there is no confusion between identifying sources in front of the antenna or behind it. Increasing the number of directors helps in further reducing the beamwidth. For instance, an eleven director (13 element) Yagi has a beamwidth of about 40°.

A parabolic dish-type of radio telescope has a potential resolving power roughly equivalent to its optical counterpart, namely:

$$R \simeq 70\lambda/D,$$

where R is the effective beamwidth (and hence resolution), in degrees, and D is the diameter of the dish, in metres. So, even at the high frequency (and short wavelength) of 300 MHz, a 10 metre diameter parabolic dish-type radio antenna cannot resolve sources less than about 7° apart in the sky. The beamwidth further increases on going to lower frequencies. The surface of the dish ought to be accurately shaped to better than about $\lambda/4$, otherwise its resolving power will fall short of that indicated by the equation.

Fortunately there is one way of achieving reasonably high resolutions without the need to build a collosal structure, such as a 10 metre parabolic dish: *interferometry*.

A radio interferometer

A radio *interferometer* can be made from two, or more, antennae well separated from each other. The idea is illustrated in Figure 16.4(a). The signals from each of the two antennae shown are conveyed back to the radio receiver where they are combined. In most amateur set-ups the antennae are pre-pointed in the correct azimuth and altitude and are then left fixed. The Earth's rotation then conveys the celestial radio source across the beam of the antennae, producing a time-varying response in the receiver. Consequently, the usual arrangement is for the antennae to be orientated east–west of each other.

Figure 16.4(b) shows how the signal would vary if only one of the antennae were connected. If we assume that each Yagi had an effective, or half-power, beamwidth of about 30°, the signal strength would rise from zero, reach a peak, and then fall back to zero in the time taken for the source to move very roughly 60° across the sky. If the source was close to the Celestial Equator this would take about four hours. The maximum response happens when the source moves directly across the 'line-of-sight' of the antenna.

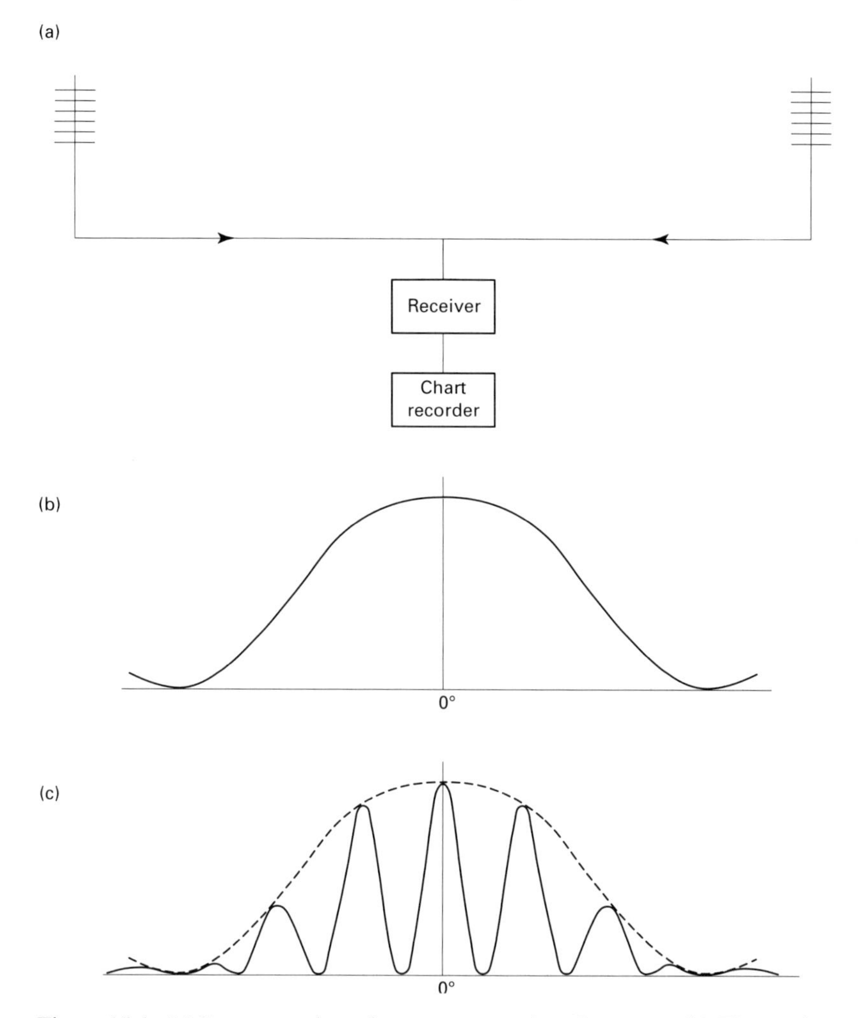

Figure 16.4. (a) Representation of a two-antenna interferometer. (b) The receiver response to a radio source passing across the 'field of view' of a single antenna (the other not connected). 0° corresponds to the direction of maximum sensitivity of the antenna. (c) The response obtained with both antennas connected.

If both antennae were brought into operation the response at the receiver would be something like that shown in Figure 16.4(c). The dotted line on the diagram illustrates the signal variation of just one antenna on its own. Note that if the central maximum of these variations can be identified, it places the position of the source with much greater accuracy. The spacing of the fringes (the minima–maxima–minima) depends upon the spacing of the antennae. If they are placed a distance of N wavelengths apart, then the angular distance, θ, that corresponds to adjacent minima on the receiver response is:

$$\theta = 57.3\ N.$$

For instance, if the antennae were set 10 wavelengths apart then θ would be just under 6°. A radio source close to the Celestial Equator would then produce one complete fringe every 24 minutes (with the antennae orientated east–west).

The response shown in Figure 15.4(c) would be that obtained if the signals from each of the antennae were simply added together. In practice this is not done, as is explained in the next section.

The radio receiver

Since the radio receiver has to deal with a very feeble input signal and produce an output which will drive a recording device, a considerable amplification factor, or *signal gain*, is needed. Electronics engineers usually deal with gains in decibels, dB, defined as follows:

$$\text{Gain in dB} = 10\ \text{Log}\ (s'/s),$$

where s is the strength of the original signal and s' is its strength after amplification. For instance, an amplification factor (s'/s) of 10 is a gain of 10 decibels. An amplification factor of 1000 is a gain of 30 decibels. A typical value for the overall gain produced by a receiver for amateur radio astronomy would be around 130 dB.

However, there are various sources of spurious signals, or *noise*, generated by the antennae and the electronics of the receiver, combined with the celestial signal. There is little use in amplifying the noise together with the celestial signal. In practice, this necessitates using a receiver system rather more elaborate than a signal *adder* (to combine the signal of the two or more antennae) coupled to a high gain amplifier. Instead a *phase-switched interferometer* is used.

A block diagram of this device is shown in Figure 16.5. After the signals from the antennae are combined by the adder, the resulting interference signal is given its first stage of amplification by the *preamplifier*. Its gain may be quite moderate, at around 20 dB, but it must be of particularly high quality and generate little noise.

The next stage coverts the signal to a lower frequency, by mixing it with a fixed frequency signal from a *local oscillator*. The subsequent electronics can then better deal with the lower freqency signal and possible oscillations due to signal feedback are avoided. The resultant *intermediate frequency*

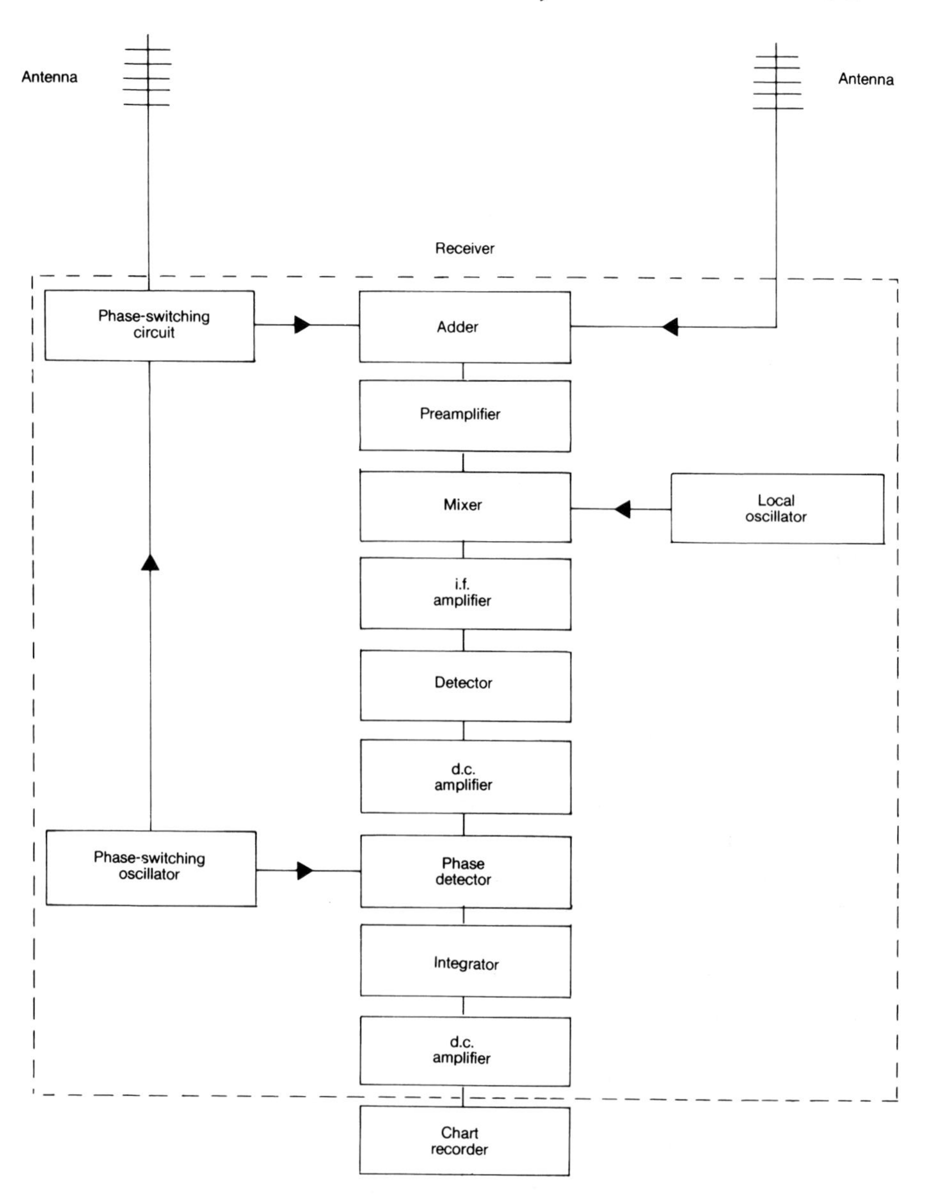

Figure 16.5. A block diagram of a two-antenna phase-switched interferometer.

(i.f) signal is then subjected to a very large amplification. A typical i.f. gain might be about 70 dB.

The *detector* rectifies the output of the i.f. amplifier and extracts the signal variations (effectively, the fringes produced by the antennae) and passes them on to the next stage, a *d.c. amplifier*, of gain value about 20 dB. If I may surreptitiously avoid dealing with the phase detector and its associated circuitry for a moment, the signal is then passed on to the *integrator*. This averages the signal over a short period known as its *time constant*. The time constant is usually set at around a few tens of seconds to a minute or two. The idea is to suppress the random fluctuations of the signal. Of course the time constant must be much shorter than the time needed to resolve individual fringes.

Finally the signal is further amplified in order that it can drive a recording device, such as an *analogue meter*, or a *chart recorder*.

Now I must go back and explain the phase-switching circuitry. The *phase-switching oscillator* produces a continuous stream of of square wave pulses. These simultaneously actuate the *phase-switching circuit* and the *phase detector*. The first of these produces a 180° phase shift in the signal in the transmission line from one of the antennae. This phase switch is repeatedly turned on and off by the pulses from the phase-switching oscillator. A 180° phase shift corresponds to advancing the fringes by half the distance between adjacent peaks. The phase detector, acting in synchronism with the phase-switching circuit, subtracts the phase-shifted signal from the non-phase-shifted signal.

The phase-switching oscillator is usually set to a frequency of several tens of Hertz and so the output of the integrator is effectively the two sets of signals superimposed. The result, the receiver output, is shown in Figure 16.6(b). If the phase-switching circuitry was not used then the output would be that shown in Figure 16.6(a). As well as the receiver output variations having twice the amplitude, this sytem makes the receiver much more immune to the effects of internal electrical instabilities. Phase-switching is almost mandatory for any sensitive radio telescope.

In these notes I have left out much of the practical details involved, such as calculating the correct lengths of the transmission lines, impedance matching, tuning, etc. These matters will be familiar to all those who already have some radio ham experience.

A radio telescope anyone can build

The December 1989 issue of *Sky & Telescope* magazine carries an article written by David Rosenthal on the construction of a radio antenna that,

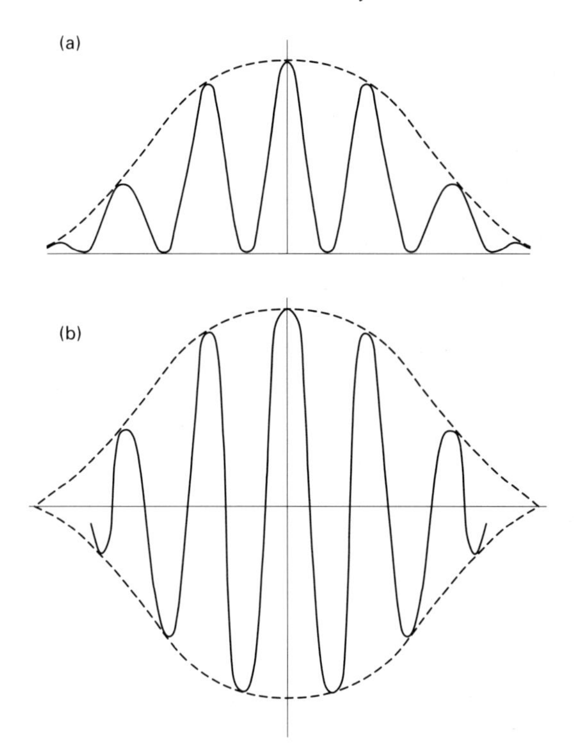

Figure 16.6. The output of an interferometer (a) without, and (b) with phase-switching.

when used with an ordinary radio set of reasonable quality, will allow one to listen for decametric radio emissions from Jupiter. These emanations arise from the passage of the satellite Io through the Jovian planet's powerful and complex magnetic field. The exact mechanisms involved are not precisely understood. The antenna design originates from that by Rob Sickles, a member of SARA.

I spent an hour or so constructing a version of this antenna myself, from scrap materials (see Figure 16.7). The diameter of the loop is 21 inches and it is positioned about 12 inches (305 mm) above the reflector, which is slightly larger than the loop. A sheet of metal, or a wire gauze, will do as the reflector but I used a sheet of aluminium foil taped (and additionally held by the four wooden posts and the earthing terminal) to a wooden board. The centre wire of a coaxial cable attaches to one end of the wire loop, the other end not being electrically connected to anything. The outer shield of the coaxial cable is attached to the reflector. The other end of the cable's inner wire is connected to the aerial feed of the radio and the outer screen to the earth connection.

(a)

Figure 16.7. (a) [above] Not exactly Jodrell Bank! However this radio antenna and domestic radio receiver combination is good enough to enable radio observations of meteors and Jupiter's emissions to be carried out. The reflector, about 23 inches (0.6 metre) square is made from aluminium foil and the wire loop is 66 inches (2 metres) of stout tinned iron wire. The hoop is held on four wooden posts by wood screws and washers. (b) [opposite, top] Close up of the hinge arrangement which allows the antenna to be pointed over a range of altitudes. (c) [opposite, bottom] The centre wire of the coaxial cable is attached to one end of the wire loop. The other is left unconnected. The outer braid (the screen) of the coaxial cable is attached, via a short wire running down the wooden post, to the aluminium foil reflector. The connection is made via a wood screw and washer, screwed, through the foil, into the wooden board.

I have found that this antenna works well over a wide range of frequencies, though it was intended for use at a frequency of 21 MHz. In his article David Rosenthal cautions that there is only about a one in six chance of successfully hearing Jupiter's emissions over a 20 minute listening period. I was quite lucky when I tried. I aimed my newly contructed antenna at Jupiter, tuned the radio to 15 metres on the shortwave band, turned up the volume control to full, and had to wait no more than about half an hour before I heard the characteristic 'wooshing' noise of the Jupiter–Io interaction.

I found that Rosenthal's description of the sound as similar to ocean waves lapping on a shore as quite accurate. The wooshes came in at a rate of about one a second. After a few minutes the sound gradually faded away, leaving the general background 'mush'. The radio beams have to be both active and sweep across the Earth in order for us to hear anything. More

(b)

(c)

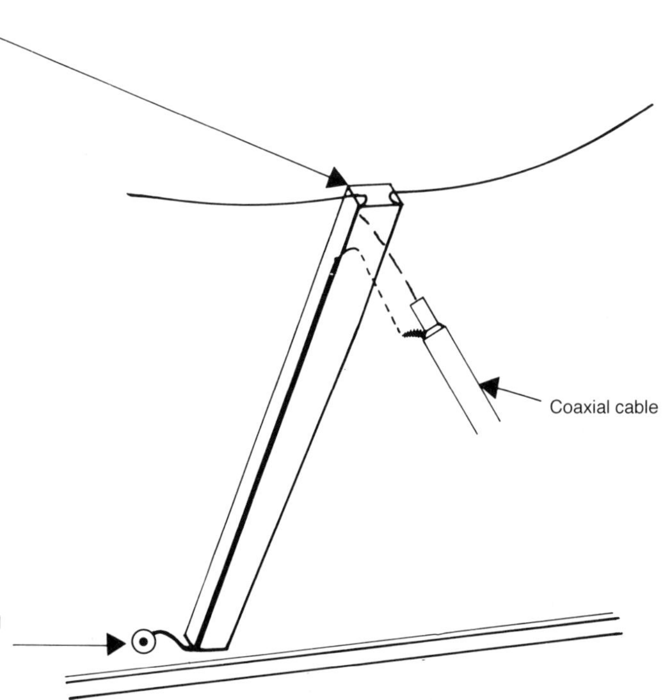
Small piece of plastic (not shown) is screwed into end of wooden post to hold ends of wire loop in position.
Coaxial cable
Earthing terminal screwed through the alluminium foil into the backing board.

often than not some man-made interference will swamp out the Jovian decametric signal. So, do have a go but also be prepared for a long wait before you hear the giant planet's wheezing signals!

As a bonus I found that the antenna was extremely sensitive to meteors entering the Earth's atmosphere. The shrill chirps – a short whistle of descending pitch – are quite unmistakable. They can be counted and recorded in the same way as for visible meteors, although one can't easily tell apart shower and sporadic meteors. Still, it is possible to plot a graph of the hourly meteor count against time in days and months. The peaks should indicate the times of the shower maxima and the relative heights of the peaks will give a good indication of the shower's activity.

In this chapter I have only scratched the surface. As well as the radio observation of meteors, amateurs currently tune in to the Sun's outbursts and provide valuable data to the solar observing and aurorae observing sections of the national astronomical societies. You would do well to contact those for information. However, I hope that in these pages I have been able to give you an idea of what is involved. If you are interested in this fascinating field then I suggest you follow up the references given in the next chapter and, above all, contact SARA.

17
Further information

This chapter provides a listing of specialist books, star atlases, magazine articles, etc., for use/further reading. These are mostly itemised under the same headings as the chapters in this book. A limited listing of equipment suppliers and national astronomical societies/associations is also included. The author and publisher wish to point out that one manufacturer's products are not recommended above another's, here. This listing is merely for information purposes and does not constitute any form of endorsement of the services or products provided by any individual company. Consequently the author and publisher cannot be held responsible for any disputes which may arise between any purchaser and any named company.

Astronomy on the internet

There is now a mind-boggling array of web-sites that may be of interest and use to the amateur astronomer. I recommend reading the article 'Astronomy on the Internet' in the August 1995 issue of *Sky & Telescope* magazine, as a start. However, it might be of some use to give one important address here. IAU circulars can be found on E-Mail:

iausubs@cfa.harvard.edu

This is the same address for reporting discoveries (see also Chapter 10).

General practical

Star atlases and catalogues

Uranometria 2000.0 (Vols. 1 & 2), Tirion, W. *et al.* Willmann – Bell, 1982.
Sky Atlas 2000.0., Tirion, W. Cambridge University Press, 1981.

Sky Catalogue 2000.0

Volume 1: Stars to magnitude 8.

Volume 2: Double stars, variable stars, and nonstellar objects.

Both the above volumes are edited by Hirshfield, A. and Sinnott, R. W. and published by Cambridge University Press (Volume 1: 1982 and Volume 2: 1985).

NGC 2000.0: the complete New General Catalogue (NGC) and Index Catalogue (IC) of nebulae and star clusters. Edited by Sinnott, R. W. and published by Cambridge University Press, 1989.

Norton's 2000.0. Edited by Ridpath, I. Longman Scientific & Technical and John Wiley, 1989.

Books

The Deep-sky Field Guide to Uranometria 2000.0., Cragin, Lucyk and Rappaport. Willmann–Bell, 1992.

The Deep Space Field Plan, Vickers, J. C. Sky Publishing Corp., 1989.

Astronomical Methods and Calculations, Acker, A and Jaschek, C. John Wiley, 1986.

Practical Astronomy with your Calculator, Duffett-Smith, P. J. Cambridge University Press, 1989.

Astronomy with your Personal Computer, Duffett-Smith, P. J. Cambridge University Press, 1985.

Compendium of Practical Astronomy:

Volume 1–Instrumentation and reduction techniques.

Volume 2–Earth and Solar System.

Volume 3–Stars and stellar systems.

Roth, G. D.(Ed.), Springer–Verlag, 1994.

The Observer's Guide to Astronomy (2 vols.), Martinez, P. Cambridge University Press, 1995.

Sky & Telescope magazine

This monthly magazine has world-wide distribution and is either obtainable direct from the publishers or through newsagents. It features up to date articles on both professional and amateur astronomy. I highly recommend it. The publishers address is:

Sky Publishing Corporation; PO BOX 9111,
Belmont, MA. 02178-9111.

CCD Astronomy magazine

A new magazine to meet the needs of practitioners of this new field. At the time of compiling this listing it is being issued quarterly. The same publishers as *Sky & Telescope*.

Telescope optics / eyepieces, seeing and magnification

Books

Seeing the Light: Optics in Nature, Photography, Colour, Vision and Holography, Falk, D R, Brill, D R and Stork, D G. John Wiley, 1985.

Sky & Telescope magazine articles

'Mounting a 6-inch Tri-Schiefspiegler'. September 1979.
'Resolution criteria for diffraction-limited telescopes'. February 1983.
'Nine low-power eyepieces'. May 1988.
'A new approach to colour correction'. October 1985.
'Analysing a Foucault test'. February 1990.
'Mirror testing for non-opticians'. February 1990.
'The Rayleigh water test'. July 1990.
'Celestron and Tele Vue tackle coma'. September 1991.
'Knife-edge testing with a video camera'. April 1994.
'Star-test your telescope'. March 1995.
'S & T test report: Your basic eyepiece set'. April 1996.

Telescope hardware

Books

How to Make a Telescope, Texereau, J. (translated by Allen Strickler). Willman–Bell, 1984.
Star Testing Astronomical Telescopes, Suiter, H. R. Willmann–Bell, 1995.
Star Ware, Harrington, Philip, John Wiley & Sons, 1994.
Unusual Telescopes, Manly, P. Cambridge University Press. Second edition, 1995.

Sky & Telescope magazine articles

'Spin-offs of the Poncet mounting'. February 1980.
'Further notes on the Poncet platform'. March 1980.

'A way to make big focusers'. January 1987.
'Crosshairs from spider web'. January 1987.
'Celestron v. Meade: an 8-inch showdown (part 1)'. December 1989.
'The top ten telescope ideas of 1988'. December 1988.
'The top 10 telescope ideas of 1989'. December 1989.
'The top 10 telescope ideas of 1990'. January 1991.
'The top 10 telescope ideas of 1991'. January 1992.
'The top 10 telescope ideas of 1992'. January 1993.
'The top 10 telescope ideas of 1993'. January 1994.
'The top 10 telescope ideas of 1994'. January 1995.
'The top 10 telescope ideas of 1995'. January 1996.
'Three collimating tools'. March 1988.
'Collimating your telescope–1'. March 1988.
'Collimating your telescope–2'. April 1988.
'The amazing stepper motor'. May 1989.
'Mirror testing for non-opticians'. February 1990.
'A handcrafted chair for viewing comfort'. March 1990
'Little observatory for a big scope'. March 1990.
'S & T test report: The next generation telescope for amateurs'. January 1991.
'A cone based polar platform for large telescopes'. December 1991.
'Rules of thumb for planetary 'scopes–I'. July 1993.
'Rules of thumb for planetary 'scopes–II'. September 1993.
'Telescopic performance on the planets'. March 1995.

Journal of the British Astronomical Association papers/articles

'Controlled frequency supply for a telescope drive'. February 1978.
'The Hill–Poncet heated observatory: a rocking type equatorial'. December 1978.
'An efficient gearless telescope drive'. December 1979.
'Making a sidereal clock'. October 1978.
'Controlling a telescope with a microcomputer and stepper motors'. October 1988.
'A home-made autoguider'. April 1989.
'Constructing grinding and polishing machines'. August 1992.
'An acclimatization fan for Newtonian reflectors'. June 1993.
'Testing mirrors for micro-ripple'. August 1993.
'A smooth sector lead-screw drive system'. October 1994.

'A run-off roof observatory'. December 1994.
'A simplified method for designing an 18 point flotation system for primary mirror cells'. December 1995
'Mirror support: 3 or 9 points?'. September 1994.
'S & T test report: Astrobeam laser collimator'. February 1996.
'Refigure your mirror while you observe'. February 1996.
'More thoughts on mirror cell design'. April 1996.

Astrophotography with the camera / through the telescope

Books

Astrophotography for the Amateur, Covington, M. Cambridge University Press. Revised edition, 1991.
A Manual of Advanced Celestial Photography, Wallis, B. and Provin, R. Cambridge University Press, 1988.
High Resolution Astrophotography, Dragesco, J. Cambridge University Press, 1995.

Sky & Telescope magazine articles

'Filters to pierce the nightime veil'. March 1989.
'Optical configurations for astronomical photography'. July 1980.
'Notes on gas-hypersensitizing'. February 1980.
'Experiments with gas-hypered film' February 1980.
'Guiding off-axis or on'. July 1989.
'Enhanced-colour astrophotography'. August 1989.
'Deep-sky photography without guiding'. November 1989.
'Temperature control of a hypering tank'. December 1989.
'S & T test report: Better ways to focus your telescope'. June 1992.

Journal of the British Astronomical Association papers / articles

'Cooled emulsion photography for amateurs'. October 1979.
'Making films more efficient for astronomical photography'. December 1981.
'Reciprocity failure of photographic emulsions before and after hypersensitization by forming gas: a densitometric study'. August 1988.

'A lensless Schmidt camera'. December 1989.
'Focusing a Schmidt–Cassegrain telescope'. December 1989.
'A 203 mm diameter Wright camera'. December 1993.
'Focusing an SLR camera using the knife edge method'. August 1995.

Electronic imaging

Books

CCD Astronomy, Buil, C. Willmann–Bell, 1991.
Choosing and Using a CCD Camera, Berry, R., Willmann–Bell, 1992.

Sky & Telescope magazine articles

'A primer for video photography'. February 1990.
'Image processing with an apple computer'. February 1984.
'Silicon eye: a CCD imaging system'. April 1986.
'Do-it-yourself image processing' and 'Practical image processing'. August 1988.
'S & T test report: a versatile CCD for amateurs'. September 1990.
'S & T test report: a powerful CCD camera for amateurs'. March 1991.
'S & T test report: the ST-6 CCD imaging camera'. October 1992.
'The Universe in colour'. May 1993.
'Image processing in astronomy'. April 1994.

Journal of the British Astronomical Association papers/articles

'Experiments with solid-state imaging devices and image processing in amateur astronomy'. June 1986.
'CCD imaging'. August 1991.
'A remotely controlled CCD camera for telescope guidance and imaging'. October 1992.
'Some applications for amateur CCD cameras'. August 1994.
'Astronomical colour imaging – notes on a new technique'. June 1995.
'Colour coding of intensity levels in CCD images'. June 1995.

The Moon

Books

The Moon Observer's Handbook, Price, F. W. Cambridge University Press, 1988. Although much of the material is outdated (and beware of obsolete nomenclature), this book provides a useful description of the Moon's surface and certain observational techniques.

Guide for Observers of the Moon, edited by Patrick Moore. Published and obtainable from the British Astronomical Association.

Atlas of the Moon, Rükl, A. Paul Hamlyn Inc., 1991.

Lunar Sourcebook – a Users Guide to the Moon, Heiken, *et. al.* Cambridge University Press, 1991.

The terrestrial planets/the gas-giant planets

Books

Introduction to Observing and Photographing the Solar System, Parker, D., Dobbins, T. and Capen, C. Willman–Bell, 1988.

Sky & Telescope magazine articles

'The art of planetary observing–1'. October 1987.
'The art of planetary observing–2'. December 1987.

Comets, asteroids, meteors and aurorae

Books

Methods of Orbit Determination for the Microcomputer, Boulet, D. Willmann–Bell, 1991.

Celestial Mechanics: a Computational Guide for the Practitioner, Taff, LG. John Wiley, 1985.

Observing Comets, Asteroids, Meteors and the Zodiacal Light, Edberg, J., and Levy, D. Cambridge University Press, 1994.

The Aurora Watchers Handbook, Davies, N. University of Alaska Press, 1992.

Sky & Telescope magazine articles

'How to observe comets'. March 1981.
'How to reduce plate measurements'. September 1982.
'Meteor observing–I'. August 1988.
'Meteor observing–II'. October 1988.
'A jam-jar magnetometer as "aurora detector"'. October 1989.

Journal of the British Astronomical Association papers/articles

'Photometry of the aurora'. April 1981.
'A fluxgate magnetometer'. February 1984.
'The photography of comets'. December 1984.
'Quick-look photo-astrometry with a linear micrometer'. February 1985.
'An amateur's computerised camera for the automatic tracking of comets'. December 1985.
'Meteor photography'. December 1992.
'Automatic photography of bright meteors and spectra'. February 1994.

The Sun

Books

Observe Eclipses, Reynolds M., and Sweetzin, R. Astronomical League Sales, 1995.
Observing the Sun, Taylor, P.O. Cambridge University Press, 1991
Guide to the Sun, Phillips, K. J. H. Cambridge University Press, 1995
Solar Astronomy Handbook, Beck, Hilbrecht, Reinch, Völker, Willmann–Bell, 1996.

Sky & Telescope magazine articles

'S & T test report: the Solaris hydrogen-alpha telescope system'. August 1992.
'A purely solar telescope'. June 1993.
'S & T test report: the Baader prominence coronagraph'. June 1994.

Journal of the British Astronomical Association papers/articles

'Solar graticules'. October 1977.
'Filter-type solar prominence telescope for amateurs'. February 1979.
'Metallized polyester film as a solar filter'. August 1979.
'Measurement of areas on the solar disc'. December 1980.
'Computer reduction of solar whole-disc drawings'. April 1983.
'The Sevenoaks spectrohelioscope'. December 1988.
'Modifications to the Sevenoaks spectrohelioscope'. October 1994.
'Carrington's method of determining sunspot positions' April 1996.

Variable stars and novae

Books

Observing Variable Stars, Levy, D. H. Cambridge University Press, 1989. Though chiefly aimed at the beginner, the author's description of a large number of variable stars, with positions and finder-charts, will make this book invaluable to the advanced observer. Highly recommended.
Variable Stars, Petit, M. John Wiley, 1987.
The AAVSO Variable Star Atlas, Scovil, C. E. Sky Publishing Corporation, 1980.

Sky & Telescope magazine articles

'Observing variable stars'. October 1980.
'Light curves and their secrets'. October 1989.
'Your telescope's limiting magnitude'. November 1989.

Methods of photometry

Books

Astronomical Photometry, Henden, A. and Kaitchuk, R. Van Nostrand Reinhold, 1982.
Getting the Measure of the Stars, Couper, W.A. and Walker E.N. Adam Hilger, 1989.
Photoelectric Photometry of Variable Stars: a Practical Guide for the Smaller Observatory, Hall, S. and Gennet, R. M. Willman–Bell, 1988.

Sky & Telescope magazine article

'A lightweight pulse-counting photometer'. September 1986.
'Adventures in photometry'. February 1996.

Journal of the British Astronomical Association papers/articles

'A photoelectric photometer'. June 1984.
'A microcomputer-assisted photoelectric photometry system.' June 1986.
'The Joint European Amateur Photometer (JEAP)'. December 1986.
'Getting started in photoelectric photometry'. December 1986.
'A simple automatic photoelectric telescope'. In three parts commencing December 1989.
'A cheap but accurate densitometer'. April 1990.
'Photoelectric photometry and the JEAP'. October 1992.
'Photoelectric photometry at Marton Green Observatory – a retrospective of a decade's work'. February 1995.

Double stars, star clusters, nebulae, galaxies and Supernovae

Books

Visual Astronomy of the Deep Sky, Clark, R. N. Cambridge University Press, 1989.
The Universe from Your Backyard, Eicher, D. J. Cambridge University Press, 1988. This is a very useful field guide to virtually every deep-sky object in the entire sky accessible to the user of a moderate telescope. Includes star maps showing the location of these, as well as many photographs taken by amateurs. Highly recommended.
Observing Visual Double Stars, Couteau, P. (translated by Alan Batten). M.I.T. Press, 1981.
Supernova Search Charts and Handbook, Thompson, G. D. and Bryan, J. T. Cambridge University Press, 1989.
Planetary Nebulae – a Practical Guide and Handbook for Amateur Astronomers. Hynes, S. J., Willmann–Bell, 1991.

Sky & Telescope magazine articles

'Measuring double stars with a grating micrometer'. June 1980.
'Visual double stars for the amateur'. November 1980.

'S & T test report: nebula filters for light-polluted skies'. July 1995.
'So you think you've made a discovery . . .'. April 1996.

Journal of the British Astronomical Association articles/papers

'An experiment in double-star photography'. October 1989.
'Nebular filters in deep sky astronomy'. October 1994.

Spectroscopy

Books

Stars and their Spectra, Kaler, J. B. Cambridge University Press, 1989.
Astrophysical and Laboratory Spectroscopy, Brown, Lang (eds.) Edinburgh University Press, 1988.

Sky & Telescope magazine articles

'An objective prism spectrograph'. May 1983.
'A simple slit spectrograph'. January 1987.
'S & T test report: Rainbow Optics star spectroscope'. October 1995.

Journal of the British Astronomical Association papers/articles

'A reflective spectroscopic slit and its applications'. April 1981.
'The construction of a small spectrograph for stellar spectroscopy and its use on some brighter stars'. February 1993.
'Automatic photography of bright meteors and spectra'. February 1994.
'Stellar spectroscopy with CCDs – some preliminary results'. February 1996.

Radio astronomy

Books

Interferometry and Synthesis in Radio Astronomy, Thompson, A R, Moran, J M, and Swenson, G W. John Wiley 1986.

Sky & Telescope magazine articles

'An amateur radio telescope'–1. May 1978.
'An amateur radio telescope'–2. June 1978.
'An amateur radio telescope'–3. July 1978.
'An amateur radio telescope'–4. August 1978.
'An amateur radio telescope'–5. September 1978.
'An amateur radio telescope'–6. October 1978.
'An r.f. converter for radio astronomy'. November, 1979.
'Jupiter on your shortwave'. December 1989.

Journal of the British Astronomical Association paper

'An easily adjusted radio telescope'. June 1979.

General (non practical) books

Cambridge Atlas of Astronomy, (Various). Cambridge University Press, 3rd edition 1994. A vast compendium of up-to-date facts, figures, and photographs in textual form.

Astronomy Explained, North, G. Springer–Verlag, due 1997 (Formerly: *Mastering Astronomy,* Macmillan, 1988). More advanced than is suggested by it's title, with the emphasis on explaining, rather than just relating, facts. Includes background physics.

Bubbles, Voids and Bumps in Time: the New Cosmology, Cornell, J. (ed.) Cambridge University Press, 1989. A fascinating exposé of our present understanding of cosmology.

Equipment suppliers (UK)

AE Optics Ltd. 28, Dry Drayton Industries, Scotland Road, Dry Drayton, Cambridge, CB3 8AT. Custom-made optics and telescopes.

Arbour Astrographic. Pennell Observatory, 29, Wrights Way, South Wonston, Winchester, Hants., SO21 3HE. Astrographic telescopes, drives and accessories, including Chesire eyepiece and 'knife edge' focuser.

Astro-Electronics. 61, Chelmsford Rd., Exeter, Devon, EX4 2LN. Telescope drives.

Astro-Promotions, 1A, Hartley Road, Luton, Bedfordshire, LU2 0HX. Custom-made optics, telescopes and accessories.

Beacon Hill Telescopes, 112, Mill Road, Cleethorpes, South Humberside, DN35 8JD. Telescopes and accesories.

Dark Star Telescopes. 61, Pinewood Drive, Ashley Heath, Market Drayton, Shropshire, TF9 4PA. Dobsonian Telescopes.

EEV Ltd., 106, Waterhouse Lane, Chelmsford Essex, CM1 2QU. CCDs and CCD cameras.

FDE Ltd., Bodalair House, Sandford Lane, Hurst, Berkshire, RG10 0SU. Manufacturers of the 'Starlight Xpress' range of CCD cameras and accessories.

Hamamatsu Photonics (UK) Ltd. Lough Point, 2, Gladbeck Way, Windmill Hill, Enfield, Middlesex, EN2 7JA. Photomultiplier tubes and equipment.

Orion Optics, Unit 12, Quakers Coppice, Crew Gates Industrial Estate, Crewe, Cheshire, CW1 1FA. Telescopes, accessories and aluminising service.

Philips Components Ltd. Mullard House, Torrington Place, London, WC1E 7HD. CCDs and CCD cameras.

Texas Instruments Ltd. Manton Lane, Bedford, MK41 7PA. CCDs.

Thomson Components Ltd. Ringway House, Bell Rd., Daneshill, Basingstoke, Hants., RG24 0QG. CCDs.

V. V. Observatories. 25, Lea Bridge Rd., Clapton, London, E5 9QB. Domes.

Equipment suppliers (USA)

Advanced Technology Instruments. P.O. Box 246, Carmel Valley, CA 93924. Computer-aided telescope hardware (including CAT).

Astro-Link, 1997, Friedship Dr., Suite F, El Cajon, CA 92020. Image intensifier and CCD cameras and accessories.

California Telescope Company. P.O. BOX 1338, Burbank, CA 91507. Telescopes and accessories, including computer-aided and computer-controlled telescopes.

Celestron International. P.O. Box 3578, 2835 Columbia St.,Torrance, CA 90503. Catadioptric telescopes and accessories. Computer Aided Telescope (CAT) devices and computer-controlled telescopes.

D & G Optical. 6490 Lemon St.,East Petersburg, PA 17520. Large refractor objectives and tube assemblies.

EEV Inc. 4, Westchester Plaza, Elmsford, NY 10523. CCDs and CCD cameras.

Hamamatsu Corporation. 420, South Avenue, Middlesex, NJ 08846. Photomultiplier tubes and equipment.

A. Jaegers. 6915, Merrick Rd., Lynbrook, NY 11563. Telescope components and accessories (including refractors).

Lumicon Corporation. 2111 Research Dr., # 58 Livermore, CA 94550. Filters and optical accesories, including 'off-axis' guiders.

Texas Instruments. P.O. Box 655303 MS 8206 Dallas, TX 75265. CCDs.

Thomas Mathis. 830, Williams St., San Leandro, CA 94577. Telescope mounts and drives.

Meade Instruments. 1675, Torronto Way, Costa Mesa, CA 92626. Telescopes, components and accessories.

Optec, Inc. 199, Smith Street, Lowell, M1 49331. Photometric equipment, particularly the SSP-3 photometer.

Orion Telescope Center, Box 1815-S Santa Cruz, CA 95061-1815. Telescopes and parts, and binoculars.

Photometrics Ltd. 2010, North Forbes Blvd. Tucson, AZ 85745. CCD camera and equipment.

Spectra. 6631, Wilbur Ave., Suite 30, Reseda, CA 91335. Accessories for astrophotography, including 'SureSharp' focuser and off-axis guiding units. CCD camera systems.

Santa Barbara Instrument Group, P.O. Box 50437, 1482, East Valley Road, #33, Santa Barbara, CA 93150. Manufacturers of SBig range of CCD astrocameras and accessories.

Roger W. Tuthill. 11, Tanglewood Lane, Dept. ST, Mountainside, NJ 07092. Telescopes, components and accessories.

Tectron Telescopes. 2021 Whitfield Park Ave., Sarasota, FL 34243. Telescopes and accessories, including Chesire eyepiece and autocollimating eyepiece.

Tele Vue. 20, Dexter Plaza, Pearl River, NY 10965. Eyepieces and refracting telescopes.

Thousand Oaks Optical. Box 4813, Thousand Oaks, CA 91359. Full-aperture solar filters.

National astronomical societes

British Astronomical Association. Burlington House, Piccadilly, London W1V 9AG. England.

Royal Astronomical Society. Burlington House, Piccadilly, London W1V ONL.

Irish Astronomical Society. c/o The Planetarium, Armagh. Northern Ireland.

British Astronomical Association (New South Wales branch). 33, Cotswold Road, Strathfield. NSW 2135. Australia.

Royal Astronomical Society of Canada. 136 Dupont St, Toronto M5R 1V2, Canada

Royal Astronomical Society of New Zealand. P.O. BOX 3181, Wellington C1. New Zealand.

American Astronomical Society. 2000 Florida Ave., NW, Washington DC 20009–1231. U.S.A.

Astronomical Society of the Pacific. 390 Ashton Ave., San Francisco CA 94112. U.S.A.

American Association of Variable Star Observers. 25 Birch St, Cambridge, MA 02138. U.S.A.

Association of Lunar and Planetary Observers. c/o The Observatory, New Mexico State Observatory, Las Cruces, New Mexico 88001. U.S.A.

Société Astronomique de France. Hôtel des Sociétés Savantes, 28 rue Serpente, Paris. France.

Koninklijk Sterrenkundig Genootschap van Antwerpen, Leeuw van Vlaanderenstraat 1, B. 2000 Antwerp. Belgium.

Vereinigung der Sternfreunde e. V., 8 München 90, Theodolindenstrasse 6. Germany.

Schweizerische Astronomische Gesellschaft, Schaffhausen, Vordergasse 57. Switzerland.

Svenska Astronomiska Sällskapet, Stockholm Observatorium, Saltsjöbaden. Sweden.

In addition there are local astronomical societies in most of the major towns and provinces of the world.

Appendix
Useful formulae

The formulae given here are those that may be useful to practical observers and are in addition to those given in the main text of this book.

Relation between hour angle, sidereal time and right ascension

$$\mathrm{HA} = \mathrm{ST} - \alpha$$

and

$$\mathrm{ST} = \mathrm{HA} + \alpha,$$

where HA = Hour angle, ST = Sidereal time and α = Right ascension. All three are measured in units of time. Distinguishing between Greenwich hour angle (GHA), Greenwich sidereal time (GST), measured at the Greenwich Meridian, and local hour angle (LHA) and local sidereal time (LST), measured at the observer's location:

$$\mathrm{LHA} = \mathrm{LST} - \alpha,$$

$$\mathrm{LST} = \mathrm{LHA} + \alpha,$$

$$\mathrm{GHA} = \mathrm{GST} - \alpha,$$

and

$$\mathrm{GST} = \mathrm{GHA} + \alpha.$$

In using any of these formulae, if the final answer is greater than $24^{h}\ 00^{m}\ 00^{s}$, then subtract this amount. The remainder is then the correct answer.

The relation between LHA, GHA, LST, GST and longitude

Right ascension increases in an eastwards direction on the sky. From any particular observation site, hour angle increases going westwards. The relation between LST and GST is:

$$\mathrm{LST} = \mathrm{GST} + L/15,$$

and

$$\mathrm{GST} = \mathrm{LST} - L/15,$$

where L is the longitude (measured eastwards from the Greenwich Meridian in degrees) of the observation site. The relation between LHA and GHA is:

$$\text{LHA} = \text{GHA} + L/15,$$

and

$$\text{GHA} = \text{LHA} - L/15.$$

Converting altitude and azimuth to declination and hour angle

In these equations:
ø = latitude of observation site,
δ = declination,
h = hour angle,
A = azimuth,
a = altitude.
All the quantities are measured in degrees. To convert hour angle to degrees multiply the numbers of hours (and decimal fractions of an hour) by 15.
Alternative formulae are given. Choose whichever provides you with the answer, given the numerical data you know:

$$\sin\delta = \sin a \sin ø + \cos a \cos A \cos ø$$

$$\cos\delta = \frac{\sin \alpha \cos ø - \cos a \cos A \sin ø}{\cos h}$$

$$\cos\delta = \frac{-\cos a \sin A}{\sin h}$$

$$\sin h = \frac{-\cos a \sin A}{\cos\delta}$$

$$\cos h = \frac{\sin a \cos ø - \cos a \cos A \sin ø}{\cos\delta}$$

Converting declination and hour angle to altitude and azimuth

The symbols are the same used as for the previous set of equations:

$$\cos a = \frac{\sin\delta \cos ø - \cos h \cos\delta \sin ø}{\cos A}$$

$$\cos a = \frac{-\cos\delta \sin h}{\sin A}$$

$$\sin A = \frac{-\cos\delta \sin h}{\cos A}$$

$$\cos A = \frac{\sin\delta \sin ø - \cos h \cos\delta \sin ø}{\cos a}$$

Index